16,50

Neoplastic Development

NEOPLASTIC DEVELOPMENT

LESLIE FOULDS
London

VOLUME 1

1969
Academic Press
London and New York

ACADEMIC PRESS INC. (LONDON) LTD
Berkeley Square House
Berkeley Square
London, W1X 6BA

U.S. Edition published by
ACADEMIC PRESS INC.
111 Fifth Avenue
New York, New York 10003

Copyright © 1969 By ACADEMIC PRESS INC. (LONDON) LTD

All Rights Reserved

No part of this book may be reproduced in any form by photostat, microfilm, or any other means, without written permission from the publishers

Library of Congress Catalog Card Number: 70-82389

PRINTED IN GREAT BRITAIN BY
W & J Mackay & Co Ltd
Chatham, Kent

Preface

Cancer has always been a "Problem". In earlier days, within my own memory, it was an Intellectual Problem which many thought to be beyond the power of the human mind to solve. Now it has become a Biological Problem, which, as is sometimes implied, biologists might solve in their spare time, if they had any. The pendulum has perhaps swung too far. Cancer is still a disease that kills people and the limited contributions of laboratory cancer research to the alleviation of human suffering proscribe facile optimism. Nevertheless it is a real advance that Burnet could write of neoplasia as ". . . a process as inevitable as evolutionary progress and of the same general nature . . ."; whether or not it is true is a matter for discussion. It is surely true that no theory of cancer—or of biology—is acceptable unless it comprehends neoplasia as one of the possible consequences of biological organization.

Neoplasia is not only an important field for the application of biological principles but a rich source of material for their study and enlargement. There is no inherent conflict between the study of neoplasia as a killing disease and its study as a biological problem, but there is a big gap in method, knowledge, language and thought. It is one of the aims of this book to provide some links between the extremes of academic biology and clinical medicine and pathology, to the possible advantage of all.

The unifying principle used in this book is the concept of neoplasia as a *developmental process* akin to normal development in some respects but differing from it in important particulars that are not yet well-defined. The emphasis is on the pathology of neoplasia as an epigenetic process and not on the pathology of tumours as lumps. This concept is not advanced as revealed truth but as a working hypothesis, which has proved useful to myself in picking my way through a substantial proportion, though far from the whole, of the accumulated wisdom and folly of "cancer research".

Part I of this volume provides an historical background of experimental cancer research and Part II a general consideration of empirical observations on neoplastic development in animals and in man and of the pathological and biological problems that emerge therefrom. Some investigators are fond of saying "What we need is more facts." The truth is that we already have more "facts" than anybody knows what to do with. Experimental analysis has produced an alarming mass of empirical facts without providing an adequate language for their communication or effective concepts for their synthesis. The orthodox pathology of overt neoplasia in man and in animals is not enough. Part III of this volume is devoted, therefore, to a consideration of the extent to which current knowledge and ideas about biological organization and development can provide a serviceable terminological and conceptual machinery for dealing with the facts and problems that emerge from Part II. It is hoped that this discussion will be a helpful introduction to readers who are not conversant with the comparatively recent impact of cybernetics and molecular biology, in their widest senses, on biological method and thought. Volume 2 deals in more detail with special cases of neoplastic development in animals and in man.

This book has been a long time on the making and it is impossible to thank individually, as I should wish, the many friends and colleagues in England and abroad from whom I have received encouragement, criticisms and ideas. I would like to express my appreciation of generous hospitality received on many occasions at the National Cancer Institute and in other institutions in the United States and at the Netherlands Cancer Institute in Amsterdam. Most of the work was done at the Chester Beatty Research Institute in London. This volume took final shape in the Division of Oncology of the Chicago Medical School and the bibliography was completed in the National Cancer Research Laboratories of New Zealand in the University of Otago Medical School, Dunedin with the assistance of a Visiting Research Fellowship provided by the W. H. Travis Trust. I thank the Directors of these institutes for their hospitality and support.

Thanks for permission to publish figures are due to Dr R. Baserga and the Editors of *Cancer Research* for Fig. 41, to Dr F. Jacob and Academic Press, Inc., for Figs 65 and 66 and to the Editors of the *British Journal of Cancer* and of the *Journal of the National Cancer Institute* for figures of my own previously published in those journals. The Publishers and especially the Production Department have been unfailingly helpful and patient in difficult circumstances.

Dunedin
February, 1969

LESLIE FOULDS

Formerly Member of the Scientific Staffs of the Imperial Cancer Research Fund, London and the Chester Beatty Research Institute of the Institute of Cancer Research, Royal Cancer Hospital, London. Lately Professor of Oncology in the Chicago Medical School Chicago, Illinois, U.S.A.

Pollards Wood Research Station, Chalfont St Giles, Bucks, England

Contents

PREFACE v

Part I: HISTORICAL INTRODUCTION

Chapter 1. Historical Introduction 3
    I. Tumour Transplantation 7
   II. The Filterable Tumours of Fowls 15
  III. Mammalian Virus-Tumours 16
  IV. Experimental Carcinogenesis 17
   V. Endocrine Neoplasia 22
  VI. Mammary Tumours of Mice 24
      Bibliographical Note 27
      Nomenclature of Carcinogenic Polyclic Hydrocarbons . 27

Part II: GENERAL AND EXPERIMENTAL PATHOLOGY OF NEOPLASTIC DEVELOPMENT

Chapter 2. Aetiological Factors 31
    I. Re-production 31
   II. Induction 31
  III. Enhancement 35
  IV. Sporadic Neoplasia 38

Chapter 3. Neoplastic Development 41
    I. Introduction 41
   II. Epidermal Carcinogenesis in Rabbits 41
  III. Epidermal Carcinogenesis in Mice 42
  IV. Inferences from Epidermal Carcinogenesis in Rabbits and Mice 44
      A. Initiation 44
      B. Conditional tumours 45
      C. Progression 45
   V. Mammary Neoplasia in Mice 46
      A. Transplanted mammary tumours 46
      B. Spontaneous mammary tumours 52
      C. Histology of responsiveness and progression . . 59
      D. Plaques and hyperplastic nodules 68

	E. Factors influencing progression	69
VI.	General Principles of Tumour Progession	69
VII.	Patterns of Neoplastic Development	75
	A. A generalized schema of neoplastic development	76
	B. Regional neoplasia	82
	C. Localized neoplasia	86

Chapter 4. Definitions, Classifications and Terminologies 91

Chapter 5. Characteristics of Neoplasms 97

I.	The Growth and Spread of Neoplasms	97
	A. Normal growth and cell proliferation	97
	B. Neoplastic growth	104
	C. Invasion	106
	D. Metastasis	112
	E. The clinical course of neoplasia	118
	F. Anomalies of behaviour of neoplasms in man	121
II.	Retrogressive Neoplasia	125
	A. The spontaneous regression of cancer in man	125
	B. Spontaneous and induced regressions of neoplasia in animals and in man	128
	C. Varieties of retrogressive neoplasia	133
	D. The significance of retrogressive neoplasia	135

Chapter 6. The Histological Analysis of Neoplasms 137

I.	The Components of Neoplasms	139
	A. The parenchyma	141
	B. The stroma	153
	C. Interactions of neoplastic and non-neoplastic tissues	155
II.	The Structural Organization of Tumours	160
	A. The significance of organoid tumours	174
III.	Complex Tumours	176
IV.	Embryonic Tumours	179
	A. Retinoblastoma	180
	B. Neuroblastoma and ganglioneuroma	181
	C. Nephroblastoma (Wilms Tumour)	184
	D. The trophoblast and its neoplastic derivatives	187
	E. Some characteristics of embryonic tumours	192
V.	Experimental Studies of Embryonal Tumours	192
	A. The chemical induction of neoplasia in transplanted embryonic tissues	192
	B. Exposure of embryonic tissues to urethane *in utero*	193
	C. Exposure of tissues to carcinogens during later pre-natal or early post-natal life	196

     D. Effect of carcinogens administered during early pregnancy 199
     E. The relationship between carcinogesis and teratogenesis 200
     F. The significance of experimental studies of carcinogenesis and teratogenesis during pre-natal and early post-natal life 203
  VI. Teratomas 205
     A. Teratomas in man 205
     B. Experimental investigations 207

Chapter 7. Biochemical Characteristics and Biological Problems of Neoplasia 219
   I. Biochemical Characteristics of Neoplasia 219
     A. Early biochemical hypotheses of cancer 219
     B. Minimal deviation tumours 220
     C. Recent biochemical hypotheses of cancer 220
     D. Biological problems of neoplasia 222

Part III: BIOLOGICAL ORGANIZATION AND DEVELOPMENTAL BIOLOGY

Chapter 8. Biological Organization 229
   I. Genetic Materials and Genetic Actions 229
     A. Genetic materials 229
     B. The genetic code 231
     C. Transcription 233
     D. Translation 236
  II. Changes in the Genetic Materials and in their Actions . 241
     A. Mutation 241
     B. Phenocopy 243
 III. Genetic Fragments and Viruses 244
     A. Bacterial transformation and transduction 244
     B. Viruses 247
  IV. Genetic Materials and Actions—Summary 260

Chapter 9. Dynamic Organization 265

Chapter 10. Cytoplasmic Organization 273
   I. The Ultrastructure of Cells 273
     A. The endoplasmic reticulum 273
     B. Mitochondria 275
     C. Chloroplasts 279

Chapter 11. Biological Organization in Unicellular Organisms ... 281
- I. Enucleation and Nuclear Transplantation ... 281
  - A. Enucleation ... 281
  - B. Nuclear transplantation ... 283
- II. The Ciliated Protozoa ... 284
  - A. Extra-nuclear genetic materials ... 284
  - B. Genetic and biotonic systems in the ciliates ... 289

Chapter 12. The Organization and Development of the Metazoa ... 305
- I. The Cell Theory ... 305
- II. Concepts of Development ... 307
  - A. Preformation and epigenesis ... 308
  - B. The genetic basis of normal development ... 310
  - C. The programme of development ... 311
- III. Early Vertebrate Development ... 314
  - A. The gametes ... 314
  - B. Fertilization ... 315
  - C. Cleavage ... 315
  - D. Gastrulation ... 316
- IV. Differentiation ... 318
- V. Competence and Capacity ... 324
- VI. Metaplasia ... 329
- VII. Embryonic Induction and Tissue Interactions ... 334
  - A. Secondary inductions and tissue interactions ... 334
  - B. Induction of the lens in the amphibian eye ... 335
  - C. Inductive tissue interactions in mice ... 336

Chapter 13. The Relationship between Cell Proliferation and Cytodifferentiation ... 343
- I. The Preparatory Role of Cell Proliferation ... 343
- II. The Inverse Relationship between Cell Proliferation and Differentiation ... 344
- III. Cell Proliferation, Differentiation and De-differentiation *in vitro* ... 349
  - A. Short-term cultures ... 350
  - B. Transformation ... 350
  - C. Established Cell Lines ... 351

Chapter 14. Differential Utilization of the Genome ... 357
- I. The Polytene Chromosomes of Diptera ... 357
- II. Lamp-brush Chromosomes of Oocytes ... 358

III. Some General Principles of Differential Utilization of the Genome 360
IV. Mechanisms of Differential Utilization of the Genome . 365
    A. The histone hypothesis 365
    B. The Jacob-Monod hypothesis 367
    C. Nuclear transplantation in amphibia 369
    D. Paramutation and treption 371
V. Conclusions 373

Chapter 15. Applications of Recent Concepts of Biological Organization to some Problems of Normal and Neoplastic Development 375
  I. Normal Development 375
    A. Genetic materials and genetic actions 375
    B. Extra-nuclear genetic mechanisms 377
    C. The dynamic component of biological organization . 378
    D. The "network" concept of biological organization . 380
 II. Neoplastic Development 381

REFERENCES 387
AUTHOR INDEX 425
SUBJECT INDEX 435

Part I

Historical Introduction

CHAPTER 1

Historical Introduction

"Cancers" and "Tumours" have been known since antiquity. Literally, "tumour" means no more than a swelling or lump but it has come to mean a special kind of lump known popularly as a "growth" or, if of the most dangerous kind, a "cancer", and medically as a "new growth" or "neoplasm". As Virchow once remarked, no man, even under torture, can say exactly what a tumour is (Ewing, 1916). Nobody has yet devised a wholly satisfactory, or universally acceptable concise definition. Nevertheless, the main characteristics of the too frequent neoplasms of man are well known as a result of clinical observations extending back for several millennia and of the study of their cellular pathology for a century. The accumulated experience of clinicians and pathologists is the basis of the modern concept of neoplasia upon which all current investigations are founded. To clinicians and pathologists the concept is a real and valuable one but it is difficult to communicate because as yet it cannot be expressed in precise biological terms that convey its elusive essence. Nearly all proposed definitions are summaries of the more common and more conspicuous properties of neoplasms as observed in man and are either too embracing or, perhaps more seriously, too restrictive. Most of them emphasize the characteristically progressive, disproportionate and seemingly purposeless *overgrowth* of a tissue that continues indefinitely after all known or suspected inciting stimuli have ceased to operate.

Nineteenth century pathologists divided neoplasms into *benign*, *simple* or *innocent* tumours and *malignant* tumours. Benign tumours were the circumscribed and often encapsulated lumps that grew, slowly as a rule, without local invasion or destruction of contiguous tissues and without more remote dissemination. Usually they were curable by local excision and, although unpleasant and troublesome, did not endanger life unless by reason of their size or position they interfered mechanically with important functions as, for example, by obstructing or compressing essential passages of the digestive, vascular or excretory systems. Malignant tumours, by contrast, usually grew more quickly and, more importantly, they grew more extensively. Their outlines were indistinct because they invaded and destroyed neighbouring tissues and they often gave origin to secondary growths of a similar kind in distant parts of the body. Malignant tumours usually recurred after local excision and ordinarily they were lethal, being curable, if at all, only by drastic and extensive surgical or radiological procedures.

Pathological classifications and descriptions of tumours are of two main varieties, the histological and the histogenetic. The histological classifications are descriptive and based on observed histological structure. The histogenetic classifications are inferential and based on the type of tissue from which particular tumours are believed to originate. Histological classification based on observation is preferable to histogenetic classification, which is commonly based on inferences from observations or, not infrequently, on guess-work. The two classifications are often in agreement because of the important circumstance that many tumours, but not all of them, retain some distinctive characteristics of the tissues from which they originate. The primary division is between tumours of connective-tissue type and tumours of epithelial type, each being further divided into benign and malignant varieties. Benign, connective tissue tumours are named by adding the suffix "oma" to the name of the particular kind of connective tissue which they most resemble or from which they are presumed to originate; they include fibroma (fibrous), myoma (muscular), lipoma (fatty), myxoma (mucinous), osteoma (bony), and chondroma (cartilaginous). The corresponding malignant tumours are sarcomas and include fibrosarcomas, myosarcomas, liposarcomas and so on. The benign epithelial tumours are wart-like papillomas or gland-like adenomas. The malignant epithelial tumours are carcinomas or epitheliomas. Usually the name of the tissue or organ in which the tumour develops is added, for example, mammary carcinoma (breast), pulmonary carcinoma (lung), vesical carcinoma (bladder) and so on. Another division is based on the disposition of the epithelial cells to make either a gland-like epithelium in *adenocarcinoma* or a stratified epithelium resembling the epidermis in *squamous* or *epidermoid carcinoma*. Some terms are merely topographical or evasive. Hepatoma can mean either a benign or malignant tumour of the liver. *Gliomas* are tumours of the supporting tissues of the nervous sytem. The *leukaemias*, of various kinds, are neoplasias of blood-forming tissues. The limitations of histological classifications will be discussed later, but it is relevant to draw attention here to the important fact that most tumours have distinctive histological *patterns* by which they can be recognized and named. Tumours, in general, are not formless, chaotic conglomerations of cells but have an organized structure which sometimes approaches in perfection that of their parent tissues.

Virchow and some of his contemporaries thought that tumours grew by the accretion of cells from surrounding tissues, which were stimulated in some way by a primary tumour nodule in them. They thought also that distant tumours resulted from the proliferation of local tissue cells in response to a product liberated from a fragment of a tumour transported to that site by way of the blood or lymph streams from a primary tumour elsewhere in the body. This opinion gradually gave way to the view embodied in Ribbert's famous dictum that tumours grow "aus sich heraus"; they grow by multiplication of their own constituent cells and not by accretion of cells from surrounding tissues. Remote tumours likewise grow by the proliferation of cells transported from the primary growth, most often by the blood or lymph streams, and not by proliferation of previously normal cells belonging to the new locality. This process of discontinuous extension by transported cells, known as *metastasis*, is the most dangerous attribute of malignant tumours and the most formidable obstacle to their cure or control. It is essentially a process of migration and colonization. The migrant cells carry the main characteristics of the primary tumour with them wherever they go and when they multiply in their new situation they breed true-to-type. Consequently, the metastic or secondary tumours ordinarily resemble the original primary tumour closely in structure and behaviour. The body of a patient dying of cancer may be riddled with secondary tumours and all of them are composed of cells that are direct descendants of migrant cells from a primary tumour, which may be small. Tumour metastasis illustrates perhaps better than anything else the fundamental difference between malignant neoplasia and a dangerous destructive disease such as tuberculosis. In fatal tuberculosis, the body may be riddled with "tubercles" but these lesions are formed by the reaction of local cells to transported tubercle bacilli and not by the multiplication of cells transported from the primary tubercular lesion. The major contribution of morbid anatomy and histology to the understanding of neoplasia was to demonstrate that the behaviour of tumour cells is attributable to their own intrinsic constitution and not to a reaction of cells to a continuing stimulus operating from outside them.

Pathological study of tumours of man provided no sound information about their aetiology and pathogenesis but stimulated much speculation. Cohnheim supposed that during embryonic development some cells of the embryo failed to keep pace and mature with their fellows but persisted as pockets or "rests" of cells with "embryonic" properties, and, in particular, with a high latent capacity for multiplication and a low degree of differentiation. The release of the latent capacity for multiplication was presumed to result in neoplasia. It was necessary to presume further that the latent capacity was held in check for many years and eventually, often in middle age, somehow released. Ribbert agreed that many tumours originated from "embryonic rests" but thought that adult cells also had a similar but less easily disclosed capacity for growth. Normal epithelium, according to this view, had a capacity for infiltrative growth that normally was held in check by the contiguous connective tissue; an epithelial cancer was believed to develop when the control exercised by the connective tissue failed. The Cohnheim-Ribbert theory has been criticized and discredited repeatedly, notably in Great Britain by Nicholson and by Willis, but it continues to re-emerge from time to time in a more or less modified form. The main alternative to it was Virchow's belief in "chronic irritation" as an essential stimulus to neoplasia. In the eighteenth century the English surgeon Percival Pott, being, it is said, temporarily immobilized by a broken leg, inferred that irritation by soot was responsible for the unduly high frequency of cancer of the scrotum in chimney sweeps and many subsequent observations, especially on workers handling tars and oils, supported the theory of chronic irritation. Bashford, in his remarkable "Draft Scheme for Enquiry into the Nature, Course, Prevention and Treatment of Cancer", prepared in 1902, and submitted as an application for the Directorship of the newly-founded Imperial Cancer Research Fund in London, specifically mentioned the systematic study of the persistent irritation of various epithelial surfaces in different species of animals, bearing in mind the possible experimental production of Petroleum Cancer and Sweep's Cancer (Bashford, 1908). More than a decade passed before experiments of the kind envisaged by Bashford bore fruit and meanwhile the experimental study of cancer began along different lines.

Bashford's "Draft Scheme" was remarkably comprehensive, as well as uncannily prophetic. He initiated critical enquiries into the validity of the statistics then available about the incidence of cancer in man and may be credited with one of the earliest attempts to establish an "epidemiology of cancer" on a sound statistical basis. Moreover, he and his colleagues at the Imperial Cancer Research Fund carried out one of the first enquiries into the "Ethnological Distribution of Cancer" or as it is now known, the "Geographical Pathology of Cancer". In this they enjoyed governmental cooperation in drawing material from the vast area signified in those days by the term "Imperial" which was part of their Institute's title and reported that cancer occurred "among all races and under all climates throughout the Empire" (Bashford *et al.*, 1905). Bashford's First Assistant, J. A. Murray, who later succeeded him as Director, carried out yet another pioneer study of the "Zoological Distribution of Cancer" and recorded numerous malignant tumours, not only in various mammals, but also in birds, amphibia and fishes. Bashford *et al.* (1905) remarked that ". . . we hesitate hastily to seek an analogy in the invertebrates, the protozoa and the vegetable kingdom, although for the formulation of a working hypothesis it is necessary to do so." This gap in knowledge has since been partially filled but not altogether satisfactorily, perhaps because investigators have sought close analogies with neoplasia as it is found in man and other mammals. It is not unlikely that neoplasia in plants, invertebrates and protozoa needs studying with different criteria and concepts.

At the beginning of the century, the demonstration that cancer developed in all races of mankind living in all tolerable circumstances of climate and nutrition, as well as in all species of vertebrate animals, was of great importance in dispelling many misconceptions and superstitions about cancer, including the widespread belief that cancer was peculiarly a disease of civilized man. Moreover, it was a necessary foundation for the experimental study of cancer in laboratory animals, which encountered a degree of scepticism and even hostility which is not now easily comprehended. Bashford and his colleagues themselves soon became engrossed in laboratory research and, before long, experimental study of neoplasia in animals became almost synonymous with "Cancer Research". It is interesting and not unimportant to reflect that, comparatively recently, increasing attention has been given to the subjects to which Bashford and his colleagues first devoted themselves; to Environmental Cancer, Geographical Pathology of Cancer, Epidemiology, and the Comparative Pathology of Cancer. These together provide an invaluable and indeed indispensable check on the increasingly complex laboratory investigations now in vogue and will receive consideration later. First, it is desirable to sketch the main lines of experimental studies and the enlargement of the materials and methods of investigation which they have provided.

I. Tumour Transplantation

In the latter part of the nineteenth century the remarkable success of bacteriologists in transmitting human infective diseases to animals encouraged attempts similarly to transmit neoplasms. Accidental transmission of human cancer from one individual to another was reported from time to time but on evidence that was almost certainly fallacious. Reports of the same period on "cancer houses" and the contagious spread of human cancer do not withstand critical examination. A few deliberate and dubiously justifiable attempts to transmit a neoplasm from one individual to another were, perhaps, successful. Numerous attempts to transmit human cancer to animals failed consistently. The first successful transplantation of malignant tumours in animals seems to have been achieved, in dogs by a Russian veterinarian, M. A. Novinsky, in 1875–1876 (Shimkin, 1955*b*). In 1889 Hanau transmitted a squamous carcinoma from rat to rat and in 1891 Moreau transferred a squamous carcinoma from one mouse to another and onwards from mouse to mouse through seventeen passages in series. The observations of Hanau and Moreau attracted little notice probably because by then the experimental transmission of tumours deservedly had fallen into disrepute, but at the beginning of the twentieth century transmission of mouse tumours was achieved independently by Borrel in France, by Leo Loeb in the United States and by Jensen in Denmark. The most convincing of the reports was

Jensen's, and his classical paper published in 1903 marks the beginning of modern experimental cancer research.

Jensen (1903) removed a carcinoma, probably a mammary carcinoma, from a mouse, broke up the tissue, and made a suspension which he inoculated under the skin of five mice. Tumours developed at the site of injection in three mice and these tumours in turn were transmitted to other mice. At the time of his report Jensen had made nineteen such transfers in series. All the growths were carcinomas of the same histological type as the original tumour. Jensen killed the mice at short intervals after the inoculations, before growths were palpable, and examined the inoculation sites histologically. He found that soon after inoculation most of the injected material was dead but some carcinoma cells survived and after a while began to multiply and by so doing produced a tumour. Jensen's especial merit was to show that the transmission of tumours was effected by a transplantation of tissues. The new tumours developed from implanted tumour cells and so differed fundamentally from infective lesions, which comprised a reaction of the hosts tissues to organisms of extrinsic origin. Jensen was not the first to transplant a tumour but he was the first to demonstrate the mechanism of transmission and to convince investigators that transplantation was likely to be a fruitful method for the study of tumours. For the next decade tumour transplantation dominated experimental cancer research. Soon after Jensen's publication many tumours of varied histological types were being transplanted, mostly in rats and mice, by investigators in various parts of the world. Transplantation within the same species (homo-transplantation) was often successful but transplantation from one species to another (hetero-transplantation) always failed. Implants will grow almost anywhere in the body, although some sites are more favourable than others. For convenience, implants are usually placed under the skin or into muscles. The technique of implantation varies somewhat from one laboratory to another. In one method of proved value, a tumour fragment of pin-head size is inoculated subcutaneously through a hollow needle. If the implant "takes" a small nodule becomes palpable after a latent period which varies greatly from one tumour to another. Thereafter the tumour grows into a visible mass, although some tumours after initial establishment or "take" fail to grow progressively and some of them regress and disappear. The various transplantable tumours, even those of similar origin, like the abundant mammary tumours of mice, are highly varied in structure and behaviour; probably no two are exactly alike in every particular. Tumours differ in histological structure, in percentage of successful "takes", in percentage of subsequent regressions, in the rate and manner of growth of established tumours and in the frequency and mode of dissemination. Some of the most vigorous tumours, starting from implants of pin-head size, become palpable after a few days and grow within two or three weeks into masses one or two inches in diameter. Other tumours are "latent" for many weeks and then grow almost imperceptibly from week to week or month to month. A few celebrated tumours have been maintained by serial transplantation for more than half a century; they include the Jensen rat sarcoma (JRS), the Walker rat tumour 256, the Bashford mouse mammary tumour (M63), the Ehrlich mouse carcinoma, the Crocker Fund mouse carcinoma 180, the Brown-Pearce rabbit tumour and others. These tumours constitute a selected sample of the most successful and useful transplantable tumours; many other transplantable tumours have died out or been abandoned. In the early days of tumour transplantation the results were erratic and failures were more common than successes.

The painstaking microscopical examination of successive stages in the development of tumours from the time of implantation onwards, especially by Bashford and his colleagues in London, confirmed and extended Jensen's observations. Much of the implanted material degenerated but, at the edge, some of the proper tumour cells, the parenchyma cells, survived, whereas all of the supporting tissue or stroma, comprising connective tissue and blood vessels, died. The surviving parenchyma cells multiplied and connective tissue cells and blood vessels grew in from the surrounding tissue to provide a new stroma for them. If the "stroma reaction" of the host failed, the implant perished. Tumour transplantation is often called tumour grafting and there is a rough analogy between grafting a tumour on a mouse and grafting a cultivated rose on a briar stock; the host mouse or the briar provides support and nourishment to the graft. A transplanted tumour is an extraordinary growth composed of tissues derived from two different animals; the parenchyma cells are lineal descendants of the cells of a tumour in a mouse that died perhaps fifty years ago whereas the stroma develops anew in each host and has changed, maybe, a thousand times. The stroma represents in Bashford's words "merely a reaction on the part of successive hosts whereby the parenchyma is nourished and supported by an artificial circulation renewed from time to time." This duality of structure is a troublesome complication which needs more attention than it usually receives in the interpretation of experiments with transplanted tumours.

The parenchyma evokes a stroma reaction that is essential for the establishment of an implant in a new host but also, as Bashford insisted, it evokes a "specific stroma reaction" by determining the quantity and quality of the stroma, which are characteristic for each strain of transplantable tumour. Characteristically, the histological structure of a transplantable tumour, depending on the association of a particular kind of parenchyma with a particular kind of stroma, remains steady during successive transfers over long periods of time. The steadiness is attributable to the stability of the properties of the parenchyma through many cell generations; the constancy of the stroma is a secondary consequence of the specific stroma reaction evoked by a constant stimulus emanating from the parenchyma cells. Transplantable tumours in general retain not only their characteristic histological structure but also their idiosyncrasies of growth and behaviour through repeated serial transplantations. With important reservations to be emphasized later, the parenchyma cells of transplantable tumours "breed true"; their characteristics are heritable in that they are transmitted from cell to daughter cell during prolonged serial transplantation. The persistence of all characters of transplantable tumours through repeated transplantations in previously normal animals shows that the capacity for behaving as a neoplasm and the idiosyncrasies of individual tumours are determined by inherent properties of the tumour cells themselves and do not depend on a continuing extrinsic stimulus provided by an abnormal environment. Transplantation, described by Bashford as "artificial metastasis", strongly reinforces the implications of natural metastasis in human beings to the effect that the structure and behaviour of tumours are attributable to heritable intrinsic qualities of tumour parenchyma cells. There is still wide disagreement about what impels tumour cells to behave as they do but wide, if implicit, agreement that the decisive mechanism is situated within the tumour cells. Those, for example, who believe that viruses provide the "driving force" of neoplasia grant that the viruses must be intra-cellular so that, in effect, cell plus virus behaves as one unit.

The pioneer investigators of transplantable tumours devoted much time to the study of *immunity* or *resistance* to transplantable tumours. They described a *natural immunity* in some animals in which tumour implants did not grow and an *acquired immunity* in others which, having had one transplanted tumour that regressed, were then refractory to implantation of the same tumour or a different one. Some tumours, without regressing, evoked a *concomitant immunity* so that the host became resistant to another transplantable tumour, which formerly it would have accepted. Finally, injections of normal tissues of another animal of the same species evoked an *artificial immunity* preventing the "take" of transplantable tumours. Haaland (1911) gave the first warning of the disillusionment to come by showing that no form of immunity, natural, acquired or artificial, was effective against a tumour transplanted within the animal in which it originated (auto-transplantation) or in any way modified the tumour's growth or dissemination in its original host. Numerous experimenters continued to use transplantable tumours without appreciating their limitations. Woglom's valuable but depressing review of 1929 shows how much labour they wasted. It gradually became evident that "immunity" did not operate against implants because they were "cancerous" but because they were "foreign" to the new host. Similar "immunity" opposes skin-grafting, blood transfusion and other transfers of normal tissue from one animal to another. Tyzzer, Leo Loeb and Little seem to have been the first to recognize that the hereditary make-up of hosts and donors were of decisive importance in tumour transplantation but the establishment of genetically-homozygous inbred strains of mice, first by Little and Strong, was a necessary prelude to a precise study of the genetics of transplantation. The inbred strains have become of great importance in many branches of cancer research. From observations on these inbred strains of mice and hybrids obtained by crossing them, Little and Strong (1924), formulated a genetic theory of transplantation which they expressed as follows: "The fate of the implanted tumour tissue when placed in a given individual (host) is brought about by a reaction between the host, determined by its genetic constitution, and the transplanted tumour cell, controlled to some extent by its genetic constitution." In a later review, Little said that the fate of transplants depends on the degree of biologic similarity between the host and the tumour tissue, the degree of similarity being determined mainly by Mendelian genes varying in number from one to twelve or fifteen.

It is now believed that tumour transplantation is limited in much the same way as blood transfusion by inherited disparities between donor and recipient animals; incompatibility evokes a reaction that is disastrous to the transferred tissue. More precisely the outcome of transplantation depends on "histocompatibility genes" numbering at least six or seven and probably fourteen or more in mice (Snell, 1953). Only three have been identified in hamsters (Palm, 1961). An implant evokes a reaction leading to its own destruction if it carries histocompatibility genes not present in the recipient host. The converse is not true; absence from the implant of histocompatibility genes present in the recipient host is not a bar to successful transplantation. A tumour originating in an inbred strain is transplantable to all other animals of the same strain (*iso-transplantation*) and into all F_1 hybrids of that strain and any other strain because the implants contain no histocompatibility genes not present, at least in single dose, in the recipients. Nothing is yet known of the activities or functions of histocompatibility genes in the intact animal, if they have any. Their existence is revealed only by the unnatural act of grafting (Palm, 1961).

The genetic theory of transplantation is applicable to normal as well as neoplastic tissues and its scope has been much enlarged by skin-grafting experiments carried out by Medawar, Billingham and others. The general validity of the theory is not disputed but some reservations may be made about its application to the transplantation of tumours. The "transplantable tumours" of Jensen, Bashford, Ehrlich and other pioneers of experimental cancer research grew from the beginning in random-bred mice, in apparent defiance of genetic differences. Tumours are apt to undergo genetic loss or simplification during transplantation and in consequence of having fewer histocompatibility genes to provoke antagonistic reactions they become able to grow in a wider range of hosts than they did at first. Nevertheless, genetic loss is not a completely satisfying explanation of the past or present behaviour of the old "transplantable tumours" (Barrett, 1958; Gorer, 1948).

The property of "transplantability" is not a simple one. It is now possible to suppress the natural resistance against homo- or hetero-transplantation of normal, as well as of neoplastic tissues, in several ways. It can be done by inducing *immunological tolerance* before or soon after birth or depressing antagonistic reactions in adult animals by irradiation or by the administration of cortisone. It should be remarked that certain sites, notably the anterior chamber of the eye and the brain, provide "immunologically privileged environments" for the growth of normal or neoplastic homografts and also that homograft reactions are very much weaker in hamsters than in mice. By combining two or more of the favouring circumstances some human tumours can now be transplanted into laboratory animals, especially successfully into the cheek pouch of an irradiated and cortisonized hamster. It has been objected that such an animal is little more than a mobile incubator. In truth, the only value of transplantation under these contrived circumstances is for demonstrating the replicability of cell characters. The method does not, without refinement, differentiate neoplastic from non-neoplastic tissues. Injected in known numbers in suspension, normal as well as neoplastic cells will proliferate in the hamster cheek pouch. In general, "normal" cells must be injected in much larger numbers than neoplastic cells to ensure proliferation, but even with this technical elaboration the distinction between normal and neoplastic cells is not nearly so sharp as could be wished.

It is a misconception to suppose that homo-transplantation is or ever has been a consistently reliable test of "malignancy". Loeb and Fleischer (1916) serially transplanted mammary fibroadenoma in rats and many others have done so since but nobody has proposed that the fibroadenoma is a malignant tumour. Homo-transplantation in random-bred animals demonstrates some difference, qualitative or quantitative, between at least some neoplasms and the generality of normal tissues. The significance of hetero-transplantation will be discussed elsewhere but the experiments with the hamster cheek pouch imply that success depends on a quantitative rather than a qualitative difference between neoplastic and normal cells. The outstanding value of tumour transplantation is for demonstrating the heritability or replicability of various characters of tumours and for studying their variability under reproducible conditions.

After half a century of study and argument, the resistance to transplantable tumours is not yet satisfactorily clarified. At least two different mechanisms seem to operate. The first, emphasized by Bashford and his colleagues, is failure of the stroma reaction and consequent failure of *establishment* so that the implant never begins to grow. The second, receiving more attention now, is temporary growth followed by regression accompanied by a cellular reaction in which lymphocytes and plasma-cells predominate. It is almost interminably arguable whether the cellular reaction or the regression starts first. The currently-favoured interpretation is in terms of an immunological reaction.

Two periods in the history of every transplanted tumour need separate consideration. In the first period an implant establishes contact with its new surroundings and acquires a new stroma and blood supply. Phenomena that depend on moving tissue from one place to another and especially from one animal to another dominate this period of establishment or "take". During the second period the established implant grows and its behaviour depends to a greater extent on the intrinsic properties that determine the general characters of a "tumour" and on the special properties that make that tumour different from other tumours. The two periods are differently regulated the first being, in the main, the more sensitive to interference. It is noteworthy that once tumours have become established in a heterologous host with the help of cortisone they will often continue to grow if cortisone is then withheld (Billingham, 1961). After establishment, when "tumour" properties come to the fore, the transplanted tumour is still not exactly comparable with a primary spontaneous tumour. Even in auto-transplantation and iso-transplantation, with genetic differences eliminated, the stroma and vascular supply, at least, are not the same in the new surroundings as they were in the original host. It is usually hazardous to transfer from transplantable tumours to spontaneous tumours any inferences about tumour-host relationships. It is well to note that any circumstance that makes an animal ill also, as a rule, impedes the growth of a transplanted tumour in it.

Somewhat paradoxically, although transplanted tumours are most esteemed for providing abundant tumour material of fairly uniform and constant quality, they are important also for demonstrating the capacity of tumours for changing their structure or behaviour. Transplantation gives the opportunity for studying the response of tumours to varied environments and it also prolongs the life of a tumour far beyond the lifetime of its original host. Some tumours are liable to conspicuous reversible changes or *modulations* in structure or behaviour when exposed to new environments; when returned to their customary environment the tumours revert to their former state. More importantly, during the course of serial transplantation tumours are liable to irreversible change or *progression*. In the general experience, transplantation usually becomes easier after the first few transfers; the percentage of "takes" increases, the latent period shortens and growth accelerates. Histological structure may be complicated and unstable for a while but then becomes more uniform. The period of variability may correspond with the selection of one or more of the variant types in an originally heterogeneous tumour but, after the early stabilization, the tumour may remain liable to modulation as well as to progression. Sometimes there is an abrupt change in transplantability attributed by Strong, Bittner and others to genetic *mutation* or the responsiveness of tumours to hormones or other extrinsic stimuli may alter conspicuously and permanently. One notable permanent change to which many but not all tumours are liable is conversion into ascites tumours. *Ascites* denotes an accumulation of fluid in the peritoneal cavity. After intraperitoneal inoculation, ascites tumours evoke an effusion of fluid in which tumour cells live and multiply. Ascites tumours are much valued for the opportunities they give for studying isolated tumour cells and applying to tumours the principles of population genetics, profitably used in studying fluid cultures of bacteria. The conversion to an ascites tumour is effected quickly or slowly by serial intraperitoneal transplantation and to that extent is a consequence of the experimental procedure. More generally, progression in transplanted tumours seems to depend chiefly on the prolongation of the life of the tumours and the consequent increase in time and opportunity for progression to occur. The capacity of transplantable tumours for unpredictable progression adds to the substantial list of reservations about their value in many kinds of investigation. Most long-maintained transplantable tumours have changed so greatly over the years that they no longer much resemble the original parent tumour from which they are descended or any other primary tumour that has ever existed. The gain in experimental convenience is often great but it is achieved at the cost of ever-increasing divergence from the natural phenomena of neoplasia.

Observations on transplantable tumours showing that the structure and behaviour of tumours are determined by intrinsic characters of the tumour cells and that these cells, with the reservations just mentioned, breed true to type, provide the main justification for some form of *mutation theory* of neoplasia. According to the mutation theory, the heritable characteristics of tumour cells result from genetic mutation in previously normal somatic cells. The earlier versions presumed changes in chromosome numbers occurring at random in dividing cells. Later versions presume changes in genetic material not necessarily entailing visible alterations in the chromosomes and not necessarily random in occurrence (Koller, 1960, 1964). Mutation theories focused attention on the dependence of tumour structure and behaviour on permanent heritable intrinsic characters of tumour cells. The most vigorously-supported alternative theory, the virus theory, emphasized the role of viruses as the "continuing cause" or "driving force" of tumours, responsible for their progressive indeterminate growth. The first substantial experimental support for this theory came from a study of sarcomas in domestic fowls as next described.

II. The Filterable Tumours of Fowls

In 1910, when the main features of tumour transplantation had been established, Peyton Rous opened up a new field of inquiry by transplanting a spontaneous sarcoma of a domestic fowl into nearly related fowls. This transplantation was unremarkable but a year later Rous (1911) found that, contrary to all previous experience with rodents, the fowl sarcoma was transmissible by cell-free extracts. Subsequently, Rous and his colleagues found four other fowl sarcomas, varied in structure and behaviour, that were similarly transmissible by cell-free extracts. The usual and most convenient method of freeing extracts from cells was to filter them through bacteria-proof filters and the tumours became widely known as "filterable tumours". Cell-free preparations that transmitted the tumours were made also by desiccation, preservation in glycerine, ultracentrifugation and other varied and ingenious methods. The first tumour studied by Rous, designated by him Chicken Tumor I, but now usually known simply as the Rous Sarcoma, was described as a spindle-celled sarcoma. The filterable tumours subsequently studied by Rous and others included other spindle-cell sarcomas, fibrosarcoma, osteochondrosarcoma and endothelioma. Fowl leukemia became recognized as a neoplastic disease and its cell-free transmission by Ellerman and Bang in 1908 gives to them priority in the discovery of filterable neoplasms but the introduction of these tumours into experimental cancer research is due to Rous.

It was questioned at first if the tumours were transmissible by extracts completely free from living cells and if they were "true tumours". Evidence soon accumulated to show that transmission in the absence of whole cells or of microscopically visible fragments of cells was a reality. Cell-free transmission is not now in doubt. It was less easy to prove that the growths were "true tumours" because nobody could say exactly what a true tumour was, but they grew invasively and metastasized, and were transplantable by cellular grafts from which, under some circumstances at least, tumours developed predominantly or exclusively by proliferation of the implanted cells. The tumours thus satisfied the criteria by which tumours were usually identified and, in addition, had the unusual property of being transmissible by cell-free extracts. Gradually the fowl sarcomas became increasingly recognized as "tumours", although perhaps tumours of an unusual kind, and controversy was re-directed to the nature of the filterable agents and the relevance of the fowl tumours to neoplasia in mammals.

Early in their investigations Rous and Murphy came to the conclusion that the properties of the filterable agents corresponded in most details with those of filterable viruses. The filterable agents, nevertheless, were remarkable because they not only evoked tumours but reproduced the particular kind of tumour from which they were derived and were recoverable in greatly increased amounts from the tumours they induced. They provided, indeed, the first example of the transmission of specific cell characters by a replicating sub-cellular agent, a phenomenon not seen again until Griffith found the transforming factor of pneumococci in 1928. Later Claude and Murphy (1933) emphasized this property of the agents and the analogy with the transforming agents and proposed to call the filterable agents *transmissible mutagens*. Most adherents of the virus theory continued to believe that the agents were living micro-organisms of extrinsic origin. The outstanding attraction claimed for this theory was that an intra-cellular virus by supplying a continuous stimulus to cell division could account for the progressive growth of tumours. The factual basis for Murphy's idea of a transmissible mutagen was insufficiently regarded, as indeed it still is, but now that the conception of a virus has so greatly changed, the two interpretations are no longer inconsistent with one another.

III. Mammalian Virus-Tumours

The first solid evidence for the participation of viruses in neoplastic disease in mammals was long delayed. In 1932, Shope found that endemic papillomas of the skin of cottontail rabbits of the Mississippi Valley were transmissible by a filterable virus. Rous and later other investigators showed that the virus evoked papillomas of the skin of domestic rabbits and that a proportion of the papillomas in domestic rabbits, as well as in cottontail rabbits, eventually became carcinomas. Remarkably, no virus was recoverable from the carcinomas. In 1936 Bittner showed that female mice of inbred strains prone to develop mammary tumours transmitted in their milk some influence or agent with the general properties of a virus that, in co-operation with other favouring circumstances, induced the development of mammary tumours in their female offspring several months later. After another long gap, Gross reported the cell-free transmission of mouse leukaemia and since then many other investigators have described viruses in leukaemias or leukaemia-like diseases in mice. Furthermore, a virus or group of viruses known as polyoma virus has been shown to initiate neoplasia of several kinds in mice, rats and hamsters. Still more recently a virus isolated from a monkey, the SV40 virus, and adenoviruses isolated from human beings have been shown to be capable of producing neoplasia in laboratory animals. The significance of these observations will be discussed elsewhere; it must suffice to notice here that the so-called "virus-tumours" are highly diverse and that the role of the virus is not always the same. The viruses themselves are also varied, some containing ribonnucleic acid and others deoxyribonucleic acid, and the mechanism of their action remains obscure. There is no evidence whatsoever for the existence of an ubiquitous "cancer virus". The "virus-tumours" of animals are being studied vigorously by many who believe that they provide a clue to the aetiology of neoplasia in man. They are likely to prove far more important, in my opinion, for giving information about intra-cellular mechanisms in neoplasia. As yet there is no factual evidence for the participation of viruses in any form of malignant neoplasia in man and the only convincing example of a virus-induced benign neoplasia is the harmless common wart, which, so far as I can learn, has never been known to give origin to carcinoma.

IV. Experimental Carcinogenesis

Bashford never put into operation his scheme for the investigation of the possible experimental production of Petroleum Cancer and Sweep's cancer in animals and the first decisive advance in this direction was made in Japan by Yamagiwa and Itchikawa who reported in 1914 that they had induced papillomas by applying coal tar repeatedly to the inner surface of rabbits' ears. A year later they reported the induction of carcinoma and in 1918 Tsutsui found that tar similarly induced papillomas and carcinomas of the skin of mice, which thenceforth, became the favourite animals for the experimental induction of tumours or, as it became known, *experimental carcinogenesis*.

The Japanese workers succeeded where others had failed because they were fortunate in their choice of an experimental animal and because they patiently continued their applications of tar for many months. Hanau, who first transplanted a rat tumour and gained little credit for it, applied tar repeatedly to the skin of rats, which are now known to be refractory to the chemical induction of skin tumours. Bayon in 1912 injected tar into the ears of rabbits but kept them under observation only for the inadequate period of one month. The experiments of Yamagiwa and Itchikawa and of later investigators showed that only rabbits and mice amongst the common laboratory animals respond well to cutaneous applications of tar and that the applications must be continued for a long time. Soon after the end of the 1914–18 war, "tar cancer" was being investigated, mainly in mice, in most cancer research institutes of the world.

The usual procedure was to apply a drop of tar, by means of a small paint brush, twice or thrice weekly to skin of the nape of the neck where a mouse can least conveniently lick or scratch itself. The hair soon fell out of the tarred area; it might regenerate once or oftener but eventually the patch became bald. Warts developed, usually after about four months, at one or more points on the tarred surface. Some warts disappeared but many of them grew progressively and became carcinomas. The carcinomas were locally invasive, they recurred after local excision, and they metastasized to regional lymph nodes and to the lungs; they were undoubtedly malignant tumours. Malignant properties were recognizable after three to eighteen months but not often in less than four months. If applications of tar were discontinued soon after warts appeared, some of the warts disappeared, promptly or after a period of growth, some persisted and grew as simple warts, and some developed into carcinomas as though tarring had been continued. Tarring of even shorter duration, discontinued before warts appeared, resulted in the growth, after considerable delay, of papillomas and carcinomas. The age of the mouse had no appreciable influence on the response to tar. The experiments discredited the prevalent assumption that cancer was a disease of senile tissues and indicated that duration of the carcinogenic stimulus was the determining factor. The belief steadily gained ground that human cancer was manifest, as a rule, in middle or later life because the inciting stimulus must operate for a long period; an induction period of about four months in mice was about one-eighth of the life span, corresponding to eight or ten years in man. The emergence of tumours long after the end of a certain minimum period of tar applications to mice was paralleled by the finding of tumours in workmen in certain industries many years after retirement. Close analogies between tar cancer in mice and occupational cancers in man implied that mice could be used to test industrial materials for carcinogenic constituents dangerous to man.

Woglom (1926), reviewing the investigations on tar cancer, wrote: "There are almost as many explanations of the way in which tar produces a malignant growth as there are investigators." The majority, it seemed, thought that the primary effect was on the epithelium though many favoured a primary alteration of the connective tissue. The views of Ribbert and his school were still influential and tumour production was often attributed to an "emancipation" of the epithelium from control by connective tissue. It became evident, however, that a fundamental change leading inevitably to cancer occurred before evidence of neoplasia was apparent. The relative importance of local and constitutional effects of tar was debated with a majority in favour of predominantly local action. The local action depended, according to some, entirely on the intensity and duration of "irritation", the nature of the irritant being unimportant; others recognized a certain specificity of action since all irritants were not carcinogenic. The solution of these and other problems was hampered by the fact that coal tar was a complex and variable mixture of ingredients, some of which were highly toxic. Chemical investigations of tar were soon in progress and some purification was attained without sacrificing carcinogenic action. Kennaway then made carcinogenic tars artificially by heating simpler substances at high temperatures (700–900°C); carcinogenic tars were made by heating isoprene or acetylene with hydrogen, thus indicating that the cancer-producing materials (or *carcinogens*) were hydrocarbons. Further progress resulted from two circumstances: organic chemists determined the chemical constitution of polycyclic aromatic hydrocarbons present in small quantities in tar, and Mayneord found that carcinogenic oils and tars had characteristic fluorescence spectra. The fluorescence spectrum seemed likely to be a serviceable guide in the search for carcinogenic substances and Hieger used it to examine a series of polycyclic aromatic hydrocarbons including one, namely 1:2-benzanthracene, which had a spectrum similar to that of the carcinogenic mixtures. This substance did not produce tumours in mice but Kennaway and Hieger (1930) extended the observations to related and somewhat more complex hydrocarbons which had recently been synthesized and found one which had the characteristic fluorescence spectrum and which proved carcinogenic when applied to the skin of mice. This substance, 1:2:5:6-dibenzanthracene, was purified and was the first pure substance of known chemical constitution to be proved carcinogenic. The search for other carcinogenic compounds was then pursued with vigour and success, notably by Kennaway, Cook, Hieger and others in London. Numerous polycyclic hydrocarbons, many of them previously unknown, were prepared and tested. Almost all those that proved carcinogenic were derivatives of 1:2-benzanthracene. Two of these compounds were of especial importance. The highly potent carcinogenic compound 3:4-benzpyrene was isolated from tar and prepared synthetically; it was probably responsible for most, though not all, of the carcinogenic activity of oils and tars. Another compound of similar high potency, methylcholanthrene, was synthesized by Fieser from cholic acid and desoxycholic acid, the two principal acids in human bile; the possible origin of carcinogenic substances from normal constituents of the body was thus indicated. Amongst numerous other compounds, 9:10-dimethyl-1:2-benzanthracene was notable as the most potent carcinogenic compound yet known.

The use of pure, defined, chemical substances introduced into carcinogenic experiments an accuracy and consistency which were impossible when highly complex and variable mixtures like tar and mineral oils were used. Elimination of the most objectionable toxic effects of tar extended the scope of experiments and permitted observation of specific carcinogenic effects less complicated by irrelevant side-effects. Some of the compounds, moreover, were exceedingly potent. It was never easily credible that tars or oils were responsible for more than a small fraction of human cancers. The discovery of methylcholanthrene as a carcinogen led to a plausible hypothesis that human cancer in general might be induced by methylcholanthrene or a similar substance produced *in situ* from normal constituents of the tissues as a result of various stimuli but more recently this has been discredited.

Non-specific "chronic-irritation" was discountenanced as the main cause of cancer; specificity of chemical action was now emphasized. An opinion gradually developed that carcinogenesis might depend on a specific chemical reaction between the carcinogen and some constituent of cells. The resulting change might still be called a "mutation" but it was understood as a permanent alteration brought about by direct chemical action rather than, as previously, a "spontaneous" change happening amongst dividing cells according to the laws of chance.

Coal tar or the carcinogenic hydrocarbons produced carcinoma when applied to the skin of mice and sarcoma when injected into the subcutaneous tissues; the hydrocarbons produced carcinoma in several glandular organs into which they were directly introduced. The effect of these carcinogens depended on the site of application and scarcely at all, except in degree, on the particular compound or the particular kind of mouse chosen for the experiment although the skin responded variously in different species of animals. Use of the pure hydrocarbons demonstrated more clearly a remote carcinogenic action which some observers had noticed in experiments with tar. When administered so as to minimize the local carcinogenic action, the hydrocarbons greatly increased the number, and accelerated the appearance, of tumours of the lung in mice. Again all the hydrocarbons produced the same kind of result but not equally in all kinds of mice. The tumours were mostly benign lung adenomas such as occur sporadically in mice. The sporadic tumours were frequent in some inbred strains of mice but rare in others; the liability of mice to develop these tumours was demonstrably a hereditary characteristic of the strain. The most conspicuous effect of remotely administered carcinogenic hydrocarbons was to increase the number and accelerate the appearance of lung adenomas in those strains of mice in which they were apt to occur spontaneously. The hydrocarbons similarly increased the incidence of leukaemia and, less decisively, of mammary cancer in strains of mice in which they occurred naturally. Experiments in which carcinogenic hydrocarbons were applied to inbred mice indicated that prolonged local action supplied, as a rule, an overwhelming stimulus to tumour formation in skin and subcutaneous tissues but that the remote effect was strongly conditioned by the hereditary make-up of the animals. Some investigators, indeed, suspected that the carcinogens did not, in the strict sense, *initiate* tumour formation but only aggravated, accelerated or *enhanced* a process that occurred, though feebly, in their absence. Aside from differences in potency the numerous chemical carcinogens produced the same result; all elicited the same kinds of tumours when similarly applied to similar mice.

One substance, 3:4:5:6-dibenzcarbazole was more specific in action for it evoked, locally, carcinoma of the skin or sarcoma of the subcutaneous tissues according to the site of application and, also, induced remote hepatomas in the liver. This compound, by virtue of its chemical structure and distinctive carcinogenic action, links the carcinogenic hydrocarbons with a group of azo-dyes whose tumour-producing properties were first noted by Japanese workers in 1931, although Fischer registered a "near-miss" in 1906 when he applied the dye Scarlet Red to the skin of mice and observed epithelial proliferations which, though difficult to distinguish microscopically from carcinoma, were impermanent. Scarlet Red achieved popularity as a constituent of ointments for promoting wound healing but not as a reagent for cancer research partly because attention was soon concentrated on the carcinogenic action of coal tar and partly as a result of the over-sensitive recoil of investigators from lesions which were not unequivocally "true" and "malignant" tumours. The Japanese workers used a derivative of Scarlet Red namely 4-orthoaminoazotoluene, administered it orally to rats for at least four and a half months and a month or two later found tumours of the liver (hepatomas) in many rats and papillomas of the bladder in a few. Many related compounds were tested and some specificity of action was disclosed; one compound, for example, induced carcinoma of the bladder but not hepatoma and another induced hepatomas, and, additionally, tumours of the stomach but none of the bladder. In contrast to experience with the carcinogenic hydrocarbons, the remote effects of azo-dyes depended substantially on the particular compound used and on the hereditary characteristics of the animals. Moreover, the results were notably dependent on the diet given to the animals. Presumably chemical change within the body preceded carcinogenic activity; the formation of a carcinogenic metabolite in the liver and its excretion in the urine might account for the predominant localization of tumours in liver and bladder.

A still wider range of carcinogenic activity was discovered accidentally in the course of toxicity tests of a substance, 2-acetylaminofluorene, under investigation for insecticidal properties. This substance had no local carcinogenic action on skin or connective tissues but when administered orally to rats or mice it elicited a wide variety of tumours, mainly epithelial, including hepatoma, papilloma and carcinoma of the bladder, carcinoma of the breast, adenoma and carcinoma of the thyroid, squamous carcinoma of the external auditory canal, tumours of the lung and of the small intestine, colon and pancreas. The localization of tumours differed according to the species and strain of the test animal, depending apparently on the inherent susceptibilities of particular tissues or on the sites of chemical transformation and routes of excretion of the compound administered. Some related compounds had a direct local carcinogenic action on skin. Similar tumour formation of wide range with differences in detail followed administration of derivatives of 4-aminostilbene. More recently tumours of varied types have been induced in large numbers by nitrosamines and aflatoxins. Much simpler chemical compounds have proved carcinogenic, although usually only in special test-systems. Urethane greatly increases and accelerates the incidence of pulmonary adenomas in mice. Moreover, in addition to this "enhancement" of neoplasia which was at first thought to be its only action, it has been shown to initiate neoplasia in the epidermis and other tissues. Carbon tetrachloride and chloroform long known as liver poisons induce hepatomas in some strains of mice, and even hydrochloric acid and distilled water have been credited with carcinogenic activity in certain limited conditions, especially in the subcutaneous tissues of rats. For a discussion of the enormous literature on chemical carcinogenesis, reference may be made to the book by Clayson (1962). It is sufficient to notice here that no clear and consistent relationship between chemical structure and carcinogenic activity has been found.

Laboratory experiments have abundantly confirmed observations on the carcinogenic action of radiations of various kinds; ultra-violet light, the radiations from the radioactive elements, the radio-isotopes and the atomic bomb are all carcinogenic in animals and in man. Metazoan parasites induce neoplasia by an unknown mechanism in animals and probably but less certainly in man. The best-studied example is the induction of sarcomas in the livers of rats by a tape-worm, Taenia crassicollis, as reported long ago by Bullock and Curtis (1920, 1924).

In several methods of inducing neoplasia experimentally, no specific carcinogenic substance has yet been identified. For example, sheets of chemically inert plastics, glass or metals induce sarcomas when implanted in the subcutaneous tissues of rats or, less consistently, of mice. The balance of opinion at present favours a physical rather than a chemical action. Another mystifying phenomenon is the "spontaneous malignant transformation" of mouse fibroblasts into sarcoma cells in the course of long-term cultivation *in vitro*. No chemical, physical, biological or viral carcinogenic agent has yet been detected (Sanford, 1965, 1967).

V. Endocrine Neoplasia

The experimental production of neoplasia by hormones deserves attention as an example of the apparent carcinogenic action of normal constituents of the body. This can hardly be classed as a special case of "chemical carcinogenesis" as ordinarily understood. Long-sustained endocrine disturbance evokes neoplasia in hormone-regulated tissues, the results being strongly conditioned by genetic factors. The tumor-inducing actions have been variously interpreted. Some investigators have inferred a direct carcinogenic action of hormones on their target tissues comparable with the action of carcinogenic hydrocarbons on the skin. Others, impressed by the chemical relationships, have thought that the steroid hormones might be converted into polycyclic hydrocarbons, which then provided the immediate carcinogenic stimulus. Neither of these hypotheses has been substantiated and the balance of evidence now indicates that the tumour-inducing effects of hormones are allied to their physiological actions. Sustained overaction, or unbalanced action, of the normal hormones, elicits first a hyperplasia of their target tissues and eventually, under appropriate conditions, neoplasia. It is still not certain that the critical transition from hyperplasia to neoplasia can be effected by hormonal stimulation alone. It is certainly strongly conditioned by genetic factors and possibly by other unidentified factors as well.

Neoplasms of the various endocrine tissues share several unusual features which led Park and Lees (1950) to propose the generic term "endocrinoma" for them. These tumours of endocrine tissues are difficult to classify by means of the usual pathological criteria. Both cytologically and histologically they resemble their parent tissues very closely, and morphological characters are not reliably correlated with clinical behaviour. Some "benign" endocrinomas are hardly distinguishable from hyperplasias and invasive metastasizing endocrinomas retain the morphological characteristics of their parent tissues to such an extent that microscopical examination gives no indication of their malignant behaviour. Malignant as well as benign tumours secrete physiologically active hormones corresponding with those produced by their parent tissues and they respond to a greater or lesser degree to other hormones that regulate the activities of those parent tissues. The prevalent idea that neoplasia, and especially malignant neoplasia, is inconsistent with specialized function is false.

The mechanism of the action of hormones on their parent tissues is poorly understood but of the "controlling forces" that are supposed to maintain the harmonious integration of the organism and that are further supposed to be ineffective in controlling neoplasia, the hormones are the most amenable to experimental analysis. Consequently, the hormonal mechanisms concerned in neoplasia of hormone-producing and hormone-regulated tissues are of much theoretical importance and specific examples will be discussed in detail later.

The multitudinous ways of inducing neoplasia in experimental animals have become something of an embarrassment. It is difficult to discern any feature common to all of them that would plausibly account for their carcinogenic action; the significant action on the target tissues of none of them is known. Some of the "occupational" or "environmental" neoplasias are reproducible in animals and a specific chemical or physical carcinogenic agent can be identified with fair confidence. It remains doubtful, nevertheless, to what extent the artificial laboratory procedures are relevant to the aetiology of the generality of neoplasias in man. An alternative experimental approach is to analyse the aetiological factors that are concerned in the "natural", "spontaneous" or "sporadic" neoplasias of animals and of man. These terms, of which "sporadic" is perhaps the least open to criticism, do not preclude the operation of a specific carcinogenic stimulus; they imply only that no such stimulus has been identified and, more particularly, that none has been deliberately introduced by the investigator. Laboratory analysis has been carried furthest with the common sporadic mammary tumours of mice, as will be next discussed.

VI. Mammary Tumours of Mice

The early laboratory investigators of cancer perforce took animal tumours as they found them and mammary tumours of mice attracted their attention most often. Their limited progress in elucidating the aetiology of these tumours was due in large measure to the immaturity of knowledge outside their own spheres. Two essential clues were in fact discovered by early workers, namely, a hereditary factor by J. A. Murray (1911) and a hormonal factor by Lathrop and Loeb (1916). Neither discovery was, or could be, satisfactorily exploited until geneticists and endocrinologists had sufficiently advanced their own inquiries.

J. A. Murray observed that mammary tumours developed more frequently in mice whose recent female ancestors had developed similar tumours than in mice whose recent ancestors were free from them. Subsequent investigations of hereditary factors were vitiated for a long time by two main faults: namely, the use of animals of mixed ancestry and the tendency of observers to group all kinds of tumours together as "cancer". The foundations for later progress were laid by Little and Strong in the United States. They began the close inbreeding of mice and developed strains which, according to genetic theory, were substantially genetically homozygous or "pure". The value of these mice for transplantation experiments and for the examination of various carcinogenic agents has been indicated in earlier paragraphs. It transpired that the incidence of each particular kind of tumour was characteristic for each inbred strain but differed greatly from one strain to another. The incidence of mammary tumours, in particular, varied amongst different strains within the widest possible limits, from 0–100% but within the same strain it maintained a characteristic level. The genetic control of tumour development was evident also in the incidence of lung adenomas, leukaemias, and other forms of neoplasia when each was studied separately. The genetic factor was obscured if all were grouped together as "cancers".

While this information about the genetic control of mammary tumours was accumulating, the study of hormonal factors advanced fitfully. Lathrop and Loeb's original observation, that ovariectomy or prevention of breeding reduced the frequency of breast tumours, pointed to the importance of ovarian secretions. The importance was shown more directly in experiments by W. S. Murray in 1928. Murray transplanted ovaries from female mice of a strain with a high natural incidence of mammary tumours into male mice of the same strain with the result that a few of the males developed mammary tumours, an extremely rare occurrence in normal male mice. This method demonstrated a principle but was not suitable for wider experimentation, which became possible only when active ovarian extracts were commercially available. Allen's introduction of the simple vaginal smear test for oestrogenic action and advances in the chemistry of steroid compounds resulted in rapid progress in the preparation of active extracts, the isolation of naturally occurring hormones of the ovary and, subsequently, the synthesis of artificial oestrogens with actions similar to those of the female sex hormones. The chemistry of the oestrogens was closely related to that of the carcinogenic hydrocarbons and it was not entirely a coincidence that the first demonstration of the carcinogenic activity of a pure hydrocarbon was made by Kennaway and Hieger at about the same time as the demonstration of the tumour-inducing action of ovarian hormones by Lacassagne (1932). Lacassagne provoked mammary tumours in male mice of a strain R III, in which most of the female but none of the normal male mice developed these tumours, by repeated injections of an ovarian extract "folliculin" starting soon after birth. Confirmation and extension of Lacassagne's observation were soon forthcoming. Broadly, oestrogenic hormones raised the incidence of mammary tumours in male mice and in virgin female mice to the level characteristic of breeding females of the same strain.

Meanwhile, the analysis of the hereditary factor was complicated by the observation that inheritance of the liability to mammary tumours in hybrids produced by crossing two inbred lines did not conform to Mendelian rules; contrary to those rules the liability was determined preponderantly by the female parent. This inheritance, it was presumed, was not determined by chromosomal hereditary units and was described as "extra-chromosomal". Eventually, Bittner (1936, 1937) traced the extrachromosomal factor to a material transmitted in the milk from female mice to their offspring. The material so transmitted was variously described as "milk factor", "milk agent", "mammary tumour agent", "Bittner's agent", "milk-borne tumour agent" or "virus". A Mendelian hereditary factor, though often overshadowed by the milk agent, was also demonstrable. Bittner suggested the hypothesis, which gained wide acceptance, that mammary tumours developed in mice as the result of the concurrent action of three factors: a genetic factor, a hormonal factor and a milk-borne tumour agent with the general properties of a virus.

Even with the most favourable combination of the three identified aetiological factors, mammary tumours do not emerge until the factors have been operating for several months. As in experimental carcinogenesis in the skin, spontaneous neoplasia in the breast is remarkable for a long period of neoplastic development before visible tumours emerge. One step in this development was recognized by Apolant (1906) and more clearly by Haaland (1911) who described what are now known as *hyperplastic nodules*. Haaland believed that the nodules were precursors of mammary tumours and wrote "Intermediate stages seem to exist in which the parenchyma shows hypertrophic changes before the development of a true tumour can be proved to have occurred." About a quarter of a century later, observations on inbred mice, notably by Gardner (1941), confirmed and greatly extended Haaland's findings, and it became evident that most mammary tumours in inbred mice probably developed from *hyperplastic nodules* or, in some kinds of mice, as later transpired, from bigger lesions called *plaques*; the emergence of carcinoma was often the result of qualitative change by progression in a nodule or plaque. It also transpired that some at least of these "intermediate stages" depended for their persistence or growth on hormonal stimulation, whereas mammary tumours were so dependent in only a minority of mice.

Two important general principles emerge from the analysis of spontaneous mammary neoplasia in mice. First, mammary neoplasia in mice results from the co-operation of several aetiological factors; there is no single or simple "cause". Second, spontaneous mammary neoplasia, like experimentally induced neoplasia of the skin, advances by progression through stages that are *qualitatively different*.

These two principles have wide general validity as will be discussed later but their application to special cases, in particular to the mammary tumours of mice, needs some further consideration. Bittner's three-factor hypothesis applies primarily to the induction of a *high incidence* of mammary tumours; the incidence aimed at would be appalling if it were approached in women. To a considerable extent a deficiency of one factor can be compensated by excess of another so as to yield an incidence of tumours that would still be intolerably "high" by human standards. The genetic factor, to a large extent at least, operates indirectly by regulating the availability of milk agent or of hormones and the capacity of mammary tissue to react to them. There is no convincing reason for identifying either the milk agent or hormonal stimulation as the primary "cause" of mammary neoplasia and the other factor as merely a secondary adjuvant. Furthermore, mammary neoplasia can be induced in mice that are genetically "resistant" to the milk agent in at least two ways: first, by administering a chemical carcinogen remotely, as, for example, intraperitoneally or orally, and, second, by whole body irradiation. In rodents other than mice these are the two most widely effective methods of inducing mammary neoplasia.

Sporadic mammary neoplasia in rats differs substantially from that in mice. The commonest sporadic mammary neoplasm in rats is benign fibroadenoma, carcinoma being rare; the reverse is true of mice. No milk agent has been demonstrated in rats. Both fibroadenoma and carcinoma have been induced in rats by prolonged, intensive administration of oestrogenic hormones but by far the most effective procedure, as shown first by Shay and his colleagues, is to administer a carcinogenic polycyclic hydrocarbon by stomach tube. As standardized by Huggins, this procedure can induce carcinoma in virtually 100% of female rats of certain strains. Other species of rodents have been much less thoroughly investigated but, as far as present experience goes, they are much less co-operative. Only one mammary carcinoma seems to have been experimentally induced in guinea pigs and that by local innoculation of radioactive thorotrast (Foulds, 1939).

The experiments that have been reviewed in this chapter have produced much illuminating information about mammary neoplasia which will be mentioned often in later sections of this book but their contribution to elucidating the aetiology of mammary neoplasia in women has been negligible.

Bibliographical Note

Those interested in the early history of cancer research, with special reference to transplantable tumours, are commended to the Second, Third and Fourth Scientific Reports of the Imperial Cancer Research Fund, London, published in 1905, 1908 and 1911 and to the admirable reviews by Woglom (1913, 1929). Woglom (1926) also wrote a valuable account of "Tar Cancer".

Kennaway (1955) told the story of the pioneer work on the carcinogenicity of pure polycyclic hydrocarbons, Foulds (1934c) reviewed the early work initiated by Peyton Rous on the filterable chicken tumours and Gross (1961) gave an account of mammalian tumour viruses. An "Index to the Literature of Experimental Cancer Research, 1900–1935" was published by the Donner Foundation of Philadelphia in 1948.

Nomenclature of Carcinogenic Polycyclic Hydrocarbons

The chemical nomenclature of the most widely-used carcinogenic polycyclic hydrocarbons has changed since their actions were first described. The corresponding old and new names and the generally-accepted abbreviations are as follows:—

1:2:5:6-dibenzanthracene	—dibenz(a,h)anthracene (DBA)
3:4-benzpyrene	—benz(a)pyrene (BP)
20-methylcholanthrene	—3-methylcholanthrene (MC)
9:10-dimethyl—1:2-benzanthracene	—7:12-dimethylbenz(a)anthracene (DMBA)

Part II

General and Experimental Pathology of Neoplastic Development

CHAPTER 2

Aetiological Factors

It is not a primary aim in this volume to analyse in detail the aetiology of neoplastic diseases in animals or in man, but, in a comprehensive study of neoplastic development, the crucial initial steps and, by implication, the initiating circumstances, cannot be ignored. The numerous and varied methods of evoking neoplasia experimentally in animals are of great value for studying certain aspects of neoplastic development, but their relevance to sporadic neoplasia in man cannot be taken for granted. It is desirable to discuss first the available experimental procedures. Most of them act in one of three ways: by *Re-production*, by *Induction* or by *Enhancement* of neoplasia. These will be considered in turn with the reservation that it is far from sure that induction and enhancement are basically different processes.

I. Re-production

This term applies to the accurate transmission of all the properties of the filterable sarcomas and leukaemias of chickens by their proper viruses, whose nucleic acids are of the RNA type. No strictly comparable phenomenon has yet been demonstrated in mammals and there is no indication of one in man. The phenomenon is mentioned here because it gives exceptional opportunities for investigating and interpreting certain problems of neoplastic development to be dealt with in Volume 2.

II. Induction

Induction implies the production of neoplasia that would not have occurred in the absence of an inducing agent. Experimental carcinogenesis has been concerned mainly with induction by chemical, physical, and viral agents. These may act directly and locally at the point of application or remotely on a distant target tissue. Their effectiveness, especially when acting remotely, depends on many factors other than the "carcinogenicity" of the inducing agent. One group of factors govern the delivery of the carcinogenic stimulus to the target tissue in appropriate intensity. As a simple example, the effectiveness of a chemical carcinogen applied to the skin depends on its absorption which may be governed decisively by the solvent in which it is administered. Another group of factors govern the metabolic fate of chemical carcinogens; a chemical agent may not provide an effective carcinogenic stimulus in the form in which it enters the body, but only after it has been chemically altered. Both groups of factors operate in carcinogenesis by aromatic amines such as β-naphthylamine, which are not carcinogenic at their point of entry but act on a remote target tissue. The high incidence of neoplasia in the urinary bladder amongst certain dye workers is attributed to the remote action of β-naphthylamine. A suggested chain of events is as follows: the amines are absorbed through the skin, digestive tract or lungs, but do not induce neoplasia in these situations. After absorption they are carried to the liver and there detoxified by oxidation to orthoamidophenols or arylhydroxylamines. These latter substances are believed to be the *effective carcinogens*, but in the liver they are immediately rendered again non-carcinogenic by conjugation with glucuronic acid, and in this inactive form they are carried to the kidneys and excreted in the urine. Finally the glucuronidide is hydrolysed by β-glucuronidase present in the urine to liberate the effective carcinogen, which induces neoplasia of the vesical mucosa.

Other aromatic amines and azo compounds, like β-naphthylamine, induce neoplasia selectively in one or more particular tissues remote from their site of entry into the body, the preferred sites differing substantially from one species of animal to another and from one strain to another within a single species. Genetically-controlled differences in biochemical processes might contribute in several ways to the species- and strain-specific localizations of neoplasia. First, diverse mechanisms of detoxication in the liver are possible; some yield carcinogenic metabolites and some do not and the metabolites may be variously conjugated. Second, if a carcinogenic metabolite is not promptly conjugated, neoplasia of the liver is to be expected; if the metabolite is conjugated quickly some may be excreted in the bile and after hydrolysis become available for carcinogenic action in the intestine, whereas the remainder, dispersed by the blood stream, is excreted in the urine and possibly by sebaceous glands and other glands as well. Third, the liberation of the effective carcinogen from the conjugated form in contact with a reactive tissue will depend on the local availability of the appropriate enzyme. There is evidence to the effect that some carcinogenic responses are regulated by genetic factors operating within the reacting tissues themselves. Furthermore, within the limits imposed by genetic factors, the biochemical processes are subject to enhancement or repression by a variety of non-specific factors such as the pH of the urine and the time it remains in the bladder. A variety of genetic and physiological factors must operate in sequence before a carcinogen can *begin* to act on a remote tissue; the factors are not carcinogenic agents but they are important or even decisive aetiological factors.

The essence of the foregoing discussion is that induction of neoplasia depends on a variety of factors that contribute to *effective exposure* of a target tissue to a carcinogenic stimulus. *Effective exposure* implies that a carcinogenic stimulus is delivered to a target tissue in an active form and in sufficient amounts and remains there for a sufficient time. Effective exposure depends, as already recounted, on a variety of factors and circumstances which need *ad hoc* investigation in all special cases of experimental carcinogenesis. Effective exposure, by itself, does not guarantee a carcinogenic effect; the sensitivity of the target tissue to the carcinogenic stimulus is decisive. The carcinogenic effect depends, in fact, on the relationship between the potency of the carcinogenic stimulus and the sensitivity of the target tissue.

It should be recognized that "carcinogenicity" cannot be defined or measured in absolute terms; it is a function of the test system in at least as great a degree as it is a function of the agents being tested. Statements to the effect that such and such an agent is "carcinogenic" or is "non-carcinogenic" have no value or validity without clear specification of the system that has been used to test "carcinogenicity". Susbtances that are plainly "carcinogenic" on mouse skin may be as plainly "non-carcinogenic" on rat or guinea pig skin. By fortunate chance, mouse and rabbit skins seem to respond in much the same way as human skin to topical applications of polycyclic hydrocarbons and probably, also, to some other potentially carcinogenic chemical or physical agents, but there is no certainty that the correspondence is exact or that it applies to all kinds of stimuli. The correspondence of responses certainly does *not* apply to all other tissues of mouse and man. Current usages of the terms "carcinogen" and "carcinogenicity" need much more critical appraisal than they now receive and so also does the wide-spread presumption that neoplasia is necessarily initiated by a specific carcinogenic stimulus whether chemical, physical or viral. The presumption is a reasonable working hypothesis but it is not established truth.

It is especially noteworthy that neoplasia in man most closely resembles neoplasia induced by potent carcinogens in animals under either of two circumstances: first, when by reason of occupation or environment, certain groups of people are exposed to known or suspected carcinogenic stimuli to a far higher degree than the general population and, second, when individuals within the general population are conspicuously more sensitive than their fellows to the action of particular kinds of potentially carcinogenic stimuli. Under either of these circumstances, the incidence of neoplasia is increased above that in the general population and a distinctive pattern of neoplastic development, resembling that seen in experimental animals, is often conspicuous. The main features and principles are instructively and conveniently illustrated by a consideration of epidermal neoplasia.

Chimney Sweep's Cancer, is the prototype of occupational neoplasias of the skin. Many other examples of chemically induced epidermoid neoplasias of the skin have since been recognized especially in men handling tars, pitches and mineral oils. The clinical manifestations of epidermal neoplasia in workers in a Scottish shale-oil plant were notably well described by Scott (1923). An unduly high incidence of epidermal neoplasia has long been recognized in out-door workers, notably farmers, sailors and fishermen and is now attributed to excessive exposure to solar radiation. A similar high liability to epidermal neoplasia is apparent in white races habitually exposed to tropical or subtropical sunlight, as are the white immigrants in Australia. Amongst these immigrants, sensitivity is graded, blondes being more liable to epidermal neoplasia than brunettes and pale-skinned redheads being the most sensitive of all. It is well known that coloured races have a very low liability to epidermal neoplasia. These variations in sensitivity are within "normal" or physiological limits; the liability to epidermal neoplasia is low amongst races living in the habitat to which they are adapted and increases when exposure to sunlight increases substantially above the ordinary level. Exceptionally, the innate sensitivity may reach a pathological degree as in the childhood disease xeroderma pigmentosum, the basic fault in which is a hypersensitivity of the whole of the skin surface to ultraviolet irradiation. The disease is heritable and is transmitted as a recessive Mendelian character. Epidermoid carcinoma develops almost inevitably in every patient sometimes during childhood and often during adolescence. Exposure to weak daylight is a sufficient carcinogenic stimulus.

The severe manifestations of epidermal neoplasia under the exceptional conditions now under consideration are remarkably similar despite wide differences in the presumed inciting agents. They are all characterized by a stage in which various kinds of neoplastic and non-neoplastic lesions are distributed over a region coextensive with the area of exposure to the inciting agent (Foulds, 1965). Within this *region* of diseased skin, epidermoid carcinoma emerges at a later time at one or a few points but not confluently over the whole extent of the region. The pathological nature of the early lesions and the course of neoplastic development will be dealt with in Chapter 3. It will be sufficient to note here that a characteristic state of altered skin has been described as Farmer's Skin, Sailor's or Seaman's Skin and Fisherman's Skin. These are all alike and they strongly resemble what used to be called Chronic Tar Skin. Even more remarkably they also resemble the skin of xeroderma pigmentosum to such a degree that Seaman's Skin has been described as "xeroderma pigmentosum of the adult" (Halberstaedter, 1923).

Other examples of industrial or environmental neoplasia, such as "aniline cancer" of dye workers, will be discussed elsewhere, but two extreme examples deserve mention here. The first is the extremely high incidence of neoplasia of the lung amongst workers in the uranium mines in Schneeberg and Joachimsthal. The second is the extremely high incidence of leukaemia in survivors of the atomic bomb explosion at Hiroshima. The effective carcinogenic agent in each is a radio-active one. To the best of my knowledge, no other group of human beings has been afflicted with nearly so high an incidence of neoplasia of any type. Clearly, these people were subjected to an overwhelming *effective exposure* to radiation which minimized individual differences in sensitivity and elicited an incidence of neoplasia which otherwise is obtainable only by optimal carcinogenic stimuli in experimental animals. The incidence of most types of neoplasia in man is "low" by laboratory standards.

Xeroderma pigmentosum presents an extreme contrast to the Schneeberg lung cancers and the Hiroshima leukaemias in that it develops in a small minority of people of extreme sensitivity subjected to minimal carcinogenic stimuli. Fortunately, it is a rare disease and a special case, but in my view it would be unwise to underrate its general significance. Some, including myself, will hesitate to accept weak daylight as "carcinogenic"; the decisive factor is in the responding tissue. The disease illustrates in an extreme form the importance of genetically determined reactivity which, in a less conspicuous degree, may be important in many other neoplastic diseases and which must be taken into account in the next section dealing with *enhancement*.

III. Enhancement

This phenomenon was described in pulmonary neoplasia, mammary neoplasia and leukaemia in inbred mice almost simultaneously in 1939 by Brues and Marble, Engelbreth-Holm and Mider and Morton. The early observations were well reviewed by Lefévre (1945) who, following the lead of the original investigators, applied the term *acceleration* to the phenomenon. The essential result of the experiments was to demonstrate the ability of remotely-administered carcinogens to increase the number and hasten the emergence of neoplastic growths of kinds that developed spontaneously in substantial numbers in particular inbred strains or, as it was said, growths to which the animals had an inherited "susceptibility". Three effects are indeed distinguishable: (1) an increase in the number of animals that develop tumours of a particular kind, (2) an increase in the number of tumours per animal and (3) a lowering of the age at which tumours emerge. The three effects are to some extent dissociable; it needs, for example, a substantially stronger carcinogenic stimulus to reduce the tumour age than to increase the tumour incidence. The term "acceleration" is therefore inappropriate. The enhanced tumours, are in the general opinion, identical in histogenesis, histology and clinical behaviour with those that develop spontaneously in animals of the same kind.

Enhancement has been studied most thoroughly in relation to pulmonary neoplasia in inbred mice. Ninety-five per cent of the spontaneous lung tumours belong to one general type as first well described by Tyzzer (1909a). The tumours are often referred to as *lung adenomas* but neither this term nor any suggested alternative is entirely satisfactory, and they will be referred to here as Tyzzer tumours. The incidences of Tyzzer tumours in different inbred strains are highly varied but the incidence in a particular strain seems to be remarkably constant in different laboratories, implying that it depends mainly on genetic constitution and is little modified by environmental factors (Shimkin, 1955a). Varied carcinogens enhance pulmonary neoplasia in suitable inbred mice when administered remotely in a variety of ways. The degree of enhancement seems to depend mainly on the amount of carcinogen that reaches the lung and stays there and this amount depends on several technical details of administration that affect the physical state of the carcinogen and its distribution in the body. All the carcinogenic polycyclic hydro-carbons are effective but in relative degrees that do not correspond with the relative potencies inferred from their actions on skin and subcutaneous tissues. Other types of chemical carcinogens have been less thoroughly tested; some, including nitrogen mustards and azo compounds, enhance pulmonary neoplasia but less effectively than the polycyclic hydrocarbons. Nettleship and Henshaw (1943) made an important addition to the list of effective agents by showing that urethane (ethyl carbamate), a simple chemical substance not then suspected of carcinogenic activity, enhanced pulmonary neoplasia as effectively as the carcinogenic hydrocarbons. Irrespective of the nature of the enhancing agent, almost all the tumours evoked by remote administration are Tyzzer tumours indistinguishable from those that develop spontaneously.

Lynch (1927) first showed in experiments with tar that the genetic constitution of the test animal decisively influences the outcome of the enhancing procedure. The magnitudes of the responses of various inbred strains to remotely administered carcinogens stand in the same order as the liabilities of those strains to develop pulmonary tumours spontaneously and seem to depend on similar multiple genetic factors. Nevertheless the genetic factors controlling the number of tumours evoked by enhancement may not be identical with those controlling the incidences of spontaneous tumours in the same strains.

Ingenious experiments have shown that the genetic factors governing "susceptibility" to enhancement of pulmonary neoplasia operate, for the most part, within the lung tissue itself. The experiments exploit the fact that tissues from each of two inbred strains X and Y survive transplantation into first generation hybrids (XYF$_1$) between the two strains. Lung tissue from a "susceptible" strain, such as strain A and lung tissue from a "resistant" strain such as C57Bl, both survive in AxC57Bl F$_1$ mice. If a carcinogen is administered to a hybrid mouse bearing grafted A lung tissue in one flank and C57Bl lung tissue in the contralateral flank, many tumours develop in the A tissue and few in the C57Bl tissue. The two tissues thus retain their differential reaction to carcinogen when transferred to a common environment in a new host (Heston and Dunn, 1951; Shapiro and Kirschbaum, 1951).

Most investigators have inferred that remotely administered carcinogens do not, strictly speaking, *induce* neoplasia but merely accelerate, accentuate or *enhance* a neoplastic process that is able to advance "spontaneously" in genetically susceptible animals. They maintain that in spontaneously developing neoplasia or in neoplasia evoked by remotely administered carcinogens, the end result, that is the Tyzzer tumour, is the same and the genetic control is the same or almost the same. Cowen (1950) held that the administered agents "hasten a process which tends to occur spontaneously and is genetically determined". According to Stewart (1958), "It is generally admitted that in strain A mice the pulmonary carcinogens act directly upon pulmonary tissue in which there resides a potential neoplastic process, as evidenced by the spontaneous incidence of these tumours. It is believed that the carcinogens act by accelerating the neoplastic tendency." Shimkin (1955a), again, wrote— "The susceptibility of various homozygous lines of mice is parallel to the spontaneous development of this tumor. This suggests that the carcinogens, whether they be polycyclic hydrocarbons or urethane are accelerators of some process that is inherent in the animals."

Engelbreth-Holm and his colleagues argued, in a contrary sense, that there is no real difference between enhancement and induction. The gist of their argument, as presented by Lefévre (1945) is that the development of tumours is based on two factors; namely, a hereditary predisposition and an extrinsic carcinogenic agent. They maintained that the bulk of experimental evidence shows that it is impossible to induce tumours other than those to which the animal has an inherited predisposition. Tumours resulting from either direct or remote action must, alike, be considered to be "accelerated" and to result from the coordinated action of hereditary predisposition and an extrinsic carcinogenic agent.

Neither of the interpretations is satisfying. "Acceleration" is not an apt description of the observed effects since the "weaker" carcinogens increase the tumour incidence and the number of tumours per animal without accelerating their emergence at all. The demonstration that urethane has initiating carcinogenic action on, for example, mouse skin favours the Engelbreth-Holm hypothesis. Enhancement by urethane is accomplished remarkably quickly; its primary effect seems to act within 24 hr after administration to young mice. As will be described in Chapter 6, this rapid action is characteristic of initiating carcinogenic action but certainly not of promoting action. In a discussion of the rapidity of action, Smith and Rous (1948) mentioned two possible mechanisms; first urethane might exert an abrupt trigger action or, second, it might stimulate cells already neoplastic to proliferate. On the basis of his own experimental analysis of enhancement by urethane, Rogers (1951) unequivocally rejected the second alternative and accepted the first but these two mechanisms are unlikely to be the only possible ones. Smith (1952) thought that urethane did not act through the same mechanism as the conventional carcinogens. It is not clear that the difference is a fundamental one. Urethane is now recognized to be a complete carcinogen in a variety of tissues (Tannenbaum, 1962, 1964).

In the mouse lung, as in other tissues, several paths of neoplastic development leading to different kinds of tumours are available but there is one preferential or most favoured pathway; namely, that leading to Tyzzer tumours, that greatly preponderates over all others. Nearly all sporadic and enhanced neoplasias follow this pathway and locally-induced neoplasia often does so too. This preferential pathway is a genetically determined *species* characteristic shared by other rodents so far as they have been examined. The genetic factor operates within the pulmonary cells. The *frequency* of pulmonary neoplasia is a *strain* characteristic also operating within pulmonary cells. Enhancement, so to speak, works *with* the genetically determined preference to evoke Tyzzer tumours in numbers greater than usual but still related to the incidence of sporadic tumours. Induction, by contrast, sometimes, although not always, works *against* the genetically determined preferences to evoke, for example, squamous carcinomas in numbers unrelated to the frequency of sporadic Tyzzer tumours. According to the evidence now available, it seems that both induction and enhancement *initiate* neoplastic development.

It would be unprofitable to try to carry this discussion any further in previously used terms such as "hereditary predisposition" or "susceptibility", "neoplastic tendencies" or "potential neoplastic processes". When they were introduced, these terms reflected serious and able attempts to interpret important empirical observations; at that time no better terms or concepts were available. The possibility of giving substance and precision to the ideas implied by them will be examined in Volume 2 of this book. Meanwhile, it must suffice to propose that the fundamental biological problem of enhancement is that of the genetically controlled responsiveness of the target tissue to enhancing stimuli.

IV. Sporadic Neoplasia

It must be evident by now that no single "cause of cancer" has been found or is likely to be. Instead a multitude of chemical, physical and viral tumour-inducing agents have been discovered, together with many other aetiological factors that influence the effective exposure and effective action of those agents. The occupational and environmental neoplasias provide satisfactory evidence that similar agents are effective carcinogenic stimuli in man but the presumption made by many laboratory investigators that the generality of neoplastic diseases in man are "induced" by unidentified carcinogenic agents of the same general types, although a reasonable working hypothesis, has no solid basis of observation or fact. It seems likely that the known carcinogenic agents concerned in the aetiology of neoplasia in man often operate by *enhancement* rather than by *induction* (Shubik, 1961). Of the two main interpretations of *enhancement*, one explicitly assumes that the neoplastic process can be initiated by intrinsic cell factors without the help of an extrinsic carcinogen and the other, although presuming the need for an extrinsic carcinogenic agent, lays great emphasis on the essential role of intrinsic "predisposition". The two interpretations, indeed, are not so far apart as they seem to be at first sight; they become almost indistinguishable if it be supposed, not unreasonably, that sometimes the "predisposition" may be so great that neoplasia develops "spontaneously", without an extrinsic stimulus or, at least without an extrinsic *specific* carcinogenic stimulus. It is appropriate to recall here that the incidence of most forms of sporadic neoplasia in man is "low" by laboratory standards and often so low that it would be considered negligible in laboratory practice. It is not unreasonable to call these neoplasias "spontaneous" and to ascribe the substantially increased incidence in special groups of people to *enhancement* of the spontaneous disease by exposure to specific carcinogenic stimuli of extrinsic origin. It might be more profitable to investigate the "predisposition" than to seek to identify the extrinsic stimulus. In the best studied example of enhancement in laboratory animals, the enhancement of the Tyzzer type of lung tumour in mice, the "predisposition" can be traced with some confidence to the operation of genetic factors within pulmonary cells.

Somewhat comparable phenomena in normal embryonic development will be discussed in detail in Part III. Hitherto, the study of embryonic induction has been focused mainly but without much success on the identification of inducing substances. It is becoming more widely recognized that the key to the problem is the *competence* of the reacting tissue, not the identity of the inducing stimulus. The competence may increase to such a peak that non-specific trivial stimuli of many kinds can trigger the inductive action. The induction may take place in the absence of any identifiable stimulus as in "autoneuralization" of embryonic ectoderm. Grobstein has attributed this phenomenon, and other similar ones, to *developmental imminence* in the competent tissue. The concept of developmental imminence may well be highly relevant to the "predisposition" of cells to neoplasia and to its enhancement.

Brief reference has been made already to at least three examples of experimentally evoked neoplasia in which the operation of a specific carcinogenic agent or carcinogen has not been demonstrated. They are first, the "spontaneous malignant transformation" of normal connective tissue cells during cultivation *in vitro*, second, the neoplastic response of rat and mouse subcutaneous tissues to sheets of such relatively inert materials as plastics, glass and metals and third, the induction of tumours by hormones. In human pathology a "predisposition" to neoplasia of certain chronic ulcers and burned and irradiated tissue has been recognized for a long time and, sometimes, has been ascribed to "frustrated repair" entailing prolonged and repeated proliferation of cells in an abnormal environment. The common feature in all these examples is prolonged disturbance of the physiological behaviour of cells in a recognizably abnormal environment, in which specific carcinogens or carcinogenic agents have not been identified.

Nicholson described neoplasia as "continued development", implying that it was no more than an extension of a normal process. The view advanced here is that neoplasia results from deflection of normal development into an abnormal pathway; it is not presumed that the deflection is always or necessarily effected by specific carcinogenic stimuli. The frequency of "latent" neoplasia in old age provides some justification for saying that neoplasia is, in a very limited sense, a "normal" event. The incidence of latent carcinoma of the prostate in old men is remarkably high (Franks, 1954*a, b*; Rich, 1935). A sufficiently thorough and systematic study of all tissues and organs of old people for latent neoplasms is hardly practicable but from the limited evidence now available, it is reasonable to suspect that nearly all human beings who reach advanced old age harbour somewhere in their bodies at least one focus of latent carcinoma, which does not kill or even inconvenience them. The "cause" of the deflection of cells into the neoplastic pathway may be no more than the decreasing powers of adaptation that characterize senescence and the consequent inability of some cells to react physiologically and successfully to the "normal" stresses of daily life.

CHAPTER 3

Neoplastic Development

I. Introduction

Until fairly recently it has been widely presumed that a tumour comes into being with all its properties and characters established in their definitive states; the subsequent behaviour of the tumour was supposedly explicable wholly by the multiplication and dissemination of tumour cells, the cells themselves being qualitatively unchanged and unchangeable throughout the tumour's history. This is not true of neoplasia in general. In man, "cancer" is a chronic disease extending back usually for many years before clinical signs are apparent; the clinically-recognizable lesions represent only a fraction of the life-history of the disease. In clinical medicine, lesions in some sense "precancerous" have been recognized for a long time but their relationship to the later-developing cancers has been, and still is, indecisively and sometimes acrimoniously debated. Pathologists, for the most part, have devoted their interest and ingenuity to identifying the tissue of origin of tumours, that is, to their *histogenesis*. Haaland (1911) was one of the first experimental pathologists to foreshadow a developmental pathology of neoplasia by recognizing successive stages of development in mammary and pulmonary tumours and leukaemia of mice. He inferred that "a kind of intermediate stage seems to exist in which the parenchyma shows hypertrophic or hyperplastic changes before the development of a real tumour can be proved to have commenced." Little further attention was given to this aspect of neoplasia until the general idea of discontinuous, stepwise development appeared in a paper by Rous and Beard (1935) on "The progression to carcinoma of virus-induced papillomas (Shope)" and in a description by H. S. N. Greene (1940) of "progressive steps in a graded evolutionary process" in spontaneous mammary neoplasia in rabbits. Later, the concept was discussed and generalized by Foulds (1949c, 1954) under the name, borrowed from Rous, of *tumour progression*.

II. Epidermal Carcinogenesis in Rabbits

A general feature of experimental carcinogenesis is a long induction period preceding the emergence of visible tumours and the earliest of these are atypical neoplasms. Peyton Rous and his colleagues working with tarred rabbits found that early emerging warts regressed when applications of tar were discontinued: the skin regained its normal appearance but did not revert to its original state because renewed tarring quickly elicited warts that had the same location and the same specific characteristics as those that had regressed. Moreover, similar true recurrences were evoked by non-specific irritants, such as turpentine, that by themselves had no carcinogenic action on normal rabbit skin. Wound-healing also was sufficient to elicit recurrence of regressed warts. Rous and his colleagues distinguished two processes in carcinogenesis: first, a process of *initiation* whereby normal skin is converted into a *subthreshold neoplastic state*, characterized by the presence of *latent tumour cells* and second, a process of *promotion* whereby the latent tumour cells are stimulated to proliferate and form visible tumours. Initiation and promotion were held to be distinct, different, consecutive processes.

The carcinomas reported by Itchikawa and Yamagiwa on tarred rabbits were probably not true carcinomas but carcinomatoids, as Rous called them, or, in the more recent terminology, keratoacanthomas. These tumours closely resemble epidermoid carcinomas histologically but they regress spontaneously. According to Rous, true epidermoid carcinoma develops late in tarred rabbits and it rarely develops suddenly but, usually, as the consequence of secondary step-like changes in the cells of benign warts. What happens is "no mere exaggeration of the previous papillomatosis but a wholly new event, the genesis of a neoplasm distinct from its predecessor. The changes do not always stop when a carcinoma is established; as a result of further successive stages the growth may go from bad to worse" (Rous and Kidd, 1941).

The early visible lesions do not reveal the full neoplastic capacity of regions of epidermis exposed to carcinogenic agents. In particular, multiple tumours may emerge consecutively or concurrently from the exposed epidermis long after withdrawal of initiating and promoting stimuli and may continue to do so throughout remaining life. To account for the consecutive emergence of tumours for up to two years after withholding carcinogens from rabbit skin, Friedewald and Rous (1950) proposed that, as well as initiating *latent tumour cells*, the carcinogen initiates *latent neoplastic potentialities* in cells which then advance at varied, often extremely slow rates towards the neoplastic state.

III. Epidermal Carcinogenesis in Mice

In mice, warts develop a little later than in rabbits but, with sustained applications of carcinogen, carcinoma develops comparatively early in almost every mouse. As the carcinogenic stimulus is reduced, the eruption of warts is delayed to the point that, after a single application of a chemical carcinogen, either no tumours at all emerge during a long period of observation or only a few after a long delay. After such a sub-optimal or sub-carcinogenic stimulus, certain agents, of which croton oil is the most effective and most widely used, greatly accelerate the emergence of warts, most of which eventually regress. Berenblum and Shubik carried out experiments of this kind and like Rous interpreted their results in terms of *initiation* and *promotion*. They inferred that the carcinogen *initiated* latent or dormant tumour cells and that these cells proliferated only in response to *promotion* by certain non-carcinogenic or very weakly carcinogenic agents such as croton oil. Like Rous, they maintained that initiation and promotion were different and distinct *consecutive* processes. These experiments supplied important new information about initiation. They showed that *initiation* was effected *quickly*, during the limited period during which a single dose of carcinogen can be presumed to act. They showed further that, as demonstrated by the reaction to subsequent applications of croton oil, *initiation* once effected persisted at least for 43 weeks, and probably permanently (Berenblum and Shubik, 1947*a*, *b*, 1949*a*, *b*).

Croton oil, used in the above experiments as a promoting agent, has been described as an *incomplete carcinogen*, one which accomplishes one step in the carcinogenic process but not both steps. Subsequently, it was shown to be weakly carcinogenic and it is now questionable if any promoting agent is completely free from initiating action. Nevertheless, the promoting action is so greatly preponderant that the validity of the inferences of Berenblum and Shubik is not seriously compromised. The simple volatile substance urethane, already mentioned as an enhancing agent, was shown to be an incomplete carcinogen in the opposite way; it initiated epidermal neoplasia in mouse skin but had no promoting action. Experiments with urethane, whose elimination from the animal must be rapid, reinforce the earlier inferences about rapid, almost instantaneous initiation. They also clarify the interpretation of histological changes that accompany or follow immediately upon initiation by carcinogenic polycyclic hydrocarbons or tar. Urethane even after prolonged application to the skin produces no visible change although applications of croton oil show that initiation has taken place. Oral administration also initiates epidermal neoplasia. Most remarkably, a single intraperitoneal injection of urethane effects *initiation* of all or most of the epidermal surface and the initiation is apparently permanent and irreversible, but it is not accompanied by any histological or cytological changes within the range of recognition by technical methods now available. It seems clear that the hyperplasia that ordinarily follows the application of chemical carcinogens to the skin and for which specific qualities have been claimed and denied, is irrelevant to initiation; it may be related to *promotion*, or as Berenblum believes, it may be a side-effect inessential to the carcinogenic action.

The evidence for a crucially important first step of *initiation* is strong and convincing, but the presumption that it affects only scattered cells in the exposed area of epidermis is questionable. The nature, significance and generality of *promotion* are obscure; the phenomena that Rous describes under this term seem to me to differ substantially from those to which Berenblum and Shubik apply it and both Berenblum and Shubik have deprecated its indiscriminate application to carcinogenesis in other tissues.

IV. Inferences from Epidermal Carcinogenesis in Rabbits and Mice

Several important concepts emerge from the observations summarized above and are consistent with other evidence which will receive detailed consideration elsewhere. They apply especially to three important phenomena, namely, *Initiation, Conditional Neoplasia*, and *Progression*.

A. INITIATION

Three characteristics of *initiation* have been reliably demonstrated: first, initiation takes place rapidly, perhaps almost instantaneously; second, initiation is long-lasting and, probably, irreversible and life-long; third, initiation is not necessarily accompanied by any distinctive histological or cytological changes or, in particular, by any proliferation of cells; it is revealed only by future behaviour or by experimental tests.

It is not obligatory to presume that the changes effected by initiation are restricted to those places where tumours emerge at a later time. Although, according to the best available evidence, the changes are irreversible and replicable they are not necessarily final; repetition of the carcinogenic stimulus can bring about further changes. The latent or dormant tumour cell hypothesis has many shortcomings. As an alternative and more elastic hypothesis, it has been proposed that initiation establishes an expanse or region of *incipient neoplasia* (meaning, literally, *the beginnings of neoplasia*) that is coextensive with the area of exposure to carcinogenic action and that has a permanent replicable new *capacity* for neoplastic development, which can be augmented by repetition of the carcinogenic stimulus (Foulds, 1960, 1961).

The crucial change effected by initiation is in the *reactivity* or *capacity* of the exposed cells and not in any of their recognizable extant characters. The term *capacity* is used here in much the same sense as Grobstein (1959) used it in relation to normal embryonic development to refer to "the range and character of the demonstrable and immediate developmental alternatives." The concept is similar to that of *competence* but more generalized. Formally, at least, *initiation* is comparable with *determination* in embryonic development; it establishes a "developmental bias without overt signs", which is the essence of the first step in differentiation as described by Grobstein, and to which the term determination is commonly, if controversially, applied.

Incipient neoplasia is recognizable only by developmental criteria and not by extant characters. It provides the basis for the phenomena attributed by Friedewald and Rous to *latent neoplastic potentialities* and for several other empirically observed features of neoplastic development that are not satisfactorily accommodated by the latent or dormant tumour cell hypothesis. In particular, the concept of incipient neoplasia extending over an expanse or *region* of exposed tissue is usefully applicable to a variety of neoplastic diseases in man and in animals, as will be discussed later.

B. Conditional Tumours

Rous emphasized another important concept in his description of *conditional tumours* that grow only in response to extrinsic stimulation, regress when the appropriate stimulus is withheld and recur "at the same places and with the same specific characteristics" when the stimulus is restored. The recurrences, as Rous insisted, are true recurrences of the regressed tumours and not new primary growths; they must stem from some persistent altered state of the epidermis despite its normal appearance. It is proposed to call this persistently altered state, *residual neoplasia*, charactertized by an abnormal reactivity or capacity but not necessarily by abnormal histological or cytological characters. The concept of *residual neoplasia* is closely related to that of *incipient neoplasia*; both states are recognized by developmental criteria and not by extant characters.

C. Progression

The concept of *progression* is one of stepwise neoplastic development through qualitatively different stages. Progression was defined as the development of a tumour by way of permanent irreversible qualitative change in one or more of the characters of its cells. (Foulds, 1949c, 1954). The limitations of the definition will be considered later. The point to be emphasized here is that progression is not the mere extension of a lesion in space and time but a revolutionary change in a portion of the old lesion establishing a new tumour having properties not formerly manifest (Foulds, 1965).

Conditional tumours and tumour progression have been dealt with somewhat summarily here because they will receive much attention in the next section devoted to mammary neoplastic development. The outstanding contribution of epidermal carcinogenesis to the study of neoplastic development is its demonstration of *initiation* as a crucial first step about which mammary neoplasia has made no significant contribution.

V. Mammary Neoplasia in Mice

Mammary tumours that developed spontaneously in certain hybrid mice (C57Bl ♀ x RIII ♂) and their inbred descendants, hereafter designated as BR mice, provided unusually favourable opportunities for studying conditional tumours and their progression. These tumours first attracted attention because they developed in a minority of the F_1 hybrids in apparent contravention of Bittner's three-factor hypothesis. The milk that the hybrids received from their mothers of the "low-incidence" C57Bl strain was supposedly free from the mammary tumour agent, and transmission to the offspring of the milk agent known to be present in the fathers of the "high-incidence" RIII strain was not at that time envisaged. It turned out that the tumours did, in fact, contain the tumour agent. They were concentrated in a few families of mice, other families being completely free and were absent from hybrids sired by males from an agent-free strain of RIII. The inference, consonant with the three-factor hypothesis, was that RIII males erratically transmitted the tumour agent to some of their offspring (Foulds, 1947, 1949a) and this inference was substantiated independently in several laboratories (Andervont, 1950; Bittner, 1952; Mühlbock, 1950). The more immediate interest of the tumours in the present context, was the unexpected behaviour of some of them on transplantation.

A. Transplanted Mammary Tumours

A minority of the tumours in F_1 hybrids, when transplanted into hosts of similar genetic constitution grew well in the females but tardily, or not at all in the males. Ovariectomy of female mice impaired their ability to sustain the growth of tumour implants. Castration of male mice did not abolish their refractoriness but prior implantation of pellets of diethylstiboestrol enabled tumours to grow as well in males as in females (Figs 1, 2, 3). The tumours were evidently *conditional, hormone-responsive* tumours whose growth after transplantation was dependent on an appropriate hormonal environment. Similar conditional tumours were found in reciprocal hybrids (RIII ♀ x C57Bl ♂), which received tumour agent from their RIII mothers; dependence of some transplantable tumours on hormones was not correlated with a deficiency of the tumour agent (Foulds 1947, 1949a, b).

It is noteworthy that the "dependence" of transplanted tumours is not absolute. Many implants in normal male mice began to grow after long dormancy; possibly all would do so given sufficient time. The essential role of the oestrogenic stimulus is to ensure *prompt* growth. Once this has been achieved, the stimulus is no longer necessary; subsequent removal of the oestrogen pellets checks the growth of tumours temporarily but tumours do not retrogress and soon all grow again at the customary rate. Moreover the refractoriness of normal male mice can be overcome, partially at least, by increasing the size of the inoculum (Foulds, 1949b). Successful transplantation depends on more than genetic histo-compatibility. The long dormancy and

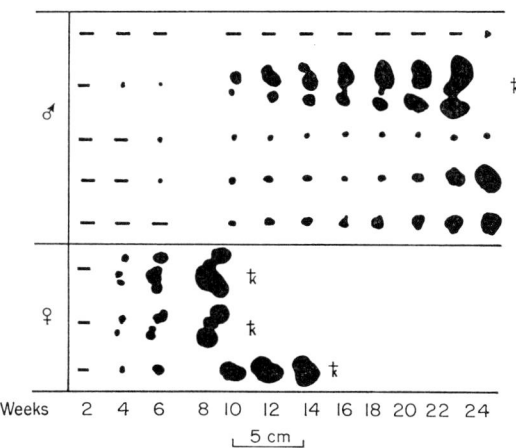

FIG. 1. First transplanted generation of Tumour BR1 showing differences in growth in males and in females.

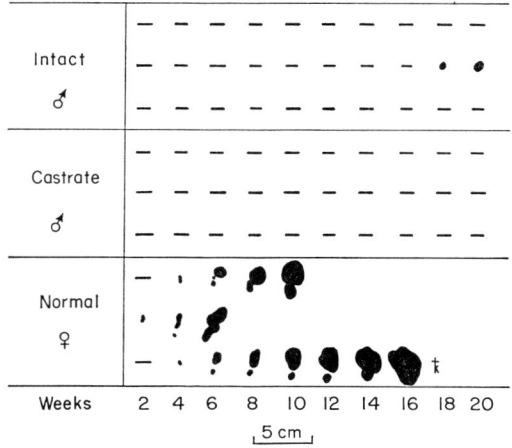

FIG. 2. Third transplanted generation of BR1. One group of males castrated fourteen days before implantation of tumour.

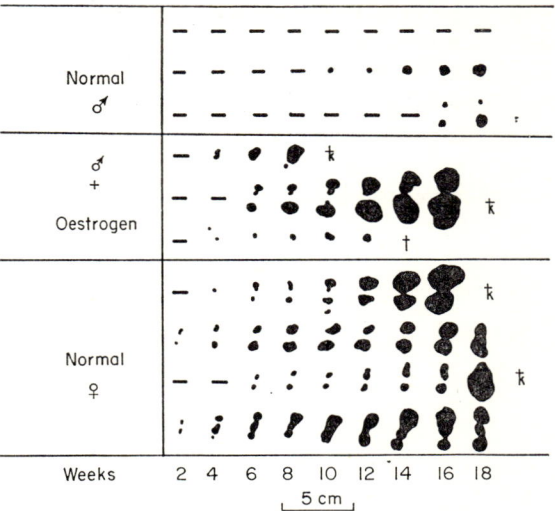

FIG. 3. Fourth transplanted generation of BR1. Stilboestrol-cholesterol pellets implanted in one group of males six days before implantation of tumour.

eventual growth of implants, as well as their state during the dormant period, present important unsolved problems but it is evident that in discussions of the "hormone-dependence" of transplanted tumours a distinction must be made between the initiation of growth and its subsequent maintenance.

Some of the tumours in one of the "dependent" transplanted strains were conspicuously milky to the naked eye. A few discharged a milky fluid through ulcerated patches of overlying skin in the living mouse. Others oozed a milky fluid when incised and slices of them turned a clear fixative milky. Large or small cysts containing milky or creamy material were evident on section and histological examination revealed accumulations of secretions with a patchy distribution. Copious milky secretion of this kind was seen only in female mice killed during late pregnancy (lactating mice were not examined) and, most conspicuously, in male mice bearing pellets of diethylstilboestrol. As mentioned above, excision of pellets after tumours had become well established checked growth only temporarily and did not stop it. By contrast none of the tumours examined after removal of pellets was milky, either macroscopically or microscopically; tumours in seventeen out of nineteen tumours from mice of a parallel series which retained their pellets were milky and the other two had apparently lost their pellets (Figs 4, 5). Milky secretion was conspicuous occasionally in other transplantable tumour strains that were not systematically studied so that its exact frequency is unknown. It is evident nonetheless that copious milky secretion is characteristic of only a minority of the tumours whose transplantation is "dependent" on appropriate hormonal stimulation provided by pregnancy in females and by an administered oestrogen in males (Foulds, 1949b, 1956).

A few observations on three other transplantable tumour strains disclosed yet another response to pregnancy; the tumours grew during pregnancy, regressed incompletely after parturition and grew again during the next pregnancy (Fig. 6). This pregnancy response was not observed in other transplantable strains including those that secreted during pregnancy. It is a specific response; the growth of transplantable tumours in general is either unaffected or inhibited by pregnancy. This form of responsiveness to pregnancy was studied in detail in *spontaneous* tumours as will be described later and the observations on transplanted tumours were not continued. It is evident, nevertheless, that the two observed responses to pregnancy in some of the transplanted tumours, namely, by milky secretion and by accelerated growth are independent of one another and each of them is manifested in only a minority of the tumours whose successful transplantation depends on the

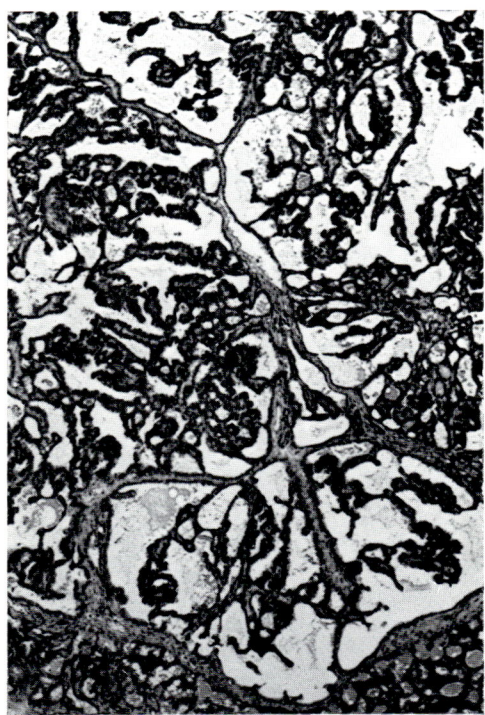

FIG. 4. Tumour of the fifth transplanted generation of BR1 growing in a male mouse which carried a stilboestrol-cholesterol pellet until death. X 60

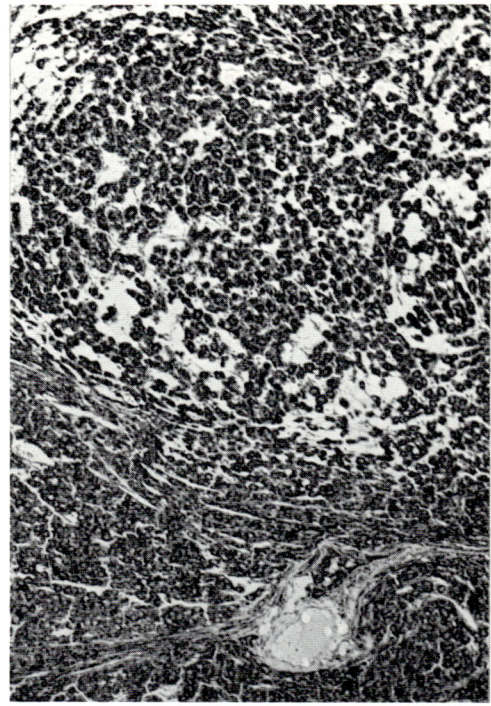

FIG. 5. Another tumour of the same series in a male mouse from which the stilboestrol-cholesterol pellet had been removed twelve weeks before death. X 60

hormonal environment provided by a normal female or an oestrogenized male mouse. Transplantability, secretory response to pregnancy and accelerated growth response to pregnancy are three *independently variable characters*, of which more will be written below. Hormone-responsiveness is

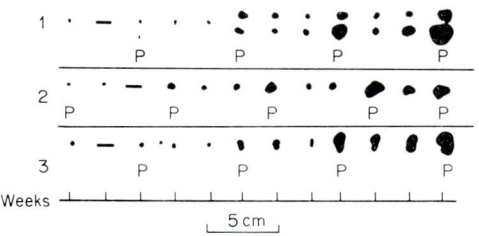

FIG. 6. Three tumours selected from first and second transplanted generations of tumour BR10 showing effect of pregnancy on growth. Pregnancy recorded at the weekly chartings marked "P".

not an all-or-none phenomenon; the responses are quantitatively graded, as well as qualitatively varied (Foulds, 1947, 1949b).

The transplantable mammary tumours of BR mice well exemplify the liability of tumours to change by *progression* during the course of serial transplantation. Sooner or later, the distinctive responsiveness to, or dependence on, hormones is lost. Figure 7 illustrates progression in the rate of growth, as

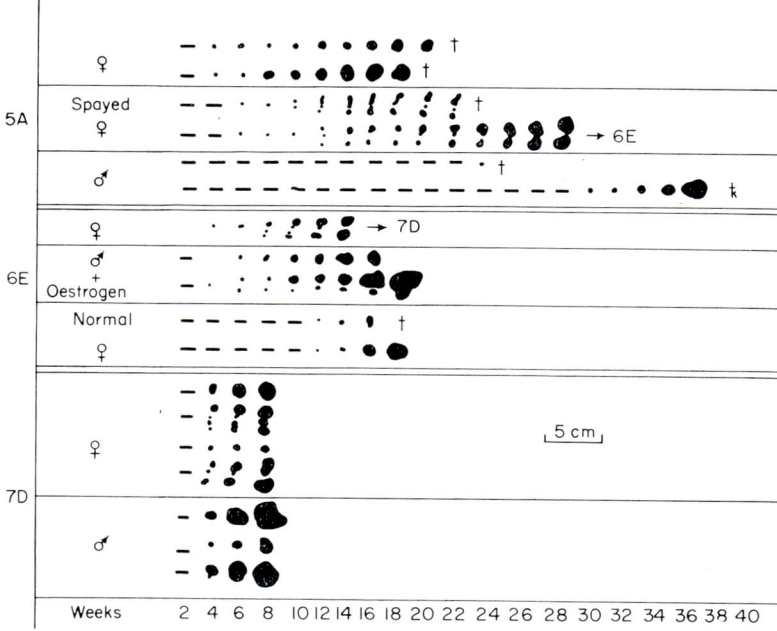

FIG. 7. Three successive transplanted generations of BR1 showing increase of growth rate and decline of sex-difference.

well as loss of hormone-dependence, in three successive generations of one transplantable BR strain; the behaviour of the same strain in three earlier passages is shown in Figs 1–3.

Whilst the foregoing observations were being made, selective inbreeding of the BRF_1 hybrids was started and attention became concentrated on the behaviour of the spontaneous tumours that eventually developed in the F_2 and subsequent generations. When first detected clinically more than half of these tumours were pregnancy-responsive or pregnancy-dependent and the frequency of progression from the responsive to the unresponsive state allowed this phenomenon to be studied in detail.

B. Spontaneous Mammary Tumours

It may be well to emphasize at the outset that pregnancy-responsiveness is highly unusual in the generality of mouse mammary tumours, which have been studied in vast numbers. Occasional examples had been noted by Gardner (1941) and by Haddow (1938) but the BR tumours gave the first opportunity for systematic study. Subsequently, Squartini and others in Severi's laboratory at Perugia reported confirmatory observations on pureline RIII mice (Squartini, 1962; Squartini and Rossi, 1962). More recently similar phenomena have been encountered in two other inbred strains. One strain designated GR is being studied in Amsterdam by Mühlbock (1965) and van Nie and Thung (1965) and the other, DD, in Bethesda by Heston and his colleagues (Heston et al., 1964). It is remarkable that all the most conspicuous examples of pregnancy-responsive mammary tumours so far reported have been in mice with at least one ancestor of European origin. The RIII strain came from Dobrovolskaia-Zavadskaia's laboratory in Paris in the nineteen-thirties. Mühlbock obtained the GR mice from Switzerland and the DD mice reached Heston by a devious route through Japan from Germany. It seems as if European mice are apt to share some common genetic or other factor that is lacking from the American inbred strains. The evidence so far obtained by Lee (1968), Mundy and Williams (1961), Squartini and Severi (1962) and in Heston's laboratory suggests that the mammary tumour agent is qualitatively different in the strains with pregnancy-responsive tumours but that a genetic factor may also be implicated. It should be remarked that chemically-induced mammary carcinomas of rats are also pregnancy-responsive as described in reviews by Noble (1964), Dao (1964) and McCormick and Moon (1965). No comparable mammary tumour agent has been detected in rats. Identification of the circumstances leading to this phenomenon would be welcome but is not essential for an analysis of progression, which is our immediate concern, and of which pregnancy-responsiveness is an especially convenient "marker". The following discussion is based primarily on personal observations made on BR mice (Foulds, 1949c) supplemented by information obtained more recently from RIII, GR, and DD mice.

A high percentage of the female BR mice of the F_2 and later generations developed mammary tumours at a relatively early age and most of them had multiple tumours. Most of the mice were subjected to forced-breeding; they were continuously in the company of males and their litters were removed as soon as possible after birth so that they had pregnancies in rapid succession. Forced-breeding is convenient for speeding up the investigations but, as Squartini (1962) showed, essentially similar results are obtained from normally breeding mice. When first recognized clinically, tumours were usually about 0·5 cm in diameter. From their first recognition onwards throughout their remaining life, they were measured frequently and their size, given as the sum of two diameters, was plotted against time on charts which showed also the dates on which a litter was recorded (Fig. 8). The litter times are subject to

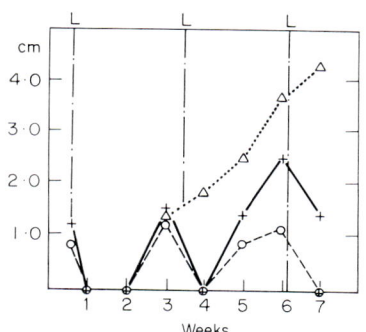

FIG. 8. Showing course of three tumours in one mouse.
+ Right axilla—pregnancy-responsive—subtype II
○ Left groin—pregnancy-responsive—subtype I
△ Right neck—pregnancy-unresponsive.

Note. In this diagram and the following diagrams in the same series, the first emergent tumour (i) or the largest of contemporaneously-emerging tumours is represented by continuous lines joining crosses which indicate the time and magnitude of the actual measurements; the second tumour (ii) is represented by a broken line and circles, and the third (iii) by a dotted line and triangles. To avoid overcrowding of the diagrams, tumours beyond the third are omitted. Scales of time are shown on the charts; the vertical lines headed "L" indicate the days on which a litter was recorded. The sites of the tumours are given in the legends.

an error of one day during the week and longer at weekends. More seriously, a Sunday litter might be eaten and be unrecorded so it was the rule to examine frequently for pregnancy and record it in order to avoid being misled by apparent discrepancies between clinical behaviour of the tumour and reproductive behaviour. Also, it was necessary to measure the tumours daily to ensure recognition of all pregnancy responses.

The chart reproduced in Fig. 8 shows the course of three tumours in one and the same mouse. The three tumours behaved differently. The two first detected grew during pregnancy and regressed after parturition but whereas one reached ascending peaks in successive pregnancies, the other showed no net increase in peak size. These tumours were designated *responsive* tumours of subtypes I and II and were sharply distinguishable from the third *unresponsive* tumour which emerged somewhat later and thereafter grew steadily without regard to reproductive activity. Amongst the several hundred tumours examined less than half were of the conventional *unresponsive* type, but in individual mice there was no regularity in the ratio between responsive and unresponsive tumours or in the order of their emergence. The responses to pregnancy were highly varied in degree; they might be represented only by slight undulations of the growth curves (Fig. 9) or by wide swings (Figs 10, 11). Regressions after parturition were sometimes dramatic in extent and in speed. Responsiveness might persist through a long series of successive pregnancies and in tumours of substantial size (Figs 10, 11). The re-emerging tumours were true recurrences located at the same site as those that had regressed after the preceding pregnancy. When regression was less than complete, the continuity between regressed and recurrent tumours could be unmistakably established by the frequent clinical examinations. When intervening regressions were complete, as judged by clinical examination, the continuity was indicated by identity of site. The tumours were *conditional*

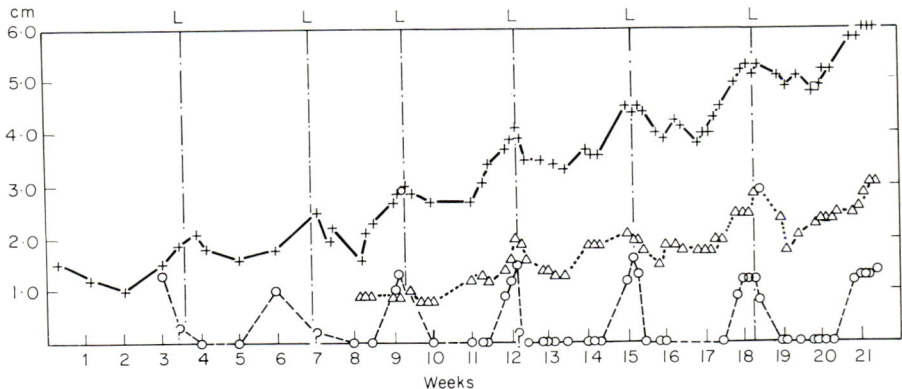

FIG. 9. (i) Right groin + (ii) Right vulva ○ (iii) Left axilla △.

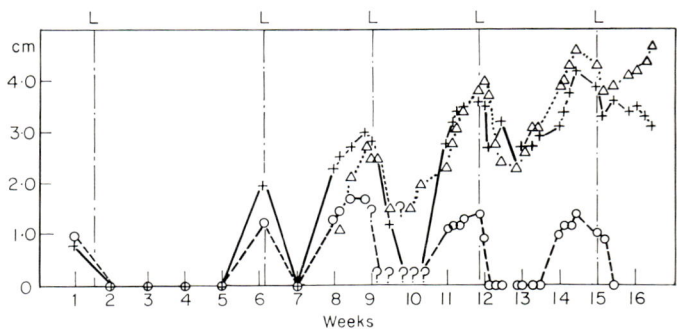

FIG. 10. (i) Left axilla + (ii) Left groin ○ (iii) Right vulva △.

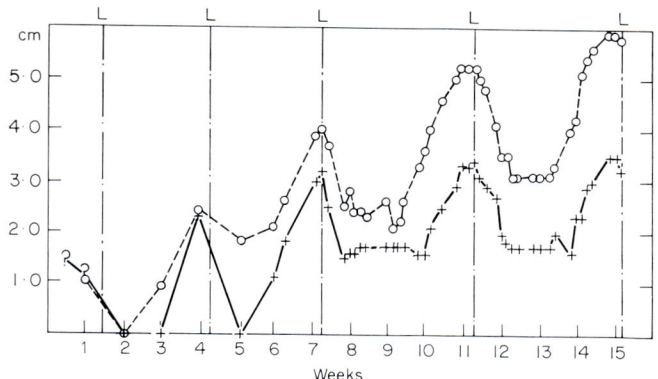

FIG. 11. (i) Left axilla + (ii) Right groin ○.

tumours. With rare and doubtful exceptions, all regressed tumours recurred at the first succeeding pregnancy however long it was delayed (Figs 12, 13, 18). In confirmation, van Nie and Thung (1965) found that all regressed

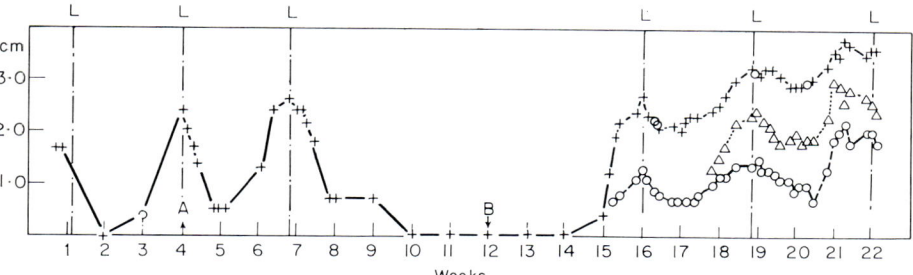

FIG. 12. (i) Right vulva + (ii) Right vulva, separate from (i) ○ (iii) Left axilla △. (One tumour omitted). Male removed at A replaced at B.

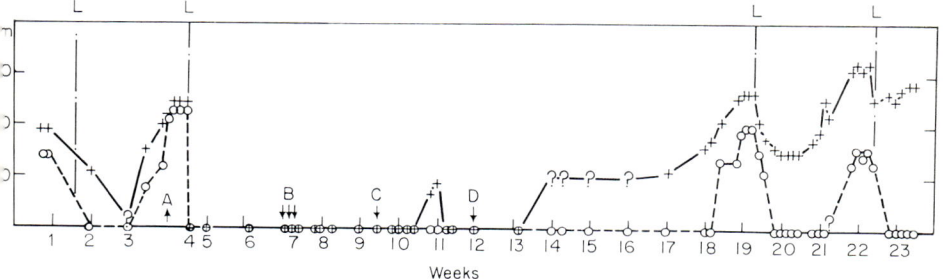

FIG. 13. (i) Left vulva + (ii) Right vulva ○. Male removed at A; cutaneous applications of stilboestrol at B; vasectomized male added at C and replaced by normal male at D.

tumours recurred after non-breeding intervals varying from 40 to 133 days. Moreover, van Nie and Thung evoked recurrences during intermissions of breeding by simultaneous administration of oestrone and progesterone. Neither hormone alone was effective, as observed previously in BR mice (Fig. 13). The hormonal treatment did not reactivate *all* regressed tumours as pregnancy did and the recurrent tumours did not reach the peak size that might have been expected during pregnancy. Nevertheless the ability to reactivate tumours experimentally, with results which might well be improved by variations in the dosage of hormones, is an important contribution to the study of the *residual neoplasia* that evidently persists between pregnancies.

During intermissions of breeding a substantial number of regressed tumours recurred without evident cause and then grew without regard to pregnancy or parturition (Figs 14, 15, 16). In addition to the recurrent tumours, new primary tumours with no spatial relationship to any previously observed lesion, emerged during intermissions of breeding, presumably from regions of incipient neoplasia. van Nie and Thung studied these phenomena in fourteen female mice kept apart from males to prevent breeding. Only one mouse, which survived for 291 days, failed to develop a mammary tumour and

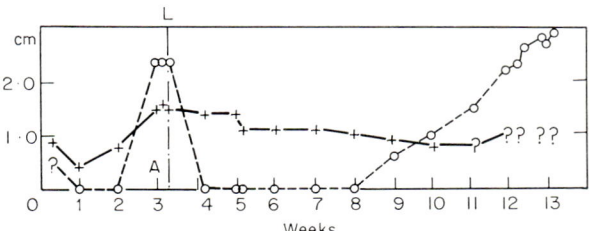

FIG. 14. (i) Right vulva + (ii) Left vulva ○. Male removed at A, not replaced.

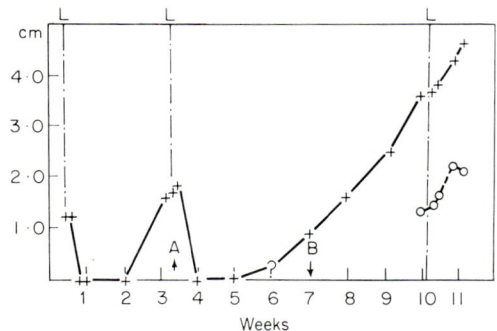

FIG. 15. (i) Right vulva + (ii) Left axilla ○. Male removed at A and replaced at B. Nursed litter from A to B.

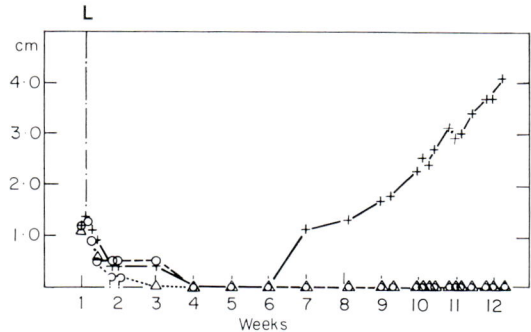

FIG. 16. (i) Right axilla + (ii) Left axilla ○ (iii) Right groin △. Male removed one day after parturition and not replaced.

one mouse developed two. Of the fourteen tumours, six emerged from hitherto unaffected glands and eight at the site of previously regressed tumours. In van Nie's and Thung's experience the recurrent tumours emerged earlier than the new primary tumours. All tumours that emerge during intermissions of breeding, whether "recurrent" or "new" are of necessity, pregnancy-independent tumours.

Some responsive tumours continued to behave in the same way throughout remaining life (Figs 9, 10, 11). Tumours of responsive sub-group I grew to the same extent during each successive pregnancy; those of sub-group II reached progressively higher peaks at each pregnancy and the intervening regressions tended to become less complete, as van Nie and Thung also found. By contrast, some tumours changed their behaviour dramatically from pregnancy-responsiveness or dependence to pregnancy-unresponsiveness or independence. Progression was often manifested and most easily recognized by a failure to regress at the end of a pregnancy (Figs 17, 18). The change was

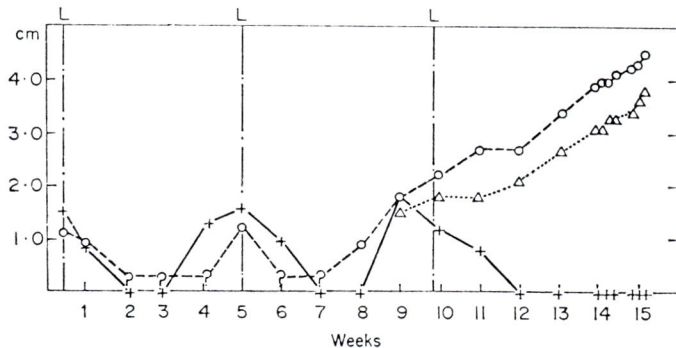

FIG. 17. (i) Left Neck +; Left groin ○; Right groin △.

persistent and, so far as could be determined, irreversible. It occurred unpredictably after short or long periods of responsive behaviour; usually it seemed to be abrupt but sometimes gradual. Only one of the multiple tumours in the same mouse changed by progression at one time. The re-emergence of regressed responsive tumours as unresponsive ones during intermissions of breeding implies that progression had taken place during the dormant phase or during the preceding pregnancy.

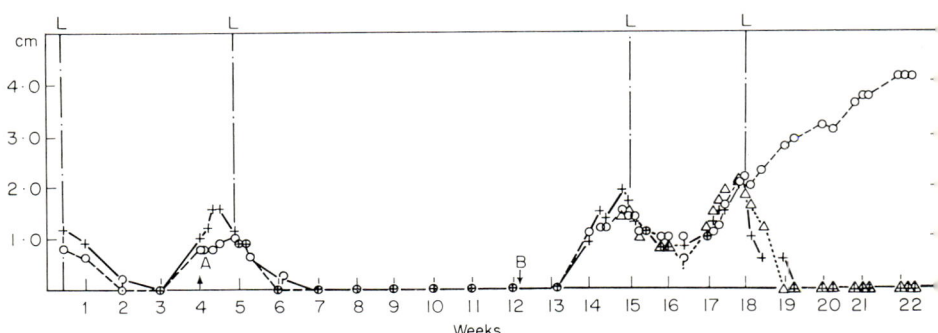

FIG. 18. (i) Left groin + (ii) Right groin ○ (iii) Right vulva △. Male removed at A, replaced at B.

Figure 18, recording the observations on a single mouse, illustrates some important general principles of tumour progression. It shows that two tumours of similar size were first detected during a pregnancy. Both regressed after parturition, recurred at the next pregnancy and again regressed at its termination. During the latter pregnancy the male mouse was removed to bring about an intermission of breeding. The tumours remained in abeyance until the male was returned whereupon another pregnancy ensued promptly. The two tumours recurred and a new responsive one emerged at a different site. The three tumours regressed incompletely after parturition and recurred at the next pregnancy which started immediately. Up to the end of this pregnancy, the three tumours followed courses so closely parallel that their growth curves are difficult to distinguish when plotted on a single diagram but after parturition two tumours regressed completely whereas the other continued to grow steadily and progressively until the mouse was killed without becoming pregnant again. The tumour that became unresponsive and independent was one of the pair first detected four and a half months previously. The abrupt change in behaviour of one of the three tumours which hitherto had behaved alike must be ascribed to progression in that tumour and not to an effect of the hormonal environment to which all three tumours were exposed.

3. NEOPLASTIC DEVELOPMENT

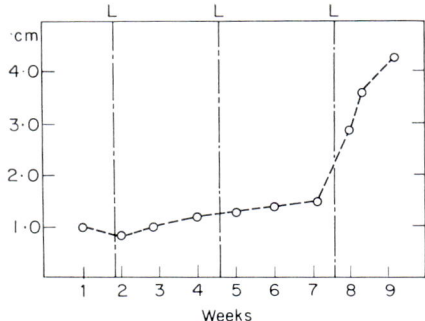

FIG. 19. (ii) Left vulva ○, showing abrupt acceleration of a pregnancy-unresponsive tumour. An earlier tumour (i)—not shown—was excised thirteen weeks previously and did not recur.

A much rarer type of progression is illustrated in Fig. 19, which show a sudden onset of rapid growth in tumours which previously had persisted with scarcely any growth at all.

C. Histology of Responsiveness and Progression

The foregoing analysis is based wholly on clinical observations on living mice subjected to no experimental interference beyond the imposition of forced-breeding and its deliberate interruption by segregation of females. At this point, the histological basis of the phenomena requires discussion. The responsive tumours when first noticed clinically are usually about 0·5 cm in diameter. Correspondingly, the most usual early stage of neoplasia found at autopsy is a disc or *plaque* measuring 0·5–1·0 cm in diameter and 0·2–0·3 cm in thickness. In sections cut vertical to the skin surface, the lesions are elliptical and smoothly outlined. Histologically they are remarkable for their symmetrical organoid structure (Figs 20, 21). The simplest of them consist of (a) a central portion or "medulla" in which small tubules are scattered sparsely and irregularly in a loose fatty stroma, and (b) a more compact peripheral zone or "cortex" in which branching tubules growing out from the medulla are disposed radially. There are many variants of this simple plaque architecture but in all of them the histological basis of responsiveness is similar (Foulds, 1956).

During pregnancy there is little change in the medulla but the cortex enlarges greatly as a result of growth and branching of tubules and massive hyperplasia of their epithelial lining cells (Figs 22, 23, 24). The radially-disposed tubules extend with repeated branchings to the periphery of the plaque where, commonly, they end in pyriform enlargements resembling the end-bulbs of *normal* mammary development. The single-layered epithelium changes, often abruptly, beyond the medulla to become multilayered. The hyperplasia is varied in degree but often extensive. At its peak, mitotic figures are numerous. Sometimes the histological picture closely simulates that of intratubular or intracanalicular mammary carcinoma.

Epithelial proliferation reaches a peak towards the end of pregnancy. Signs of regression are discernible about the time of parturition and within a day or two regression is far advanced. The earliest sign of regression is degeneration of epithelial cells with karyorrhexis of their nuclei and vacuolation of their cytoplasm. As a rule, peripheral cells of the epithelial masses seem to degenerate first (Figs 25, 26). The appearance after one or two days is distinctive. There is a characteristic "ballooning" of epithelial cells forming a clear space between the stroma and the surviving epithelium, usually single layered and notably irregular, that lines shrunken, distorted tubules (Fig. 25). It is not clear how the necrotic material is removed; there is no conspicuous phagocytosis. Meanwhile the vascularity of the stroma decreases, the connective tissue cells attenuate and collagen fibrils begin to form (Figs 25, 26).

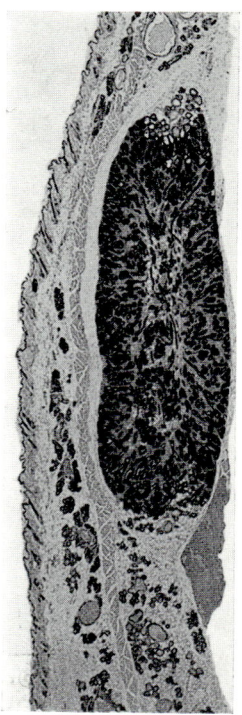

FIG. 20. Plaque from a late-pregnant mouse. Tubules radiate from a central "medulla" and branch to make a more compact "cortex". X 15

Fig. 21. A similar plaque from a late-pregnant mouse with more adipose tissue in the stroma. X 15

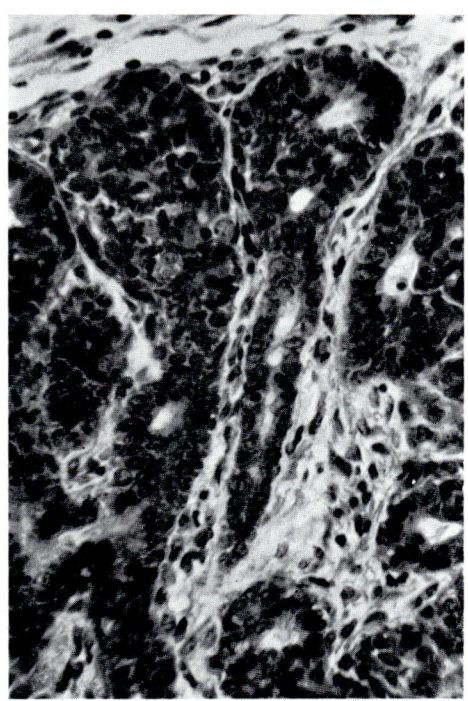

Fig. 22. Tubules and end-bulbs at the edge of a plaque from a late-pregnant mouse. X 320

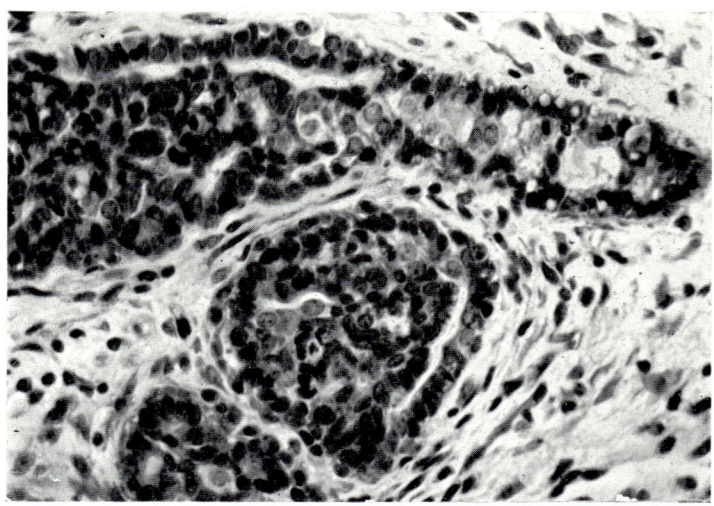

FIG. 23. Tubules in a plaque from another late-pregnant mouse with dark and pale cells and dividing cells in the hyperplastic epithelium. X 320

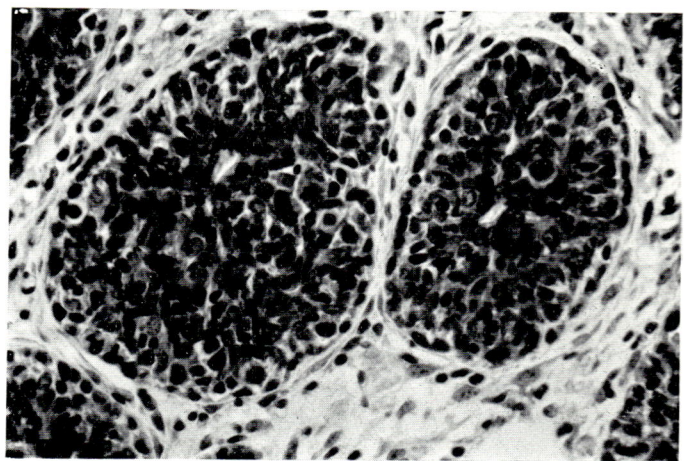

FIG. 24. Two hyperplastic tubules cut transversely in a plaque from a late-pregnant mouse. The distinct enveloping layer of smaller cells is possibly myoepithelial. X 320

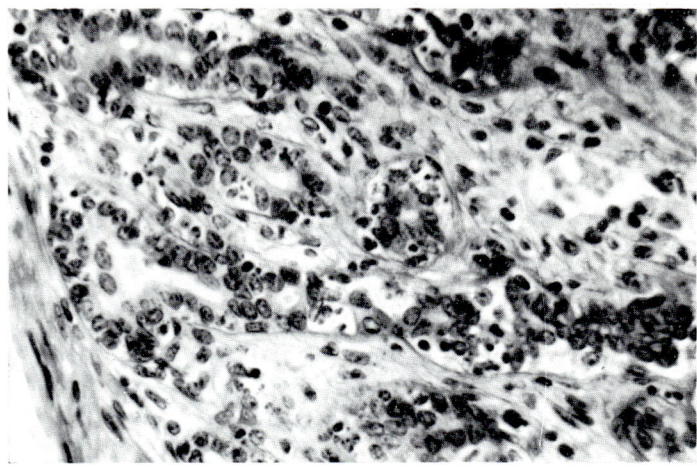

FIG. 25. Regressing plaque from a mouse killed soon after parturition with karyorrhexis and vacuolation of cytoplasm in the tubule epithelium. The stroma is loose and oedematous. X 320

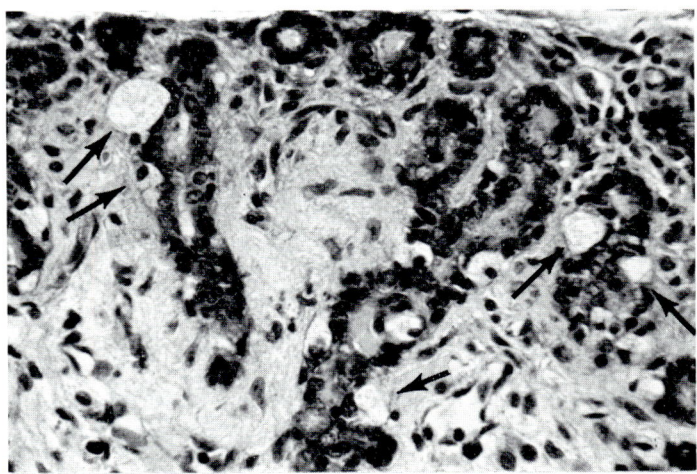

FIG. 26. Regressing plaque from a mouse killed soon after parturition. There is less nuclear debris than in Fig. 25 but greater vacuolation of cytoplasm with a characteristic "ballooning" of the peripheral cells of the tubules as indicated by the arrows; the tubules are more widely separated by stroma in which there is early fibrosis. X 320

In plaques which shrink but do not entirely disappear between pregnancies, the fibrosis in the peripheral zone is often intense (Figs 27, 28). Scanty irregular fibrosis and a few irregular tubules mark the site of some regressed plaques which have remained dormant through a long intermission of breeding (Fig. 29). These plaques, some of which are just palpable or detectable only with difficulty on post-mortem examination of the mammary glands

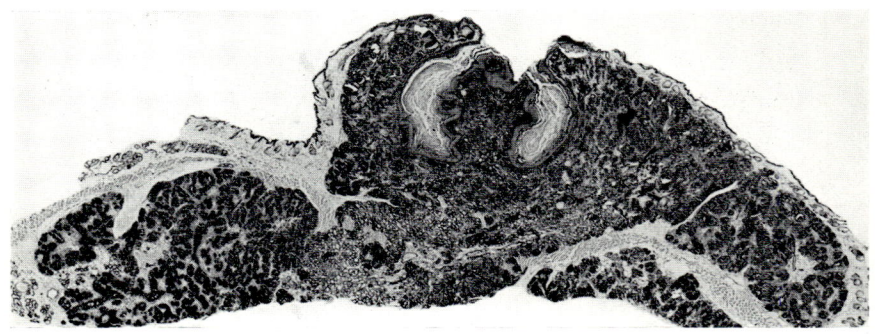

FIG. 27. A plaque near the peak of growth from a mouse that had littered three hours before the tumour was excised. × 12.75

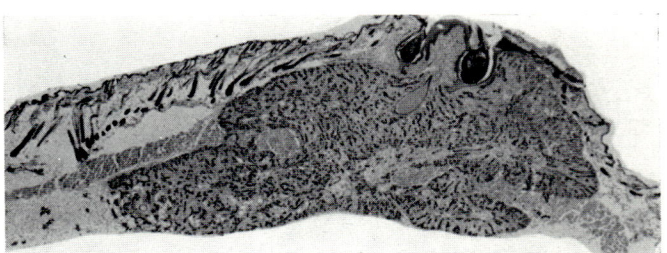

FIG. 28. A regressing growth in a similar position to that in Fig. 27 from a mouse that had littered five days previously. × 15

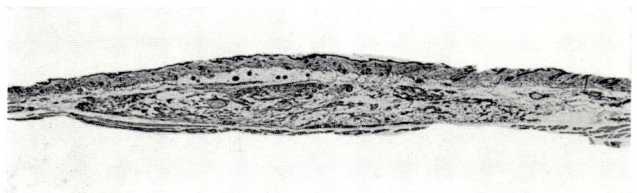

FIG. 29. A regressed plaque from a mouse that had not been pregnant for over eight months. The lesion was noticed only at autopsy. A tumour at the same site had reached a size of 1·0 × 0·5 cm during an earlier pregnancy and had then regressed and had not again been palpable. × 15

form a link between the plaques that persist at a diminished size between pregnancies and those that disappear altogether. The ragged, atrophic epithelium characteristic of regressed plaques presumably represents the *residual neoplasia* that endures between pregnancies although it is not "neoplastic" in appearance.

Histological structure is not a reliable indicator of responsiveness. It is impossible to be sure from histological examination of a growth removed during pregnancy that the growth would have regressed after parturition if left *in situ*. Soon after parturition, decisive indications of regression are visible before the architecture existing at the peak of growth has been greatly altered. From observations made at this time it is evident that highly complex variants of plaques, from which, during pregnancy a diagnosis of carcinoma could hardly be withheld, are fully responsive and liable to regression. Conversely, some gross tumours including certain "giant" plaques retain the basic histological characteristics of plaques although they may be transplantable, hormone-independent, malignant tumours. In short, hormone-dependence or responsiveness is not reliably correlated with histological structure.

The considerations advanced in the preceding paragraph impose some evident restrictions on the histological recognition of progression especially in tumours removed from pregnant mice. In partially-regressed tumours from non-pregnant mice signs of progression are more apparent and two main types are distinguishable, namely, a *focal* or *multifocal* type and a *lobular* or *diffuse* type. Focal progression is most conspicuous in partially-regressed plaques from non-pregnant mice and is evidenced by a focus of active growth within the sclerotic tissue of the remainder of the plaque. The changed focus usually has the histological structure of a commonplace cystic haemorrhagic mammary carcinoma of mice (Figs 30–32). Multiple foci of this kind are often present in a single plaque and may be of different histological types (Figs 33, 34, 35). A series of about 200 specimens with evidence of focal progression contains examples of all stages of development from a minute focus of carcinoma within an otherwise typical regressed plaque to a large carcinoma with vestiges of plaque-like material at its periphery. Lobular or diffuse progression

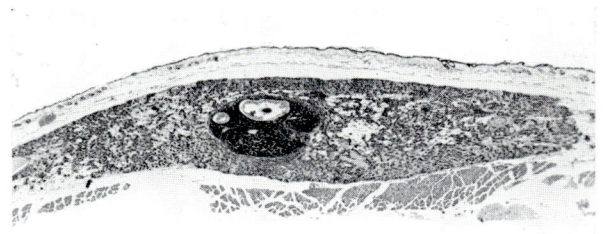

FIG. 30. Centrally-placed progression to carcinoma within a regressing plaque. X15

is more difficult to identify because often it does not entail any immediate or conspicuous change in histological structure. It is recognizable by the persistent activity of one small or large portion of a plaque from a non-pregnant

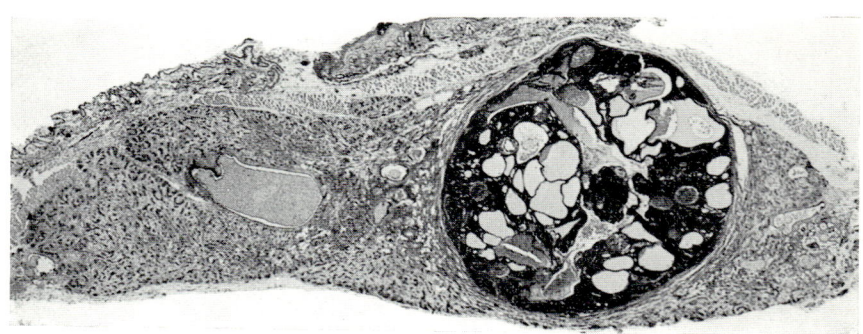

FIG. 31. Carcinoma with circular outline within a large regressed plaque. X 13

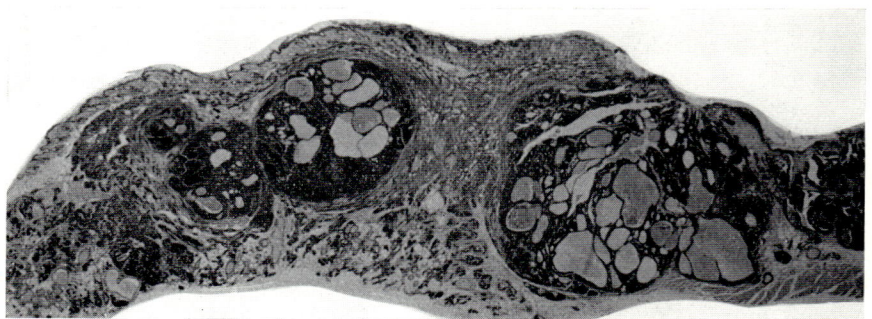

FIG. 32. Multifocal progression within a regressed plaque. X 12·75

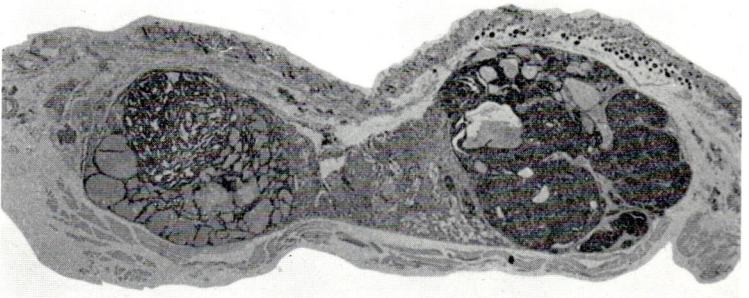

FIG 33. Two histologically-dissimilar carcinomas with regressing plaque tissue between them. X15

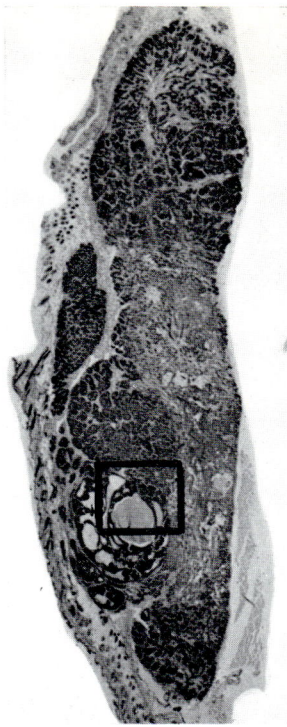

FIG. 34. Varied types of progression within a plaque. X15

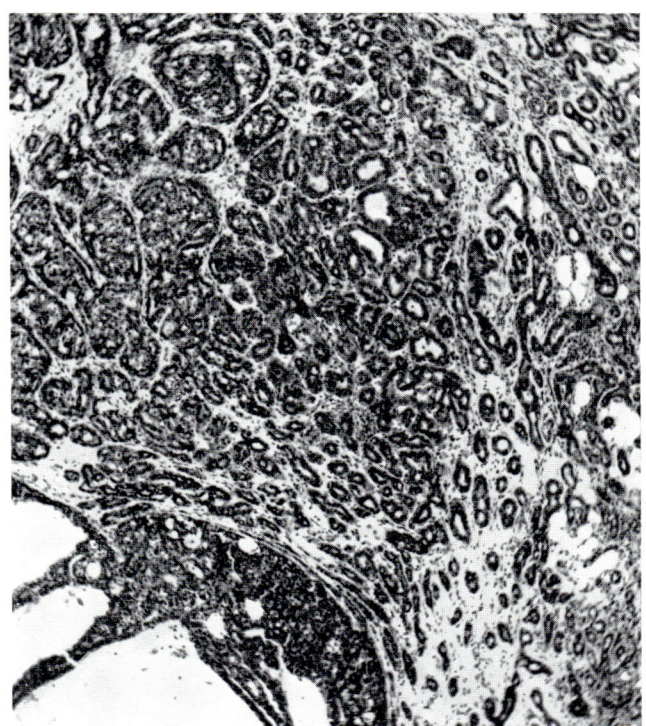

FIG. 35. Higher magnification of the marked area of Fig. 34 showing at the lower left corner the edge of a cystic carcinoma and, above it, an uncommon type of intratubular carcinoma. X60

mouse when the remainder has the characteristic appearance of a regressed plaque; one portion of the plaque has become "independent" without conspicuous histological changes. The theoretical limiting case of *total progression* is possibly seen in certain of the "giant plaques" already mentioned and also in some small tumours of the shape and size characteristic of plaques but which, histologically are carcinomas although some of them retain the radial architecture of plaques to some degree. The histological findings were discussed in greater detail by Foulds (1956).

D. Plaques and Hyperplastic Nodules

The majority of mammary carcinomas in BR mice and probably in RIII, DD and GR mice, originate by progression in pregnancy-responsive plaques. In most or all of the best known "high-incidence" inbred strains, the majority of carcinomas originate in hyperplastic nodules. The commonest form of nodule consists of tightly-packed clusters of acini applied to a duct without an appreciable stalk so as to resemble a bunch of grapes. The morphological distinction between plaque and nodule may depend on the length of duct that is implicated in neoplastic development; nodules seem to develop from a small focus whereas plaques seem to grow from a substantial length of duct (Foulds, 1956). Plaques, at their first recognition, are roughly ten times as big as nodules, which are undetectable clinically and only a little above the limit of visibility to the naked eye at autopsy, but nodules do not grow into plaques. Hyperplastic nodules and plaques are *alternative* stages in mammary neoplastic development, not consecutive ones (Foulds, 1956; Severi *et al.*, 1959).

The "high-incidence" inbred strains in which pregnancy-responsive tumours are not observed do not have plaques. By contrast the strains with pregnancy-responsive tumours, together with a small proportion of pregnancy-unresponsive tumours, have both plaques and nodules. It seems extremely probable that in these mice the pregnancy-unresponsive tumours habitually originate in nodules whereas the pregnancy-responsive ones demonstrably originate in plaques. Heston *et al.* (1964) have noted, also, that the definitive mammary carcinomas in the plaque-bearing strains mostly belong to Dunn's histological Group B, whereas the great majority in C3H mice and, apparently, in other American strains without plaques, belong to her Group A. It would be going too far to infer that *all* Type A tumours originate in nodules and *all* Type B tumours in plaques. Both types of tumour develop on occasion in all kinds of inbred mice; the difference is in their relative frequencies. The inference to be stressed here is that mammary neoplastic development in mice can advance along at least two distinctly different paths, even in one and the same mouse, to reach similar but perhaps not identical end-points. Intermediate steps along the two paths are distinctly different but the end-points are not consistently different.

E. FACTORS INFLUENCING PROGRESSION

Attempts to encourage progression to carcinoma in experimentally-induced epidermal papillomas have been indecisive. Progression has been reported as a result of cauterizing the base of a papilloma or of applying croton resin to it (Berenblum, 1944; Cramer, 1929; Des Ligneris, 1940). Wounding and subsequent scarring have been thought to encourage progression (Deelman, 1924; Lacassagne and Latarjet, 1946; Linnell, 1950; Linnell and Norden, 1950; Lipschutz 1924; Roe and Glendenning, 1956) and so also have surface applications of oleic acid and other chemicals (Roe, 1956; Twort and Twort, 1939). Unfortunately, there seems to be no widespread conviction that progression can be decisively and consistently induced by any of these procedures. *Promotion*, as studied in the "two-stage" experiments on rodent skin does not seem to have any effect on progression. In man, the disastrous effects of injudicious surgical interference with melanomas and with the more advanced carcinomas of the urinary bladder and lung might, but need not necessarily, be attributed to progression.

The likelihood of progression seems to depend primarily on the *capacity* of the neoplastic tissue and this in turn, as implied by experimental investigations, seems to depend on the intensity and duration of the carcinogenic stimulus, where one is identifiable. The *capacity* can increase progressively as a consequence of summation of carcinogenic stimuli and possibly also autonomously with time. Conceivably, the *capacity* may increase locally to a state of developmental imminence. If so, the apparent "inductions" of progression noted in the preceding paragraph might be attributed plausibly to non-specific stimulation of a tissue in a state of *developmental imminence*, whose nature will be discussed later.

VI. General Principles of Tumour Progression

The six general principles or rules of tumour progression proposed by Foulds in 1949c and based primarily on the study of mammary neoplasia in several hundred BR mice were as follows:

Rule I: *Independent progression of multiple tumours.* Progression occurs independently in different tumours in the same animal.

Rule II: *Independent progression of characters.* Progression occurs independently in different characters in the same tumour.

Rule III: *Progression is independent of growth.* Progression occurs in latent

tumour cells and in tumours whose growth is arrested. Two notable corollaries of Rule III are:
(i) At its first clinical manifestation a tumour may be at any stage of progression.
(ii) Progression is independent of the size or clinical duration of a tumour.

Rule IV: Progression is *continuous* or *discontinuous* by gradual change or by abrupt steps.

Rule V: Progression follows one of *alternative paths of development*.

Rule VI: Progression does not always reach an end-point within the lifetime of the host.

These general principles have proved widely applicable to various sorts of neoplasia in animals and in man (Foulds, 1954, 1958a, 1964b, 1965) but in the light of experience of their wider application, some have assumed greater immediate importance than others and some have needed qualification or revision. The rules as originally proposed therefore, will be reconsidered here in turn.

Rule I. Independent Progression of Tumours

This rule applied originally to, and remains valid for multiple primary tumours present simultaneously in a single host. Its applicability to multiple metastatic tumours deserves more consideration than it has yet received. Its most important application to neoplastic disease in man is to the multiple lesions typified by the mammary hyperplastic nodules and plaques of mice that are commonly designated "precancerous" and which will require repeated and detailed consideration elsewhere. Meanwhile, it may be noted that as a general rule only one of a considerable number of the early emerging epidermal lesions on human or animal skin exposed to strong carcinogenic agents undergo progression to carcinoma at any one time or, for that matter, at all. This applies equally to the distressing childhood disease xeroderma pigmentosum. Similarly only one or a small number of the abundant papillomas in familial intestinal polyposis give origin to carcinoma by progression at one time—or at all. Moreover it is the general, if not invariable rule, that progression in a single lesion seems to be usually focal or multifocal, affecting only a fraction, often a small fraction, of the available tissue. It is perhaps desirable to indicate that in this context "focal" does *not* necessarily imply restriction to a single cell or even to a minute group of cells (Foulds, 1958a).

Rule II. Independent Progression of Characters

Haaland (1911), writing primarily about mammary tumours of mice, remarked on the extraordinary variety of the tumours developing from cells of one organ; the neoplastic change, he said, obviously affects the different functions of the cell in different degrees and the relative independence of these functions can be inferred. Although evidence for the relative independence of characters has been accumulating ever since, it received inadequate attention until the rule of independent progression of characters was formulated. The associations of characters such as growth rate, invasiveness, powers of metastasis, responsiveness to hormones or chemotherapeutic agents and histological and cytological structure in tumours of the same provenance are extremely varied. Text-book descriptions and classifications refer to certain common or "average" associations of clinical and pathological characters but the distinctions are blurred by many "atypical" or "anomalous" associations of characters, as a result it has been suggested, of "out-of-step" progression of characters (Foulds, 1951). The "dissociation" of characters, as Hamperl (1957) called it, is exemplified by the "locally-malignant" tumours that invade vigorously but rarely metastasize, by the "metastasizing benign tumours" that disseminate widely despite minimal local invasion and by the highly malignant carcinomas of the human prostate and breast that are "dependent" on, or "responsive" to, hormonal stimulation.

The rule of the Independent Progression of Characters led to the more general proposition that the *structure and behaviour of tumours are determined by numerous characters that, within wide limits, are independently variable, capable of highly varied combinations and assortments and liable to independent progression* (Foulds, 1954). The concept of independently variable characters is crucially important in the study of neoplastic development. The need for particularization, or factorial analysis as some prefer to call it, can scarcely be over-emphasized. The recognition and analysis of progression depend on the study of identifiable characters that can be *separately* specified and described or measured. Some properties are dependent or consequential on others but even such correlations as those between growth rate and "dedifferentiation" are far less consistent than is commonly supposed, and the general principle of the independent variability of characters has proved widely applicable and useful in diverse studies of neoplasia in animals and in man (e.g., Foulds, 1958a; Hamperl, 1957; Horava and Skoryna, 1955; Klein and Klein, 1957; Leighton, 1957; Potter, 1962; Yoshida, 1962).

The description of independently variable characters in earlier publications as "unit" characters has been reasonably criticized on the ground that those cited were complex characters and not unitary. In truth, the generalizations were based on relatively coarse methods of observation and analysis but they hold good to a surprising degree when tested by the most delicate methods available, revealing characters that are more nearly unitary than those formerly studied. The general principle seems to be applicable at all levels of analysis, remaining valid, for example, when analysis is pushed to the level of histocompatibility genes and individual enzymes (Klein and Klein 1958; Potter, 1962). These more penetrating investigations substantiate and extend much earlier opinions about the individuality of tumours; probably no two tumours are exactly alike in every respect even when they originate from the same tissue, have the same general properties and have been induced experimentally in the same way (Foulds, 1940). Cytogenetic and biochemical individuality is now being recognized amongst tumours of the same general type. It is becoming increasingly evident that many wide generalizations about "cancer" have broken down, as most of them eventually do, because they have not taken into account the independent variability and progression of characters and the consequent individuality of tumours. Moreover, it is also becoming increasingly apparent that many prominent characters are not essential components of the neoplastic process but only incidental consequences of it. These characters, as Potter remarks, may be exploitable in therapy without being essential or relevant to the neoplastic process itself (Klein and Klein, 1958; Potter, 1962; Weinhouse, 1960).

The orthodox pathology of tumours is gravely lacking in a terminology for the specification of lesions with atypical or anomalous associations or "dissociations" of characters. In particular, there is an urgent need to distinguish those numerous lesions including the generality of "precancerous" lesions to which the orthodox terminology is inapplicable. The mammary hyperplastic nodules and plaques to which reference has been made already, are difficult to fit into orthodox classifications. Attempts to force them into existing named categories that were not designed for them do more harm than good. In general, the contentious lesions possess some but not all of the characters that ordinarily are required for a confident diagnosis of either benign tumour or malignant tumour. It has been proposed to describe such tumours as *imperfect* (Foulds, 1960, 1961). This description has not been kindly received, which is not surprising since it is manifestly unsatisfactory, but no more desirable alternative has yet been advanced. Whatever the name, the phenomenon is one of great importance as illustrated by carcinoma *in situ* of the human uterine cervix, which has the cytological features of carcinoma but fails to satisfy the criteria of local invasiveness or remote dissemination. To call it a "cancer" is dangerously misleading. Its designation as an *imperfect carcinoma* indicates, in harmony with the best available evidence, that the quality of invasiveness is not merely inapparent or "latent" but is absent and can be acquired only by progression (Foulds, 1958*a*).

Many independently variable characters enter into the complex phenomena of "malignancy". The associations of characters that contribute to malignant behaviour are highly diverse and the characters need individual specification and study (*vide infra*). It is at least arguable that the fundamental processes of neoplasia may be revealed most clearly in *benign* tumours, which represent neoplasia in its simplest form "without the frills" and uncomplicated by the numerous characters that result from or contribute to malignant behaviour (Foulds, 1954). Some of these added features may be of great interest and convenience in laboratories without being essential to the neoplastic process or at all important in determining the course and outcome of neoplastic disease in animals or in man.

Rule III. Progression is Independent of Growth

This rule is based chiefly on the re-emergence of a regressed responsive tumour as an unresponsive one during intermissions of breeding when growth of responsive tumours is repressed. It may be suspected that some of the recurrent tumours originated by focal progression during the preceding pregnancy and not during the non-breeding period and that the delay in their emergence is attributable to the time needed for the altered foci to grow to a clinically appreciable size. This seems the likely interpretation of the early recurrences but is less convincing for the late ones. Progression without manifest growth possibly accounts for the long-delayed recurrences and metastatic growth after apparent "cure" of a primary tumour in man by surgery or irradiation and for some at least of the relapses that terminate a long period of supression effected by endocrine therapy or chemotherapy. There is little experimental evidence or none that sustained proliferation *per se* enhances progression and some evidence to the contrary (Andervont *et al*, 1957; Millar and Noble, 1954). There is a disturbing possibility that therapy, by suppressing or retarding growth, may favour progression from the responsive to the unresponsive, independent state. This important matter, closely relevant to the management of human patients by endocrine therapy or chemotherapy deserves thorough investigation.

The corollaries to Rule III summarize the empirical observations in BR mice. Their generality has not been closely investigated but they are consistent with clinical experience.

Rule IV. Progression is Continuous or Discontinuous by Gradual Change or by Abrupt Steps

It is arguable that gradual change might be the consequence of a large number of successive small abrupt steps or that a seemingly abrupt step may be but the culmination of a long-lasting, slow, continuous change. There is insufficient evidence to allow these arguments to be pursued profitably. The observations on BR mice suggest that slow gradual progression is apt to result in sluggish, irregular or dubiously unresponsive growth and that rapid-growing, unequivocally unresponsive tumours are either unresponsive at their emergence or are the result of abrupt progression in fully responsive and, often, strictly "conditional" tumours.

Rule V. Progression Follows One of Alternative Paths of Development

All tumours of the same general type do not attain their definitive state by traversing the same developmental pathway. The distinction made originally was between *direct* and *indirect* paths of development of mammary tumours of mice. The *direct* path led to the emergence of a conventional primarily unresponsive tumour; the tumour acquired its definitive properties early without traversing the intermediate stages of plaque or nodule which are conspicuous on the *indirect* path or "responsive detour". Eventually the two paths lead to much the same end-point. It has been suggested that neoplastic development is always *indirect* in the sense that it entails stepwise progression through qualitatively different stages, which are traversed so early or so quickly that they are not noticeable (Hamperl, 1957; Horava and Skoryna, 1955; Shabad, 1962). Be that as it may, it is important to emphasize that the distinction between direct and indirect development is a real one in clinical practice if not in theory. It seems to be a general rule, to which no clear exception has been demonstrated, that in every tissue in which a particular kind of malignant tumour *usually* develops along an indirect path through visible intermediate lesions, a similar kind of tumour *sometimes* emerges with its definitive characters established from the beginning and without traversing any intermediate lesion during its clinically-evident course. It is believed that invasive carcinoma of the human uterine cervix usually develops from carcinoma *in situ* but sometimes it emerges as an invasive carcinoma from the beginning. The relative frequency of the direct and indirect paths of development of carcinoma of the uterine cervix is still being disputed. It is not part of the theory of progression to affirm that malignant tumours in general habitually develop by progression in a "benign" or in some sense "intermediate" or "precursor" lesion, and it has been pointed out repeatedly that, on the basis of clinical observation some, perhaps the majority, do *not*. The frequency of indirect development differs widely in various types of neoplasia and can be assessed only by empirical observation. The outstanding interest and importance of indirect development, when it can be recognized, is the opportunity it provides for therapeutic intervention in the neoplastic process before it is too late.

It should be added to the above account, that within the *indirect* group, alternative paths leading to much the same end-point can be recognized. As an example, mammary neoplasia in mice can advance along at least two indirect paths, the one through hyperplastic nodules and the other through plaques; the two paths eventually yield tumours that differ only in minor respects, if at all. When intermediate or precursor lesions are recognizable in clinical practice, it cannot be taken for granted that all malignant tumours of the same general type will develop from one and the same type of precursor nor that progression in one particular type of precursor will yield always the same type of malignant tumour.

Rule VI. Progression Does Not Always Reach an End-Point Within the Life-Time of the Host

This rule is incontrovertible in the sense that progression advances further through successive stages in the transplanted progeny of primary tumours but for wider application it needs elaboration and qualification. The negative statement should be balanced by a positive statement that some tumours *do* reach an end-point of progression in their original hosts. Neoplasia tends to "go from bad to worse", as Peyton Rous has emphasized, but this "tendency" should not be elevated into a rigorous law or definition of neoplasia. It was pointed out in a previous section dealing with the Independent Progression of Tumours that only one or a few of the multiple intermediate or precursor lesions in neoplasia of several kinds undergo progression at any one time or at all. Of the remainder some may grow progressively without qualitative change, some may persist indolently for a long time and some may retrogress and ultimately disappear. Even a few unequivocal "cancers" have been known to regress and many more persist for a long time without conspicuous change; relentless progressive growth to a fatal ending is by no means an invariable characteristic of malignant neoplasia. On the other hand, qualitative change by progression can take place in advanced stages of primary neoplasia in the original hosts, as well as still later in transplanted descendants.

VII. Patterns of Neoplastic Development

The main objective of the present section is to derive from previous ones a generalized concept of sequential stages of neoplastic development that will be serviceable as a working hypothesis to guide further studies of neoplastic development in man as well as in animals. The argument of the preceding section was based mainly on two especially favourable experimental models, namely, epidermal neoplasia and mammary neoplasia in mice. As a matter of experience, epidermal neoplasia has provided better opportunities for studying the earliest developmental stages in animals and in man, whereas mammary neoplasia has been the more convenient for the experimental analysis of tumour progression. To obtain a comprehensive view of the known patterns of neoplastic development, both must be taken into account together with information obtainable from other systems. Some other instructive special cases will be dealt with briefly in this chapter; they will be analysed in more detail in Volume 2. The quantity and quality of the relevant observations depend to a considerable extent on the accessibility of the neoplastic tissues to direct examination during life. It is important to recognize that extremely valuable observations have been made on neoplastic development at accessible sites in man and some of these will be mentioned.

A. A Generalized Schema of Neoplastic Development

Certain difficulties and pitfalls in studying sequential neoplastic development, more especially when it is not readily accessible to examination during life, need preliminary attention. As a consequence of the diverse associations and dissociations of independently variable characters that are possible in lesions of one general kind it is often possible to collect from a large number of individuals a variety of lesions that can be arranged in an apparently continuous series. To illustrate this graphically, in Fig. 36(a) a lesion, Y, in some

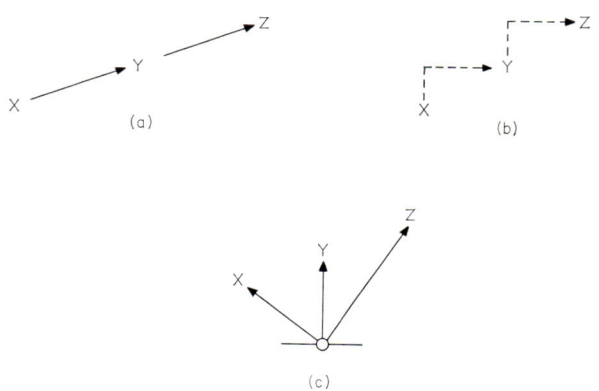

Fig. 36. Possible relationships of lesions on ascending scale of gravity produced by (a) smooth transition (b) abrupt progression and (c) concurrent but diverse development from a common source.

sense intermediate on an ascending scale between two other lesions X and Z from different individuals is apt to be interpreted as one stage of a smooth transition from X to Z. It is not uncommonly reported that "transitions have been seen" which is inaccurate and misleading and does not necessarily represent what actually happens in any individual lesion. Two other interpretations are possible and considerably more probable. First, the development of a particular tumour is more often discontinuous, by stepwise progression than smoothly continuous [Fig. 36(b)]. Second, even in a single individual, multiple lesions that can be arranged in a series of ascending gravity are not necessarily consecutive stages; they are more probably independent lesions that have developed concurrently from a common source represented in Fig. 36(c) by the small circle which is used to denote a *region* of incipient neoplasia. Clinical observations on xeroderma pigmentosum and certain occupational epidermal neoplasias of the human skin have shown beyond much doubt that the numerous and varied early neoplastic lesions do *not*, in general, turn one into another. It may be noted in passing that this is one of the kinds of observation that can be made more easily and accurately on human skin than on mouse skin. The lesions X, Y and Z may emerge concurrently or consecutively in time, but in either event they are independent of one another and different from the beginning as are the hyperplastic nodules and the plaques in mammary neoplasia in mice.

The diagrams on Figs 36(b) and 36(c) can be combined to give, with slight elaboration, Fig. 37 which shows the *possible* developmental relationships of

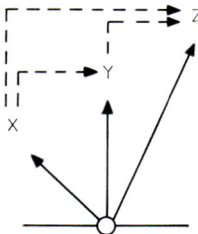

FIG. 37. Possible relationships of similar lesions resulting from consecutive or concurrent development.

three lesions X, Y and Z of increasing gravity. The actual developmental relationships can be determined only by empirical observation. In a later paragraph it will be shown that this diagram is applicable in general and in detail to early neoplasia of the human uterine cervix. In Fig. 37 and in subsequent diagrams the solid lines indicate *direct* development from a region of incipient neoplasia and the broken lines indicate *indirect* development by progression. Lesion Y can develop either directly from the incipient neoplasia or indirectly by progression in X. Lesion Z can develop in at least three ways, directly from the incipient neoplasia, by progression in Y or by progression in X by-passing Y.

Progression as studied in clinically evident lesions is an important mechanism in the clinical course of many neoplastic diseases but it is not apparent in all of them and it is not applicable without re-definition to the crucial but mysterious events of the pre-clinical or induction period. Visible lesions, even those of microscopic size do not reveal the whole complexity or extent of neoplastic diseases. The concept of incipient neoplasia, or something like it, is needed to deal with the clinically "silent" period. The lesions X, Y and Z have different assortments of independently variable characters varied within a range set by the neoplastic *capacity* of the incipient neoplasia and that capacity seems to depend on the intensity and, probably still more, on the duration and repetition of the carcinogenic stimulus, where one is identifiable. Repetition of the carcinogenic stimulus certainly increases the capacity for malignant neoplasia and in some systems may be an essential condition for it. More controversially, the neoplastic capacity may increase autonomously with time and without repeated carcinogenic stimulation. This is represented in Fig. 38 by sloping the base-line of incipient neoplasia upwards. The three

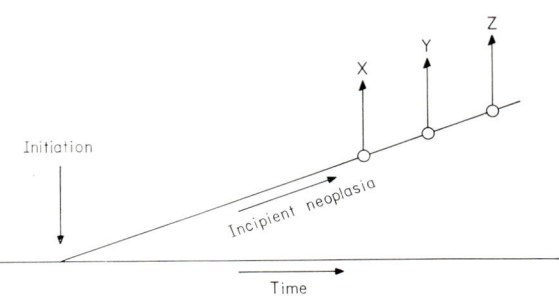

FIG. 38. Diagramatic representation of the increase of neoplastic capacity with time.

lesions X, Y and Z are shown emerging directly at successive times from the incipient neoplasia; on account of the rising neoplastic capacity of the incipient neoplasia, the three lesions may be expected to be of increasing gravity *when they first emerge*.

The composite diagram of Fig. 39 represents a generalized concept of patterns of neoplastic development. It is based on the proposition that *initia-*

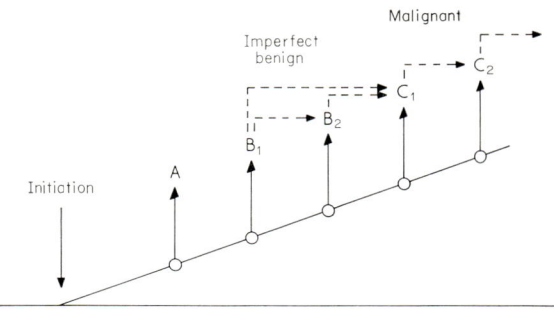

FIG. 39. Composite generalized diagram of possible paths of neoplastic development.

tion quickly establishes a region of *incipient neoplasia* whose neoplastic capacity increases with time, with or without prolongation of the initiating stimulus, and from which varied lesions emerge concurrently or consecutively over a substantial period of time. At present, incipient neoplasia is not consistently recognizable by overt signs but only by subsequent behaviour or experimental tests. The emergent lesions are divided into three main groups, A, B and C, most conveniently illustrated by reference to epidermal neoplasia in its most severe manifestations in experimental animals and in man.

1. Group A Lesions

Group A lesions comprise the trivial ones exemplified by freckles and other familiar consequences of over-exposure of the human skin to sunlight. Most of the lesions are transient; none is "neoplastic" as judged by the usual clinical or pathological criteria and none has a demonstrated capacity for neoplastic development. Whether or not they should be considered as minimal manifestations of neoplastic disease, neoplasia without a future, is arguable. Many, perhaps all, of them are probably side-effects attributable to non-specific irritation by the carcinogenic agency but they deserve notice because they give warning of substantial exposure to a potentially carcinogenic stimulus.

2. Group B Lesions

The earliest emerging lesions that are recognizably neoplastic by pathological criteria include those that are often called "precancerous" or "preneoplastic". These terms have been troublemakers for half a century or more and are best avoided. Without being accurately descriptive, they are unreliably prophetic. The Group B lesions of human and animal skin are mostly benign or "imperfect" neoplasms that, on account of anomalous associations or dissociations of independently variable characters do not permit an unequivocal diagnosis of either benign papilloma or malignant carcinoma. They include the following varieties:

(1) Conditional tumours that grow only whilst they are stimulated by extrinsic factors, regress when those factors are withdrawn and recur from a persistent focus of residual neoplasia when the stimulus is restored.
(2) Papillomas that regress spontaneously and do not recur.
(3) Indolent papillomas that persist for a long time but grow scarcely at all.
(4) Papillomas that grow progressively and remain papillomas without changing their habit of growth.
(5) Papillomas in which carcinoma originates by progression.

(6) Kerato-acanthomas or carcinomatoids, that resemble carcinomas histologically but regress spontaneously.

(7) Non-invasive carcinomas *in situ* in which invasive carcinoma may, or may not, originate by progression.

In Fig. 39 the B lesions are divided arbitrarily into two sub-groups, B_1 and B_2 of which B_2 is credited with the greater neoplastic capacity. Spontaneously regressing papillomas may be assigned reasonably enough to the B_1 subgroup and carcinomas *in situ* to B_2 but some of the other lesions cannot be so easily assigned owing to the variety of the associations of independently variable characters. At each period of time lesions of considerably varied neoplastic performances and capacities can emerge from the same region of incipient neoplasia as shown diagrammatically in Fig. 37. The situation would probably be better represented by numerous lines fanning out from the regions of incipient neoplasia so as to overlap and represent lesions intermediate between B_1 and B_2. The apparently sharp distinctions between B_1 and B_2 lesions, which the diagram might unfortunately suggest, are probably illusory, as they are in most biological phenomena, and are attributable to an insufficiency of observations.

The study of Group B lesions has been clouded for years by semantics which I have tried to avoid. Figure 40 illustrates important characteristics of

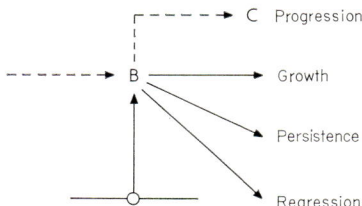

FIG. 40. Possible fates of Group B ("precancerous") lesions.

the group. First, B lesions may originate directly from the incipient neoplasia or indirectly by progression in another B lesion. Second, B lesions in general are subject to at least four different fates: (1) *Progression* to another B lesion of higher neoplastic capacity or to carcinoma; (2) *Progressive growth without qualitative change* in structure or behaviour; (3) *Indolent persistence* without qualitative change and with minimal growth or none; or (4) *Regression* and permanent disappearance. Of these four possible fates, progression to carcinoma is, in most systems, the *least* likely outcome. Progression, when it occurs is usually a focal change affecting only a small fraction of the available tissue. When the lesions are multiple, progression takes place as a rule in only one or a small proportion of them. If the capacity for progression to carcinoma is high, as it is in xeroderma pigmentosum for example, the eventual emergence of a carcinoma either directly from the incipient neoplasia or indirectly by progression in a B lesion is predictable but the time and site of emergence are not predictable. The probability that progression will occur in a particular kind of B lesion can be estimated statistically but its occurrence in a particular lesion or a particular patient cannot be foreseen. One of the basic objections to the use of the predictive labels, precancerous, preneoplastic and the like, for B lesions is that the majority of them are not "pre" anything; they are dead-ends of neoplastic development, terminal products with no neoplastic future ahead of them.

3. *Group C Lesions*

The Group C lesions comprises growths having the cardinal feature of malignant tumours. The arbitrary division into two sub-groups C_1 and C_2 is made to draw attention to the differences in clinical gravity and other characteristics within the groups; the basal cell carcinomas may be assigned to the C_1 sub-group and the epidermoid carcinomas to the C_2 sub-groups. Some of the differences are present from the beginning; others are established by progression. Further steps of progression may occur after a carcinoma has been established so that the tumours "go from bad to worse". Owing to the wide scatter of the associations of independently variable characters the diversity of tumours within the C group is far greater than the division into two sub-groups implies.

The diagrams are applicable primarily to regional neoplasia characterized by a wide expanse of incipient neoplasia from which multiple qualitatively varied lesions emerge concurrently and consecutively over a long period of time (Foulds, 1965). The region is coextensive with the area of exposure to the carcinogenic stimulus where one is identifiable; it can include both of paired organs, notably both breasts, and it can extend from one tissue to neighbouring related ones as from uterine cervix to vagina but it does not extend to unlike, unrelated tissue, as from epidermis to mammary gland. The circumstances most conducive to the induction of regional neoplasia were considered in Chapter 2. It has been widely presumed that most sporadic neoplasias of man and also of animals are, at their origin localized to a small primary focus and that the wider extension of the neoplastic disease depends on continuous or discontinuous extension from the primary focus; the multiple growths which may ultimately develop in this way are secondary tumours whereas the multiple neoplasms of regional neoplasia are independent primary growths. In the following pages the regional and localized types of neoplasia, which are probably not so sharply distinct as might appear, will be further considered with especial reference to neoplasia in man.

B. Regional Neoplasia

The whole gamut of lesions represented in the diagrams is not often easily recognizable in the neoplastic diseases in man. The diagrams are most clearly applicable to environmentally-induced epidermal neoplasias evoked by long exposure to potent chemical carcinogens and to xeroderma pigmentosum where the sensitivity of the skin is extremely high. The skin of a patient with advanced xeroderma pigmentosum could serve as an almost complete museum of epidermal neoplasms. Regional neoplasia is conspicuous also in the uterine cervix, in the urinary bladder and, sometimes at least, in the mammary glands. Some of the information about these neoplasias has been reviewed previously (Foulds, 1958a, 1965) and they will be considered in more detail in Volume 2. They will be discussed in summary fashion here mainly to illustrate the close relevance of the general principles of regional neoplasia to clinical practice.

1. Neoplasia of the Uterine Cervix in Women

There is substantial evidence to show that neoplasia of the uterine cervix is a regional disease based on widespread incipient neoplasia, lining much or all of the cervical canal and extending into the vagina. Varied lesions classifiable as A, B and C lesions develop over a long period of time. Those described by some authors as "minor atypias" together with the lesser degrees of dysplasia probably belong to Group A and have little if any neoplastic capacity. The important "early" or "precancerous" lesions are the dysplasias and carcinomas *in situ* which may be assigned to Groups B_1 and B_2 respectively, with the reservation that there is no sharp demarcation between them. The invasive tumours clearly belong to Group C. Clinical and pathological studies have demonstrated that the developmental relationships between the diverse lesions correspond well with those predictable from Figs 39 and 40. The main uncertainties concern the relative frequencies of the various possible pathways. It is widely believed, although disputed by a few pathologists, that the majority of invasive carcinomas originate by progression in Group B lesions and most often in carcinoma *in situ*. It is well established, nevertheless, that some of them do not so originate from a precursor lesion but emerge from the incipient neoplasia with their invasive qualities already established. The prospective fates of the Group B lesions are those characteristic of B lesions in general as shown in Fig. 40. Here again, the relative frequencies of the several fates are controversial. It should be emphasized that carcinoma *in situ* does not shade into invasive carcinoma and does not grow into it by mere extension in space and time; invasive carcinoma is established by a decisive change effected by focal progression within the carcinoma *in situ* whereby neoplastic cells acquire invasive properties, which they did not manifest before. To describe carcinoma *in situ* as a "cancer" or, without qualification, as a "carcinoma" is erroneous and in clinical practice, dangerous. An understanding of the natural history of this disease and of the nature and significance of the developmental processes and lesions that are involved in it is essential for the rational and humane management of patients who present themselves with Group B lesions.

It is noteworthy that no inducing carcinogen has been convincingly implicated in neoplasia of the uterine cervix in women and that the developmental history has been elucidated in considerable detail entirely by observations on human subjects. Experimental studies have made no important direct contribution to the understanding of the disease; their valuable indirect contribution has been in providing general principles of neoplastic development that greatly facilitate the direction and interpretation of clinical observations.

2. Neoplasia in the Urinary Bladder

Neoplasia of the mucosa of the urinary bladder is of two main types. One type is characterized by *solid* tumours, usually solitary, that are carcinomas *ab initio*. The other type is characterized by a multiplicity of *papillary tumours*. The industrial disease is entirely similar to the sporadic disease in the general population and so also is neoplasia experimentally induced in appropriate animals. Reference will be made later to solitary tumours. The present concern is with the papillary type of disease.

It has been inferred from both clinical and pathological observations that the papillary disease in man and in animals is a regional one affecting a wide expanse of vesical mucosa and, perhaps, all of it. Owing partly no doubt to the greater difficulties of observation, the identification of Group A lesions is more doubtful than in epidermal neoplasia and Group B lesions also are less conspicuous. They include benign or imperfect papillomas with a low capacity for progression to carcinoma. The origin of carcinoma by progression in a papilloma has been described in the industrial disease in man and in the experimentally induced disease in animals but some surgeons consider it to be rare. The disease can often be controlled for many years by fulgurating these early lesions but new primary tumours are apt to emerge consecutively at different sites. As time passes the newly emerging tumours tend to be increasingly dangerous *ab initio* and sooner or later one, or more, of them grows invasively, disseminates and kills the patient. This pattern of neoplastic development, distinguished by the consecutive emergence of new primary tumours with ascending degrees of gravity as represented diagrammatically in Fig. 38, is consistent with the supposition that the neoplastic capacity of the incipient neoplasia increases progressively with time.

3. Mammary Neoplasia

Mammary neoplasia in the "high-incidence" inbred strains of mice is clearly a regional disease. All, or nearly all the mammary epithelium in an adult female seems to be in a state of incipient neoplasia; if tumours are excised as soon as they are detected, new ones continue to emerge at other sites throughout remaining life. The usual course of neoplastic development traverses intermediate Group B lesions, which are usually multiple but some malignant tumours first erupt as such without relationship to a preceding B lesion. In women the regional pattern is most evident, as described by Cheatle and Cutler (1931), when carcinoma develops on the basis of chronic cystic disease. Cutler (1962) estimated that about 20% of all mammary carcinomas originate in this way; some authors believe that the percentage is much higher (e.g. Willis, 1967). There are several indications that in the absence of cystic disease the mammary neoplastic disease is more extensive than the clinical observations suggest. Pathological observations indicate that many or most mammary carcinomas are of multifocal origin as Willis maintains. Furthermore the extent of the incipient neoplasia is probably much greater than the visible lesions imply. Statistical analysis indicates that the presence of an apparently solitary carcinoma in one breast entails a higher than average probability of a second carcinoma in the contralateral breast. When, following radical mastectomy, a carcinoma developed in the contralateral breast after a longer or shorter interval, it used to be customary to interpret the second tumour as a metastatic tumour derived from the excised one. The probability that the second tumour is a new primary emerging from an incipient neoplasia of the previously tumour-free breast is now being increasingly recognized. Similar considerations apply to local "recurrences" after surgical or radiological treatment of a primary tumour; they may be attributable, not to a failure to remove or destroy the whole of the primary tumour, but to a failure to remove the whole region of incipient neoplasia, from which new independent primary tumours may develop.

4. Other Examples of Regional Neoplasia

Regional neoplasia is most conspicuous in surface epithelia that are available, in greater or less degree, to clinical observation during life. It is much more difficult to recognize and study in less accessible sites but evidence that it is more common than formerly supposed is accumulating. It is evident in familial polyposis of the colon in man; progression to carcinoma occurs almost inevitably in every patient but in only one or a small fraction of the numerous polyps, which are Group B lesions with the usual Group B properties. The regional pattern with progression and development through intermediate Group B lesions seems to be the rule in experimentally-induced neoplasia of endocrine tissues in animals (Bielschowsky, 1955) and good examples have been described in neoplasia of the thyroid gland in man (Frazell and Foote, 1958). In general, regional neoplasia with consecutive emergence of multiple, varied neoplastic lesions is most evident in tissues recognized by clinicians as "precancerous" (Smithers, 1960).

5. *Multifocal Origin of Tumours*

The multifocal origin of tumours of one general type within a single tissue or group of tissues is not uncommon. The tumours may be distributed for example through the epithelia of the oral cavity and pharynx or vagina and vulva. These associations of sites have been attributed to the exposure of related tissues to a common carcinogenic environment. Whilst this may be correct, the common responsiveness of the related tissues to a particular carcinogenic stimulus might be stressed; contiguous tissues of diverse type do not similarly respond. Tumours may occur in unrelated organs and tissues of the same patients but, apparently no more frequently than can be accounted for by chance association. There is no evidence in animals or in man of a generalized "neoplastic tendency" common to tissues of many or all kinds. Multiple tumours are particularly common on the skin and may be extremely numerous. The remarkable "turban tumour" of man consists of a large number of primarily independent primary basal cell carcinomas that have coalesced secondarily to form a mass of neoplastic tissue covering the whole scalp. (Moertel, 1966; Moertel *et al.*, 1961, *a*, *b*, *c*; Newman and Cromer, 1959; Smithers, 1960; Willis, 1967).

The tumours discussed in the previous paragraph have a regional distribution; they have been discussed separately from the other regional neoplasias already described because, so far as can be gathered from the descriptions, all the tumours in a particular lesion are of the same type and classifiable as Group C lesions; a progression through B lesions either does not occur or if it does, has not been recorded.

Willis (1967), Smithers (1960) and others have stressed the multicentric as opposed to the unicentric origin of tumours and evidence given in this chapter supports their view. It may be well to remark that "focal" whether applied to the origin or to the progression of tumours does not mean or imply that the initial change is in a single cell; it means only that the changes are limited to discrete groups, masses or sheets of cells within a wider expanse of tissue. It implies that the changes do not affect the whole of the available tissue uniformly and diffusely.

In practice it would be difficult to prove decisively that a tumour can originate by some change in a single cell and even more difficult to prove the negative proposition that none ever does. Nevertheless, it is possible to say with considerable assurance that many of the early lesions, comprising the Group B lesions do not stem from a single altered cell; the histological structure of carcinoma *in situ* is incompatible with an origin from a single abnormal cell (Orr, 1958; Willis, 1967). Reasons were given earlier for thinking that the mammary plaques of mice originate by neoplastic outgrowths from a substantial length of mammary duct. On the other hand, the multiple Group B lesions of regional neoplasia at their origin are focal and discrete and I am inclined to doubt the reality of simultaneous conversion of a normal tissue *en masse* into neoplastic tissue. Even the "turban tumours" are, at their onset, focal and discrete.

C. LOCALIZED NEOPLASIA

Regional neoplasia, characterized by a multiplicity of lesions emerging concurrently or consecutively over a long period of time seems, to be a consequence of strong carcinogenic action or a high degree of sensitivity of the target tissue or, perhaps of both factors acting cooperatively. The reverse conditions, weak carcinogenic action or low sensitivity of the target tissue or both, seem to favour a localized manifestation of neoplasia characterized by a long induction period terminated by the emergence of tumours or, often, a solitary tumour that is a carcinoma from the first and not preceded by any precursor lesions. The two contrasted situations can be arranged experimentally by appropriate choices of animal species and carcinogenic procedure.

Rat and guinea pig skins are notoriously refractory to the action of chemical carcinogens but long continued applications of 9:10 dimethyl-1:2 benzanthracene, one of the most potent available carcinogens, induces epidermal neoplasia where the first or only manifestation is usually a solitary carcinoma (Berenblum, 1949). A similar result can be produced on sensitive mouse skin by limiting carcinogenic action to a single exposure. Under these circumstances, also, tumours that are often solitary and that are carcinomas from the beginning, develop after a long delay. If croton oil is applied when carcinomas erupt, it elicits a crop of multiple benign or imperfect neoplasias of the same kind as those that could have been evoked by applying croton oil soon after the single application of carcinogen (Shubik, 1967*a*). These results indicate that the single dose of carcinogen quickly established a region of incipient neoplasia of limited neoplastic capacity which was disclosed only by experimental interference and, further that the capacity for benign or imperfect neoplasia was retained without substantial modification throughout the long induction period of the carcinomas. The probable interpretation of these experiments is that carcinoma originated by focal progression in a persistent region of incipient neoplasia established by the initiating carcinogenic stimulus. The eventual growth of a malignant neoplasm after a prolonged period during which no sign of neoplasia of any kind was discernible is, at present, inexplicable. It may be that progression requires a long "build up" or that it must be "triggered" by circumstances that develop very slowly.

There is some circumstantial clinical evidence to support the idea that solitary carcinomas develop by progression within a region of incipient neoplasia and can thus be accommodated in the generalized scheme on the assumption that the lesions of Groups A and B are suppressed. There is no reason to believe that the regional and localized types of neoplasia in the urinary bladder are evoked by different aetiological factors. Both types of neoplasia are found in the sporadic as well as in the industrial diseases and, furthermore, in the experimentally-induced neoplasia in animals. Possibly individual differences in sensitivity to carcinogenic stimuli decide which form of neoplasia shall be manifested.

A tumour that presents clinically as a solitary growth is not necessarily of unifocal origin; it may have been formed by coalescence of several neighbouring independent foci of neoplasia. Furthermore the existence of a substantial "field" of neoplastic tissue surrounding the visible tumours has been inferred from the observation that neoplastic epithelium at the periphery of a tumour of a surface epithelium often merges imperceptibly with the surrounding "normal" epithelium so that it is impossible to tell where the one ends and the other begins. Brunschwig and Tschetter (1938) studied the transitions at the periphery of epidermal tumours of human and mouse skin and came to the conclusion that the appearance of gradual merging of normal and neoplastic epithelium was attributable to a "progressive cancerization" of normal epithelium extending centrifugally from the periphery of the tumour. Willis (1944, 1945, 1967) confirmed and extended the observations on epidermal tumours of man and elaborated the hypothesis of "spreading cancerization" of normal epithelium at the periphery of the neoplasms. Willis proposed that epidermal tumours of man develop by the gradual and general transformation of a whole "field" of epidermis, that corresponds in extent with an area subjected to repeated carcinogenic stimulation. The carcinogenic stimuli evoke slow progressive changes in the dermis as well as in the epidermis and eventually, evoke a "precancerous hyperplasia" which persists for a short or a long time until, usually at a central focus or at several "high-potential foci", the hyperplasia passes into irreversible neoplasia. The carcinogenic stimuli, it is supposed, act similarly but not equally on all the epithelium of the exposed area and carcinoma begins where the stimuli have been maximal. As cancerous proliferation and invasion progress at these foci, cancerous change of the surrounding unstable epidermis takes place in a steadily enlarging area around them and continues until the whole of the "field" has become cancerous. Early carcinomas of the skin become available for study before this process is complete and the surrounding epidermis often thickens gradually towards the margin of the tumour where the hyperplastic epithelium is continuous with the neoplastic epithelium with no abrupt change from one to the other. When the whole field is cancerized, there is an abrupt break between normal and neoplastic epithelium at the periphery of the tumour. Thereafter, the tumour grows solely by proliferation of its own neoplastic cells.

Willis insists that the histological findings are incompatible with an invasion of normal epidermis and he rejects suggestion that some infective agent or "malignant influence" may pass from the tumour and induce neoplasia of the neighbouring epidermis. He also rejects, perhaps prematurely, suggestions that the gradual transition might be due to a secondary healing together of normal and neoplastic epithelium. Rous (1913) described "False transitions between normal and cancerous epithelium" obtained by applying finely ground Flexner-Jobling adenocarcinoma to the wound made by removing a disc of skin from rats. He found that the regenerating epidermis often united with the carcinoma tissue and that there was often a gradual transition from stratified squamous epithelium to the epithelium of adenocarcinoma. Rous concluded that the histological appearance of transitions can result from a secondary union of normal and cancerous epithelium and probably do so result frequently. I have seen secondary unions between the epithelium of mammary plaque tubules and the epidermis of mice, as described elsewhere (p. 157). It is admittedly difficult to understand how this transition is established but this is not sufficient reason for denying that it can happen.

The Field Theory of Willis corresponds in some respects with the concept of regional neoplasia presented here but differs from it sharply in others. Willis does not name any neoplastic states other than "precancerous" and "cancerous" ones. The inadequacy of these terms for describing the diverse and complex happenings that are seen during epidermal carcinogenesis becomes most evident when the field theory is extended to cover solar and senile keratosis and xeroderma pigmentosum. (Willis, 1967) writes that in fair-skinned people with widespread keratoses, the whole of the exposed skin is "ready to become cancerous" and the multiple tumours that actually develop merely denote "foci of maximal cancer potential in these areas". The statement takes no account of the great variety of epidermal lesions most of which are not "cancerous" and do not become "cancerous". It cannot be agreed that the whole region is "ready to become cancerous"; so far as I am aware it never does. Even in xeroderma pigmentosum, only a fraction of the exposed skin becomes "cancerous". To attribute the focal emergence of tumours, most of which are not "cancers", to the existence of foci of maximal cancer potential is unconvincing and essentially tautological; this asserts merely that tumours are more likely to emerge at some places than others. In another place, Willis proposes that tumours are localized to areas of maximal carcinogenic stimulation. Since in most carcinogenic experiments and in environmental neoplasias of man the exposures to the carcinogenic circumstances are repeated frequently over a long period of time, it seems likely that in the long run the degree of exposure would be approximately uniform throughout the exposed regions. If this be true the cause of the inequalities of neoplastic response must be sought in the responding tissues instead of in the carcinogenic stimulus. In truth, the focal origin of tumours and the focal progression in tumours have not yet received a plausible explanation.

CHAPTER 4

Definitions, Classifications and Terminologies

One of the greatest evils in modern cancer research is the abuse of definitions that are based on the preconceived ideas of individual investigators highly expert over an extremely narrow range of neoplastic phenomena. Statements such as "Cancer is, by definition, somatic mutation" are indefensible in logic and disastrous in application; they presume as axiomatic what remains to be proved or, sometimes, what has already been disproved. Somatic mutation supplies a reasonable working hypothesis for the study of neoplasia; as a *definition* of neoplasia it is intolerable. No satisfactory definition of neoplasia is available or will be until a great deal more is known about its nature and properties; cancer research will have reached an outstanding landmark when it becomes possible to define neoplasia in biological terms. Meanwhile, the concept of neoplasia is essentially descriptive and based on the collective experience of many generations of clinicians, pathologists and laboratory investigators but primarily on the empirical observation of neoplasia in man.

Early pathologists recognized two main kinds of neoplasm or new growth: namely, benign neoplasms and malignant neoplasms. It seems necessary to insist that benign tumours were understood to be neoplasms as much as were malignant tumours. Terms like "benign conditions" are becoming increasingly frequent in the literature; judging from the contexts, "benign conditions" may include boils and bunions as well as benign neoplasms but it is often impossible to tell which. It is tempting, if unkind, to guess that the authors don't know either. Willis has correctly affirmed that there is no sharp division between benign and malignant tumours. Smithers, in agreement, writes of a "spectrum" of tumours. Undoubtedly benign neoplasms stand at one end of the spectrum and unquestionably malignant ones are at the other end but intermediately there are tumours that are not plainly either one or the other, and for these there is no agreed terminology. Moreover, the pathology of neoplasia in man has provided no terminology for the stage of neoplastic development that precedes the first emergence of a "tumour" or "growth" recognizable by either clinical or microscopical procedures.

Ewing entitled his classical book on the pathology of neoplasia. *Neoplastic Diseases*, thereby implying a wide variety of manifestations superimposed on a basic unity. It seems to me that most of his successors have taken a retrograde step by using the much less comprehensive and adaptable title *The Pathology of Tumours*. In this book, neoplasia is used comprehensively to include all stages of neoplastic development and all types of neoplasm, benign, malignant and unclassifiable. The first stage of neoplastic development is designated *incipient neoplasia*, "the beginnings of neoplasia"; it is not accompanied by microscopically-demonstrable changes in the tissues. Incipient neoplasia is presumed to begin with the inception of neoplasia following the first exposure to carcinogenic circumstances. The term *pre-clinical* is applied to the period preceding the emergence of a clinically-recognizable tumour of growth; it includes the stage of incipient neoplasia but also the stage during which a tumour is growing to clinically-detectable dimensions. The unequivocally benign and malignant tumours at the two ends of the "spectrum" of emergent growths are so designated. Some of the "imperfect" or "anomalous" neoplasms, in some sense intermediate, but not always in the same sense intermediate, between the two extremes have received special names and will be mentioned later. The most frustrating gap in the terminology of the pathology of tumours in man is the lack of a satisfactory name for the so-called pre-cancerous lesions, which are "intermediate" in a special sense. Reasons were given earlier for rejecting "precancerous". "Pre-neoplastic", sometimes now applied to lesions of the same general kind, is even more objectionable; it can only mean that the lesions are not neoplastic whereas I maintain strongly that they are neoplastic and that this should be recognized in their designation. Provisionally they will be called Group B lesions to indicate their position in the schema of neoplastic development discussed in the previous chapter. Less seriously, the pathology of tumours provides no terminology for neoplasia artificially and indefinitely prolonged by transplantation *in vivo* or cultivation *in vitro*; sometimes, at least, it has special features which will be mentioned in due course. Reference is often made in the literature to "experimental tumours" which may apply to tumours whose existence has been prolonged by transplantation or tissue culture but also to primary experimentally-induced tumours and in most contexts it is essential to know in which sense it is being used. For the sake of clarity, therefore, "experimental tumours" should be further specified as experimentally-induced, transplanted or cultured tumours.

It is natural and proper that the most dangerous tumours of man should loom large in the minds of research workers as well as clinicians. The pioneer laboratory investigators focused their experiments on malignant tumours for at least two reasons. In the first place they had to justify their existence by convincing sceptical surgeons that they were studying true, malignant tumours, similar in all respects to those found in man and that, on this account, their experiments were relevant to the study of human cancer. This was more

4. DEFINITIONS, CLASSIFICATIONS AND TERMINOLOGIES

difficult than may be generally understood at the present time. In the second place, broadly speaking, malignant tumours were more amenable to experimental investigation than benign tumours. The first of these reasons no longer operates, which may be a mixed blessing. The discipline imposed by a need to check the artificialities of the laboratory against the realities of human life and death is salutary and becomes the more necessary with the tendency of "cancer research" to range even more widely and deeply into chemistry, physics, biology and their various hybrids. The second reason is still operative and may be responsible, in part, for the continuing and even increasing neglect of benign tumours to such an extent that many investigators seem to use the term "malignancy" as synonymous with neoplasia. This is unfortunate because it is easily credible that benign tumours, being the simplest manifestations of neoplasia, might yield the more insight into its essential nature (Foulds, 1954, 1964b; Yoshida, 1966).

Cancer and *Malignancy* are primarily clinical concepts and are not amenable to precise definition. Smithers (1960, 1964) maintained that cancer is just a shortened way of saying something that cannot be simply defined. By common usage the word has come to represent the tumours at the more dangerous end of the spectrum. The diagnosis of cancer in man is made on evidence of behaviour and neoplasms are arbitrarily classified as cancerous or not by the number and degree of behavioural characteristics that happen to be observed, none of which is peculiar to cancer but all of which, when they occur together, may form a characteristic picture. Smithers considers "cancer" a bad term for two main reasons: first, it has acquired undesirable emotional overtones of a terrifying destructive disease of cells and second, the phenomena to which it applies are terminal events in a long progressive chain of circumstances. Despite these disadvantages, as Smithers recognizes, the word does convey some idea of the clinical and pathological events that are under consideration. In medical practice, with clinicians and pathologists working close together and familiar with one anothers' habits of mind, the use of the word *cancer* probably does not imperil adequate communication seriously and *cancer* is certainly convenient for colloquial reference to the most dangerous forms of neoplasia.

In laboratory practice, the position is far different and *cancer* does not, in fact, allow adequate communication within or between the various disciplines. To quote Smithers (1964), "Since the meaning of the word can have no precise boundaries this side death, it would be better if we were to abandon its use in communications having some pretence to scientific reporting." I agree with this statement. I am in further agreement with Smithers in continuing to use the word in certain circumstances. If a publication dealing with "cancer" is to be quoted, as many such publications ought to be, probably it had best be done as a rule in the author's terminology; to translate "cancer" into some more precise terms without deep insight into the author's mind is to risk serious misinterpretation of his observations and misrepresentation of his intentions.

In laboratory research, the troublesome word is "malignancy", to which many of Smithers' criticisms of "cancer" are applicable. Malignancy is an abstract concept. To speak of it or to study it as if it were a biological entity or unitary property of neoplastic cells is futile; to do so is to concentrate on the abstraction and ignore the realities. Malignant behaviour depends on certain combinations of a number of independently variable characters none of which is peculiar to malignant neoplasms. Consistently with this statement, Yoshida (1966) has recently noted that it is the *totality* of the properties of a tumour that determines its malignancy. The characters that are usually listed as indicators of "malignancy" are rapidity of growth, invasiveness, metastasis, autonomy and anaplasia of which invasiveness and metastasis are commonly presumed to be decisive and pathognomic. In fact, none of them is pathognomic. In a normal pregnancy, the trophoblast of the early embryo invades the maternal tissues. Blair Bell (1925) inferred from this that the trophoblastic cells "are of sheer necessity malignant in character" and said in a later paper (1926) that "We regard the chorionic epithelium as being a normally malignant tissue that comes under somatic control". The argument seems to be that, since malignant neoplastic tissue is invasive, all invasive tissue is malignant.

Nicholson (1950) and Stewart (1952) have expressed similar opinions. The basic fallacy and inherent danger in this sort of word-play is that on the basis of the observation of one or two of a group of characters that commonly are found in association with one another, the presence of the remaining members of the group is inferred or implied without any basis of observation at all. The abundant evidence for the dissociability and independent variability of characters shows that inferences with implications of this sort are not valid in neoplastic diseases. Invasiveness is a good but not infallible sign of "malignancy" only if it is present in a tissue that for other reasons is demonstrably neoplastic. The acquisition of the quality of invasiveness by cells of a benign tumour, as can occur by progression, establishes a malignant tumour but the manifestation of invasiveness in normal or hyperplastic tissues does not constitute neoplasia.

The properties listed above as more or less characteristic of malignant tumours will be considered separately in the following pages with the warning that this entails abstraction. Some of these properties are disclosed by clinical behaviour and others by histological and cytological structure. Both kinds of observation have to be taken into account in assessing the "totality" of properties that is acceptable evidence of malignancy. The "clinical" or "behavioural" properties will be discussed first but some preliminary attention

4. DEFINITIONS, CLASSIFICATIONS AND TERMINOLOGIES 95

to the histo-pathological properties, with special reference to terminology, may be useful here.

Histopathological classification is primarily descriptive but has acquired undertones of retrospection in histogenetic classifications and overtones of prediction when used in assessing prognosis. The descriptions, being based on direct observation, can be precise but the retrospections and predictions are inferential and prone to error. It has been pointed out already that the descriptive terminology is defective in several respects. Adenoma and carcinoma are not adequate for describing the whole range or "spectrum" of epithelial neoplasms. I cannot accept the view, which seems to be held in some quarters, that an epithelial tumour must be either an adenoma or a carcinoma; some epithelial tumours are not plainly either one or the other. Apart from defectiveness in this and similar ways, there is a good deal of inconsistency in application of the terms adenoma and carcinoma; sometimes the terms are used, as I consider they should be, in a strictly descriptive sense based solely on histological structure but sometimes clinical behaviour is taken into account to modify or even reverse the descriptive terminology. As an example, one of the commonplace mammary tumours of mice (Dunn's Type A) is structurally an adenoma but it is denied that name because it is clinically "malignant". On the other hand the chemically-induced mammary tumours of rats mostly have the structure of carcinomas and are usually so designated but some writers, familiar with their habits, call them adenomas because they are not clinically "malignant".

These "adenomas" and "carcinomas" have been cited here as good examples of, and in support of Smithers' (1960) complaint about, the use of terminologies conveying "a mixture of descriptive, evaluative and predictive meanings". They are, as Smithers' says, potent sources of ambiguity; they lead to unnecessary and futile argumentation due not so much to a disagreement about facts as to a breakdown of verbal communication amongst the disputants. Increasing knowledge and experience necessitate modifications in the usages of words and, whatever may have been in the minds of the nineteenth century pathologists, I maintain that words like adenoma and carcinoma should be stripped of their secondary undertones and overtones of meaning and used in a strictly descriptive sense. In the examples cited I see no objection to using the terms "malignant adenoma", or "benign carcinoma" whenever structural pattern and clinical behaviour are at variance.

The undertones of retrospection inherent in histogenetic classifications of tumours are distasteful to myself; if strictly applied the classifications can be misleading or worse. Fortunately, they are not always applied strictly and when used moderately and flexibly, as by Willis for example, they are less objectionable although, in my view, not particularly helpful. The overtones of prediction are far more dangerous and are well illustrated by carcinoma *in situ* of the human uterine cervix.

Carcinoma *in situ*, as others have pointed out, is a contradiction in terms, but no more so than "aleukaemic leukaemia". Some authors have claimed to have cured, say, 95% of their patients with "cancer" of the uterine cervix. Admiration of this remarkable achievement abates when closer analysis of the reports indicates that some 95% of the patients did not have "cancer" but only carcinoma *in situ*. The term carcinoma *in situ* seems to have started an insidious chain of mental processes which may be summarized as follows: carcinoma *in situ*-carcinoma-cancer-death. As a result of this chain of "reasoning", the appropriate treatment for carcinoma *in situ* was judged at one time to be panhysterectomy. Many women were subjected to this procedure until eventually it was realized that removal of the ovaries was not imperative; hysterectomy was then considered sufficient. More recently it has become apparent that a considerable number of women with carcinoma *in situ* of the uterine cervix, perhaps more than half, will cure themselves if left alone and that most of the others can be "cured" by a limited surgical procedure that does not impair their reproductive capacity. Meanwhile many young women have suffered unnecessary mutilation and loss of reproductive function and still more have suffered incalculable psychological trauma by being told that they have been, or are going to be, cured of "cancer". The objections to the use of the word "cancer" in this context cannot be brushed aside as mere pedantry; the usage has been a bar to accurate communication and the cause of severe, avoidable human suffering.

Much of the gist of the foregoing discussion was anticipated in a memorandum prepared by a committee of Scottish physicians from which the following quotation is taken: "It is much to be wished that we had an exact definition of cancer, those of the nosologists being very imperfect and insufficient. . . . If a just and exact definition of cancer cannot yet be formed, we must be satisfied with such a description as a correct history of the disease will afford. This, it appears has never yet been judiciously and accurately done: . . . It is much to be wished that we may no longer be deceived by ambiguous words or phrases or consider them as conveying to us any essential or practical knowledge."

The memorandum was written in 1802, published in 1806 in the *Edinb. med. Surg. J.* 2, 382–389 and reprinted, with full justification, in 1967, in the *Int. J. Cancer*, 2, 281–285. The last of the quoted sentences is perhaps even more apposite today than when it was written 167 years ago.

CHAPTER 5

Characteristics of Neoplasms

This chapter deals with the characteristics of established neoplasms both benign and malignant under three headings: I. The growth and spread of neoplasms; II. The histopathology and experimental pathology of neoplasms and III. Biochemical characteristics.

I. The Growth and Spread of Neoplasms

A. NORMAL GROWTH AND CELL PROLIFERATION

For the sake of clarity in exposition and in thought it is desirable to distinguish between *cell proliferation*, which applies to an increase in cell numbers, and *growth*, which applies to an increase of protoplasmic mass irrespective of cell division.

1. Growth

Weiss and Kavanau (1957) defined growth as the net balance of mass produced and retained over mass destroyed or otherwise lost. Somewhat similarly, von Bertalanffy (1960) defined it as the quantitative increase of a living system which results from prevalence of anabolism of building materials over catabolism. Growth entails the assimilation of non-living nutritive materials, the "building materials", into living protoplasm and its derivatives with the production of materials that are specific for the living system in question. Von Bertalanffy notes that although the definitions are clear their applications to particular cases are not. He insists that "growth" is an abstraction and a definition is meaningful only if it is clearly understood that, in general, growth is not a discrete phenomenon but more a certain aspect of the process of life. A definition can be only an "operational" one having a value exactly equal to its contribution to a specific problem (von Bertalanffy, 1960). For the present purpose "increase in protoplasmic mass" is adequate and has the merit of brevity. Comprehensive and detailed reviews of various aspects of growth by various authors may be found in the book edited by Nowinski (1960) and in discussions by Bullough (1967), Goss, (1964) and A. E. Needham

(1964). The main objective here is to distinguish it from cell proliferation. Although the term *growth* appears conspicuously in almost every mention of neoplastic disease, the real problem under discussion is usually one of cell proliferation.

Growth and cell proliferation usually run parallel courses but they are distinct, dissociable processes. It is usually thought that cells grow to a limiting size and then divide but Mazia (1961) believes, on the contrary, that cell division is a cause of cell growth and not a consequence of it. In general, there is no simple relationship between mitotic activity and cell size or mass (Holtzer, 1963). X-irradiation acts differently on growth and on the division of cells and nuclei. Irradiation may stop cell division without impeding growth; it may inhibit nuclear division but allow "unbalanced growth" of the cytoplasm so as to produce a mononucleated giant cell; or it may suppress cell division without inhibiting nuclear division to yield a multinucleated giant cell (Ducoff and Ehret, 1959; Swann, 1957, 1958).

Although a few organisms seem to go on growing indefinitely, the vast majority stop when they have reached the size characteristic of the species, this size being genetically determined. An individual cell has only a limited capacity for growth, related perhaps to the inability of a nucleus to control more than a limited amount of cytoplasm (Mazia, 1961).

In the full-grown animal, the cells of different tissues behave variously to maintain the adult size and form. The diversity of behaviour is recognized in Cowdry's well-known classification of vertebrate cells. Cowdry (1942) distinguished a class of *vegetative intermitotic cells* including the basal cells of the epidermis, the primordial blood cells and the spermatogonia, which form a permanent germinal layer or zone; the cells are not visibly specialized but, as Weiss has often insisted, they are "differentiated" in the sense that the cells of different tissues have different capacities for differentiation. The basal cells have short lives ended by cell division. Some of the daughter cells remain in the basal layer and eventually divide again; the others leave the basal layer and become *vegetative mitotic cells* which divide and then differentiate progressively until they reach a state of terminal specialization when they can no longer divide again. Cowdry calls the terminally specialized cells *fixed post-mitotic cells*. The neurones of the central nervous system are fixed post-mitotic cells; they have long life spans and their nuclei are highly active in transcription but, after foetal life, they scarcely ever divide; if they did, memory and learning might be severely jeopardized. It is noteworthy that these cells do not form neoplastic growths in man and cannot be induced to do so in animals by any known experimental procedure. The loss of the ability to multiply seems to be complete.

The cells of many adult organs and tissues, including liver, kidney, capillary endothelium and connective tissues are differentiated cells that ordinarily become senescent and die without dividing but that can multiply, sometimes at a great rate, in abnormal circumstances. Cowdry calls them *reverting post-mitotic cells*.

Although Cowdry's terminology is still in use, a different classification based largely on the observations of Leblond and his colleagues is now being increasingly used. The observations depend on the technical innovation of labelling dividing cells with a radioactive isotope, preferably by administering tritiated thymidine ($[^3H]$ thymidine), which is incorporated selectively into DNA during the preparations for mitosis. The label being permanent, except insofar as it is diluted during successive cell divisions, the fate of the daughter cells can be traced in autoradiographs of tissues obtained at various times after labelling. Using a labelling method, Messier and Leblond (1960) distinguished three kinds of cell populations namely *static*, *expanding* and *renewing* cell populations. Most tissues fall clearly into one or another of these categories, although the static and expanding populations are not sharply separable (Goss, 1964).

(A) STATIC CELL POPULATIONS

These are recognized by the complete absence of labelled nuclei. They comprise cells that scarcely ever divide in the adult animal. Neurones, neural retinal cells and, perhaps, muscle cells are assigned to this category.

(B) EXPANDING CELL POPULATIONS

These are recognized by the presence of a small number of labelled nuclei that persist for an indefinite length of time. The cells have long lives and rarely divide under normal circumstances although they can proliferate rapidly in pathological conditions. They are represented by the tissues of liver, kidney, exocrine glands, endocrine glands and lens and by the connective tissues proper and some skeletal tissues.

(C) RENEWING CELL POPULATIONS

These are characterized by large numbers of labelled nuclei indicating active cell division and subsequently by a rapid decrease in the numbers of labelled nuclei and their eventual disappearance indicating cell loss. They comprise the cells of the epidermis, endodermal epithelium, transitional epithelium, endometrial cells, germinal cells of the gonads, haematopoietic tissues and some skeletal tissues.

Surface epithelia are continually being renewed by loss of cells from the surface and their replacement by the multiplication of basal cells, so regulated that a *steady-state* is maintained. Leblond and his colleagues maintained that *only* the basal cells divide; mitosis and the preceding DNA synthesis are restricted to the basal layer and are never found in cells identifiable as belonging to the spinous or higher layers of the epithelium. The basal cells are, in fact, the *stem cells* of the stratified squamous epithelium. It should be noted that Leblond and his colleagues consider that the basal cells need not be restricted to a single layer in contact with the dermis but can form a multi-layered epithelium. They prefer to identify basal cells by their histological appearances than solely by their topography.

It has been widely presumed that the two daughters of a basal cell are different *ab initio*; the one, maintaining contact with the basement membrane, being destined to divide again and the other, separated from the basement membrane, being destined to form a specialized product and not to divide again. The differential behaviour has been attributed to *differential mitosis*, *asymmetric mitosis*, or *unequal division*, of which the first term is the best. Asymmetric mitosis (Mazia, 1961) auggests asymmetry of the mitotic figure for which there is no evidence. Unequal division (Foulds, 1963) suggests that the daughter cells are unequal in size, which is true of the cells of some tissues of plants and invertebrate animals but not generally true of the tissues of higher animals (Holtzer, 1963; Jensen, 1963).

Leblond and his colleagues have disputed the reality of differential mitosis. They studied in particular the oesophogeal squamous epithelium of rats and mice and inferred from their observations that the presumption that one of the daughter cells of a basal cell remains a basal cell whereas the other becomes a differentiating spinous cell, is true only in a statistical sense. To maintain the steady state, the addition of one cell to the basal layer by mitosis must be balanced by the migration of one cell to the differentiating spinous layer but the migrating cell is not necessarily one of a pair of new daughter cells. Autoradiography, following the labelling of dividing cells with [^{3}H] thymidine showed that some pairs of daughter cells comprised one basal cell and one spinous cell, as would be predicted by the differential mitosis hypothesis, but more often the pairs consisted of two basal cells or two spinous cells. The results, indicating that a basal cell might divide to yield two basal cells, two spinous cells or one of each, suggested that to maintain the steady-state, cells are squeezed out of the basal layer in a random way to balance those added by mitosis. The cell squeezed out may or may not be one or a pair of newly-produced daughter cells; migration is not necessarily an immediate consequence of the mitosis. Histological evidence of such extrusion was found. The orientation of mitotic spindles also discredited the concept of differential mitosis; only 4% were perpendicular to the basement membrane as they would be expected to be differential mitosis. Bullough and Laurence (1964) similarly found that the two daughter cells of a mouse basal epidermal cell usually lay side by side in contact with the basement membrane.

Other discussions of tissue renewal may be found in Bertalanffy and Lau

(1962), Bullough (1967), Goss (1964), Greulich (1964), Leblond (1965), Leblond et al. (1964) and Leblond and Walker (1956).

2. Cell Division

The stages of mitosis proper, namely prophase, metaphase, anaphase and telophase are well known and do not need description here but little has been known until comparatively recently about the activities of the cell contents during the interphase between successive divisions. It was recognized by the 1950s that essential preparations for cell division were completed well in advance of visible mitosis. These preparations include the replication of DNA and the synthesis of the proteins needed for building the mitotic apparatus, especially the mitotic spindle, which during division occupies the major part of the cell and may account for from 20 to 50% of its dry weight. More controversially, it was thought that the cell used carbohydrate during the interphase to build up an energy store to be drawn on during actual mitosis. Mazia suggested that energy might be built in to the mitotic apparatus itself and not necessarily stored in the form of identifiable high energy fuels. The state of knowledge and opinion about these matters was well presented by various contributors to a Symposium on Mitogenesis held in 1956 (Ducoff and Ehret, 1959) and in reviews by Swann (1957, 1958) and by Mazia (1961). New information about the mitotic cycle and in particular about the replication of DNA was next obtained by administering tritiated thymidine, a deoxyribonucleoside that is incorporated exclusively into replicating DNA, and by using high-resolution autoradiography to trace its fate. The methods and their application to the elucidation of the mitotic cycle were discussed in an instructive Symposium on Cell Proliferation (Lamerton and Fry, 1963) to which reference may be made for details.

3. The Mitotic Cycle or Cell Cycle

It is now usual to divide the cell cycle into four phases which in sequence are: (1) the mitotic or M phase (2) immediately following mitosis, the G_1 ("first gap") or post-mitotic phase (3) the S or synthetic phase and (4) the G_2 ("second gap") or pre-mitotic phase leading to the next mitosis. The sequence is illustrated diagrammatically in Fig. 41, from Baserga (1965). As a specific example of a short cycle, the total cycle time of epithelial cells of the crypts of the mouse intestine is about 12 hr; the S phase accounts from about 7 hr and mitosis and the G_2 phase each for about one hour, leaving roughly 3 hr for the G_1 phase. It should be noted that the determination of these periods is beset by severe difficulties and that some estimates must be accepted with reserve (Mendelsohn, 1963; Quastler, 1963).

Bullough (1965) has proposed an elaboration of the scheme illustrated in Fig. 41 from which it differs mainly in the subdivision of the G_1 phases into three named portions so that the whole cycle is divided into six phases instead of four. For many practical purposes the four-phase scheme may continue to be adequate and the more convenient in use but Bullough's division is valuable in providing a basis for the discussion of certain poorly-understood but highly important phenomena.

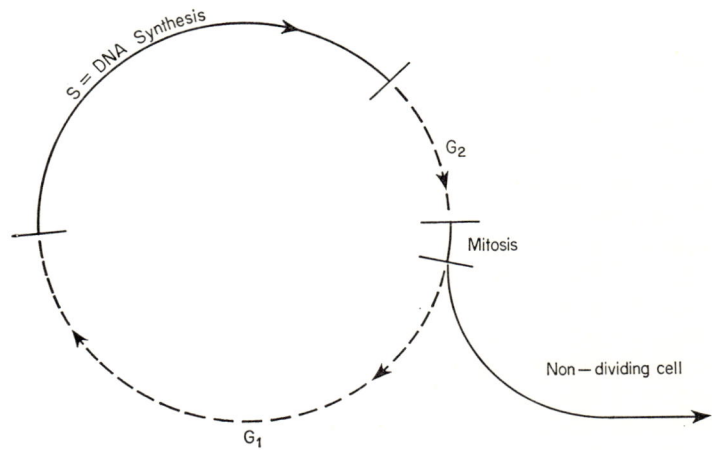

FIG. 41. The life cycle of cells. At completion of mitosis, daughter cells may go through another cycle (G_1, S and G_2) or become non-dividing cells, destined to die without any further division. (Slightly modified from Baserga, 1965 *Cancer Res.* **25**, 581.).

The six phases proposed by Bullough arranged in sequence are Apophase, Dichophase, and Prosophase, which itself is divisible into three phases (early prosphase, phase of DNA synthesis and antephase) and mitosis itself.

1. Apophase (from Gr. απο, implying "moving away from") is applied to the first portion of the G_1 phase immediately following mitosis. Little is known about this phase but Bullough suggests that it may be regarded as the final stage of the mitotic cycle during which the cell returns to its normal size and state.

2. Dichophase (from Gr. διχο, which expresses doubt about two ways ahead) is the period during which cell become committed either to specialize for tissue function (or, as usually expressed, to "differentiate") or to specialize for mitosis. The dichophase is a phase of decision. Its duration is extremely variable. It is very short in tissues having a high mitotic rate, the decision being made promptly, but more commonly it is protracted and in tissues with a long cycle time it seems as if cells remain for long periods in the dichophase in a state of indecision. Bullough estimates that even under extreme condi-

tions the sequence prosphase-mitosis-apophase is unlikely to last more than two or three days whereas it is well known that in tissues with moderate or low mitotic rates the time elapsing between mitoses is far greater. The basal epidermal cells of mice divide only once in about twenty-five days and liver cells of the rat, once in 200–400 days, but it is not clear that the cells remain all the time in dichophase. Bullough notes that cells may leave the mitotic cycle during the dichophase and differentiate and, further, that functioning "mature" cells that have not differentiated to the point of irreversible advance towards death may return to the cycle in dichophase and complete the mitotic cycle. The original "decision" of these cells to differentiate instead of dividing is clearly not irrevocable.

3. Early Prosphase (from Gr. προσ, expressing "movement towards" is usually included in the G_1 phase but is regarded by Bullough as the first subdivision of prosphase, which applies to the period during which cells carry out all the necessary preparations for mitosis to begin. Following Bullough's account, early prosphase begins with the synthesis of the first of the series of enzymes that underlie the mitotic process. Two enzymes, or groups of enzymes, are likely to be critically important. The first of these comprises the enzymes necessary for the syntheses of nuclear histones which begin to increase in quantity in the nucleus before the onset of DNA synthesis. The second essential enzyme is DNA polymerase without which the replication of DNA cannot begin.

4. The phase of DNA synthesis corresponds with the S phase of the usual terminology. The replication of DNA requires the simultaneous presence, as triphosphates, of all the nucleosides that enter into its composition and it requires, also, DNA polymerase. The DNA must be converted first into the "primer" state, which probably entails unwinding of the double helix as well as changes in the associated RNA and histone protein. Prior synthesis of RNA and protein is essential for initiating DNA synthesis but not necessarily for sustaining it. Synthesis of RNA, nevertheless, continues throughout the S phase and protein synthesis is increased. The synthesis of histones seems to continue simultaneously with DNA replication. It is worth noting, as Mazia remarks, that more substances than DNA and histones are needed to make two chromosomes out of one (Lark, 1963; Mazia, 1961; Taylor, 1963).

It is remarkable that the replication of DNA once initiated seems always to advance to completion; it is extraordinarily difficult to stop it once it has started (Mazia 1961; Swann, 1957). It is not yet certain whether the "point of no return" is at the beginning of DNA synthesis or at an earlier time in early prosphase.

5. The antephase, equivalent to the G_2 phase, extends from the end of DNA synthesis to the beginning of mitosis. Little is known about what happens during this phase. Probably the final synthesis of necessary

cytoplasmic materials are completed, notably those needed for the mitotic spindle. Possibly the energy store, too, is built up at this time.

6. Mitosis. There is still no general agreement about what, if anything, is required to start mitosis. Various "trigger" mechanisms and "realization factors" have been suggested but none has been substantiated. Protein synthesis falls to a minimum level during mitosis and late telophase. Once initiated mitosis automatically goes on to completion, even in cells removed from dead animals and then held *in vitro*. (Bullough, 1950; Ducoff and Ehret, 1959; Firket, 1965; Mazia, 1961).

B. Neoplastic Growth

Growth, or better *overgrowth*, has for long been considered a cardinal feature of neoplasia and included in all definitions of neoplasms. The central problem of neoplasia is still believed by many to be the regulation of growth and cell proliferation or, most recently and at a deeper level, the regulation of DNA synthesis. It should be recalled in this context that the earliest decisive steps in experimentally-induced neoplasia are accomplished without any overgrowth at all. Growth entails the formation of new tissue which is something much more complicated than the mere formation of new cells; it includes a considerable element of morphogenesis (Abercrombie, 1957). In man, the most important clinical problem is not so much growth as dissemination (Wallace, 1961). The distinctive characteristics usually attributed to neoplastic growth are that it is steadily progressive and indeterminate; malignant neoplastic growth is credited with the additional characteristics of rapidity and invasiveness.

The rate of growth of tumours is not extraordinary. At its maximum in some repeatedly-transplanted tumours it approaches, but rarely if ever exceeds, that of embryonic tissues or of regenerating adult mammalian liver. Most tumours fall far short of that rate and some, over long periods of time, grow scarcely at all. In general malignant tumours grow more quickly than benign ones but the correlation between rapid growth and malignant behaviour is not close or consistent. The growth rate of transplantable tumours is a stable replicable character of each tumour strain but varies widely from one strain to another and to a substantial degree independently of variation in other stable, replicable characters.

It is difficult to measure the growth rate of tumours in man and fair accuracy has been attained only by serial radiographic examinations of tumours in the lungs. The growth curves of most of these tumours seem to approximate to the exponential form. The remarkably slow growth of some tumours in man, as also in animals, is at present inexplicable (Steele and Lamerton, 1966).

Many factors may vitiate estimates of intrinsic growth rates based on external measurements of gross tumours *in situ*. Haemorrhage and oedema or the accumulation of products such as keratin, cartilage or bone may account for a large proportion of the bulk of a tumour; the amount of neoplastic tissue that is actually growing may be small and restricted to a peripheral shell. Growth may be influenced to a considerable but undetermined extent by the state of the surrounding tissues and especially by the connective tissue forming the tumour bed. Thomlinson and Gray (1955) estimated that cells could not be expected to proliferate if they are more than 150 microns distant from a capillary blood vessel. If this be true, it is conceivable that the growth of benign tumours might be self-limiting. Many years ago Dr. Peyton Rous drew my attention to the remarkable parallelism of the growth-curves of multiple mammary plaques growing in a single mouse; this applied to both pregnancy-responsive and pregnancy-unresponsive plaques. Unfortunately, this parallelism was not further investigated. It suggests either an ubiquitous regulation of neoplastic growth or genetic factors common to all neoplastic mammary epithelial cells in the same animal.

Efforts to compare the mitotic cycles of normal and neoplastic cells have been hampered by several complications in the application of the new methods of analysis to neoplastic tissues. It has been found, for example, that apparently only a minority of the cells in a tumour are engaged in the cell cycle. The clearest outcome of the experiments has been to show that neoplastic cells in general do not proliferate more rapidly than the cells of all normal tissues. The length of the cell-cycle has been found to be shorter in a non-growing normal tissue than in a rapidly growing transplantable Ehrlich tumour. The crucial comparison between a tumour and its parent tissue has not been satisfactorily made. In one experiment in which it was attempted the tumour cells apparently had shorter mitotic cycles than the cells of the normal parent tissue but reservations had to be made about the validity of this result. It has been suggested that the crucial difference between normal and neoplastic tissue is that the mitotic cycles of normal cells are subject to modulation by extrinsic agencies whereas neoplastic cells, if they proliferate at all, proliferate always at a maximum rate so that the duration of the mitotic rate is invariant. Local and systemic restraints continue to operate on the tumour as a whole and determines its growth mode and the number of cells that proliferate but individual cells proliferate always at the same rate or not at all (Baserga, 1965; Mendelsohn, 1963, 1965; Reiskin and Mendelsohn, 1964).

The distinction between growth of the tumour as a whole and the rate of division of its constituent cells should be emphasized as also the finding that only a small fraction of the neoplastic cells are proliferating at one time. It should be noted also that some tumours resemble renewing tissues except that no steady-state is maintained; the proliferation of stem cells more than keeps pace with the loss of cells by lethal differentiation or some other process. The death rate of malignant neoplastic cells is high. Most of the discussion that follows will deal with the growth of whole tumours, which depends not only on how quickly cells are proliferating but also on how many of them are proliferating and how many of them are dying (cf. Mendelsohn and Dethlefsen, 1968).

According to the foregoing analysis, the rate of cell multiplication in neoplasms is not extraordinary except that it seems not to be subject to the feed-back control mechanisms that regulate the proliferation of normal tissue cells. Even the "escape" from feed-back control now seems less crucial than was formerly presumed in the light of the recent indications that the rate of growth of the tumour as a whole depends less on the rate of proliferation of individual cells than on the proportion of cells that are dividing and the proportion that are dying and that these proportions are possibly subject to extrinsic modifying factors.

The proposition that neoplastic growth, referring here to the growth of intact tumours, is distinctive in being relentlessly progressive and in having no determinate end-point is widely accepted but is subject to so many exceptions and qualifications that it has little operational value; in particular it does not provide a reliable definition of neoplasia. Evidence in the contrary sense will be given in later chapters. It will be convenient to deal first with the most distinctive features of malignant tumours, namely, invasion and metastasis.

C. Invasion

Progressive multiplication of neoplastic cells yields an increasing bulk of tissue which must somehow be accommodated within the tissues of the animal. Benign tumours grow expansively and are circumscribed from the surrounding tissues often by a distinct capsule; the neoplastic tissue does not intermingle with neighbouring tissues, but extends by protruding into body cavities or, if superficial, by stretching of the overlying skin. Malignant tumours by contrast penetrate into surrounding tissues, which they destroy. Invasiveness is considered an obligatory criterion of "malignancy" but it is not by itself a sufficient one. Invasion is a commonplace phenomenon in normal embryonic development and certain normal cells and tissues are notably invasive in some pathological but non-neoplastic lesions. The "spurious malignancy" of certain neoplasms which are not ordinarily accepted as "true" malignant tumours must also be taken into account. These two complications will be considered first because they provide useful information about the nature of invasiness in general.

1. The Invasiveness of Non-Neoplastic Tissues

Vasiliev (1958) summarized the observations of a number of observers on what they called "inflammatory proliferations of epithelium" which were characterized by invasive growth of non-neoplastic epithelium into the connective tissue at certain stages of inflammation and complete regression of the invading epithelium at the end of inflammation. The lesions are impressively illustrated by the carcinoma-like proliferations that Fischer (1906) produced by injecting scarlet red intradermally into the ears of rabbits. Lesions conforming with the same criteria could be induced by introducing a variety of irritants into various organs of laboratory animals and into the skin of rat embryos. Moreover, similar epithelial proliferation and invasion could be provoked experimentally in lower vertebrate and invertebrate animals indicating, as Vasiliev noted, that this type of morphogenetic reaction was developed at a relatively early stage of evolution. Invasive growth of a similar type is an important mechanism of morphogenesis in embryos and is apparent also in adults in the development of the mammary gland during pregnancy and in the growth of the trophoblast to which reference is made in Chapter 6.

Vasiliev observed that the common feature in all these examples of nonneoplastic invasion is the close association of epithelial proliferation and connective tissue proliferation. The formation of a bed of young connective tissue precedes epithelial invasion. In all the examples, the formation of this "bed" or "matrix" seems to be essential for the invasive growth of nonneoplastic epithelium and apparently the growing connective tissue somehow attracts the epithelium.

The important inference from this analysis is that the invaded tissues are not passive during invasion; they are active and essential participants. It may be remarked that an independent study of the formation of the placenta indicated that the endometrium actively "accepts" or "receives" the invading trophoblast (Chapter 6).

2. Spurious Malignancy of Benign Tumours

Rous and his colleagues have reported several studies of what they called *spurious* or *factitious malignancy*. The terms apply in the main to extrinsically-evoked invasiveness of neoplasms that ordinarily are non-invasive and benign; they are not considered to be truly "malignant" but unlike the invasive growths described in the previous section they are neoplastic. The earliest and most extensive observations were made on papillomas of the rabbit skin evoked by the Shope papilloma virus (Beard and Rous, 1934; Rous and Beard, 1934a, b).

Ordinarily, the papillomas evoked by the Shope virus grow rapidly for a while and then remain stationary for a long time. Rous and Beard thought that their growth was checked by the dense layer of connective tissue, which gradually formed around their bases. Injections of the dyes Scharlach R (Scarlet Red) or Sudan III into the skin around papillomas stimulated the tumours to grow and also caused them to invade underlying tissues and form large fleshy masses below the surface. These growths had the histological characteristics of malignant carcinomas and they often invaded blood vessels but, although the experimental procedure stimulated the habitually benign papilloma to grow progressively, it effected no enduring change in the papilloma cells; the procedure did not change the benign papilloma into a malignant carcinoma because, as the injected material was absorbed, the aggressive behaviour subsided. The papillomas, indeed, reacted to the dyes in much the same way as did normal rabbit skin in the old experiments of Fischer; in both systems the spurious malignancy was induced by, and dependent for its persistence on, the continuation of an extrinsic stimulus which operated, almost certainly, on the ambient connective tissues.

A similar spurious malignacy was brought about by autotransplanting fragments of papillomas into the subcutaneous tissues, muscle, liver, spleen and stomach of the rabbit providing the papilloma. The transplants yielded actively-growing tumours which were often conspicuously invasive and which tended to recur after excision and to disseminate as an accident of surgical interference, distribution through the peritoneal cavity being especially frequent. The transplanted papillomas often penetrated into blood and lymph channels but true metastasis leading to the growth of secondary tumours was not certainly observed. Otherwise, the tumours looked and behaved like highly malignant neoplasms; they grew progressively, directly invading and destroying normal tissues and killed their hosts. In the most susceptible animals the growths, often fatal, had the histological structure of an epidermoid carcinoma. Nevertheless the "malignancy" of these tumours differed from that of the true epidermoid carcinomas that can develop by progression in long-standing, quiescent Shope papillomas or that can be induced in rabbit skin by chemical carcinogens. Rous and Beard remarked that spurious malignancy could be sustained by renewing the inciting stimulus and thought that continuous stimulation might result eventually in true malignancy. To my knowledge, this outcome has not been demonstrated but the possibility that progression to carcinoma might occur under these circumstances, as it can occur without continued stimulation, cannot be excluded.

In later studies of spurious malignancy, Rous and his colleagues were concerned with the conditional tumours induced in rabbit skin by chemical carcinogens (MacKenzie and Rous, 1941; Rous and Kidd, 1939, 1941). The spurious malignancy was most evident when tar was used as the inducing carcinogen and especially in the carcinomatoids that are among the early-emerging conditional tumours. The "carcinomas" first described by the pioneer investigators of the carcinogenic action of coal tar on rabbits' ears were almost certainly carcinomatoids, the diagnosis of carcinoma being based only on histological evidence of invasion. The more recent observations have shown that the invasive and "malignant" properties are attributable to continuing action of tar. When tar was withheld, the carcinomatoids became ordinary benign papillomas or they rounded up to form cysts or disappeared. Even if application of tar were continued they became harmless sooner or later in one of those ways.

Mackenzie and Rous noted that most carcinogenic agents not only cause cells to become neoplastic but also have adverse or encouraging effects on their subsequent growth. Tar is a notable example; as well as initiating neoplastic change it conspicuously encourages tumour growth. Other agents, highly various in kind, act only by encouraging cells already neoplastic to multiply. Mackenzie and Rous, after discussing various examples, came to the conclusion that the action of agents that stimulate the proliferation of cells already neoplastic was, almost always, essentially non-specific; the generality of such "encouraging" agents seemed to act merely by producing tissue conditions favourable for the multiplication of cells, irrespective of their being neoplastic.

3. *Invasiveness of Malignant Tumours*

In general, tumours grow along paths of least resistance through pre-existing tissue clefts and spaces and destroy normal tissue as they go. Dense tissues like the walls of arteries and the capsules of organs resist destruction for a long time and there is an effective barrier to the spread of tumours between the central nervous system and the rest of the body. Tumours often destroy the parenchyma of an organ whilst the connective tissue framework persists and sometimes tumour cells use it for their own support and creep along it, progressively replacing the normal epithelial or endothelial investment. Tumours that have entered a muscle fibre are often confined within its sheath but destroy its contents. Similarly tumours may extend for long distances within arteries, lympatics or ducts without eroding their walls. The invasive growth of tumours, therefore, is not entirely unconfined; it is subject to important checks and restraints.

As a rule, invasive growth is recognizable, clinically, by fixation of the tumour and impaired function of the invaded tissue. Histological evidence of invasion is usually decisive but, as Wallace (1961) remarks, it is not distinctively different from the invasive growth of certain normal tissues and cells. It has been affirmed and denied that the most conspicuously invasive normal cells yield the most highly invasive and "malignant" tumours.

The mechanism of invasive growth is still obscure. One of the elementary questions remaining unanswered is the nature of the essential difference between an invasive tumour and a non-invasive one (Wallace, 1961). Progressive multiplication of cells is probably an important factor in invasion because with increasing bulk of neoplastic tissue and increased hydrostatic pressure within it, tumour tissue is liable to thrust mechanically into all available spaces in the contiguous tissues. Young (1959) has supported this interpretation, which nevertheless does not alone explain comprehensively and convincingly the invasiveness of the generality of malignant tumours. The advance of a tumour does not always follow paths of least resistance and mechanical factors alone cannot explain the invasion of one tissue by another in tissue cultures *in vitro* where no head of pressure can build up. Other suggested factors include the amoeboid motility of neoplastic cells and their ability to produce lytic enzymes but although these mechanisms may favour the invasive growth of some tumours their importance in invasive growth of the majority of malignant tumour is doubtful (Hamperl, 1967).

Coman and his colleagues, who were amongst the first to study invasiveness experimentally, reported that the mutual adhesiveness of carcinoma cells was less than that of their normal present cells and proposed that the decreased adhesiveness facilitated dissociation of the tumour into single cells whose amoeboid motility enabled them to wander into the surrounding tissues and establish new colonies. More recently Abercrombie (1967*b*) has proposed that the loss of contact inhibition by neoplastic cells accounts, in part at least, for their invasiveness but it is far from being demonstrated that the cells of invasive neoplasms in general have, in fact, lost the property of contact inhibition. The phenomena described by Abercrombie as well as by Coman depend on the properties of cell membranes, which are now receiving close attention in this and other contexts. For discussions of their relevance to invasiveness, reference may be made to reviews by Abercrombie (1967*b*), Abercrombie and Ambrose (1962), Coman (1953), Leighton (1967) and to the most comprehensive account in the recent book by L. Weiss (1967).

It may be noted that both Coman's and Abercrombie's interpretations imply a dissociation of tumours into single cells and Coman explicitly writes of cells wandering out of the primary tumour to establish new colonies in the surrounding tissue. According to this analysis invasion is a discontinuous extension of the tumour and strictly speaking might be described as metastasis, whose particular quality is discontinuous extension. Recently, Kellner (1967) inferred from studies of human material a similar discontinuous extension but attributed the dissociation of cells to necrobiotic processes and not to decreased mutual adhesiveness. He demonstrated clumps of cells in the close vicinity of primary tumours but, as shown by the study of serial sections, they were not in continuity with the tumours. He named these small foci of tumour "pericarcinomatous metastases" and thought they were probably the sources of more widespread dissemination. On the other hand, invasion can be accomplished by extension in continuity without dissociation of cells. Lymphatic permeation, whose frequency and importance have probably been overestimated, is due to a continuous extension of tumour within lymph vessels, although obliteration of a portion of the vessel and its contents may give a misleading impression of discontinuous spread. The invasiveness of certain organoid tumours of the mouse mammary gland described and illustrated in Chapter 6 is also due to extension in continuity. The extension in some of these tumours is extremely rapid but histological studies reveal no trace of cell dissociation and, significantly, the tumours do not metastasize. It seems desirable to distinguish between two kinds of "invasion"; the one discontinuous and accomplished by dissociation of the tumour into single cells or small groups of cells and the other accomplished by extension in continuity and without dissociation of cells. The important clinical distinction is that invasion of the former type is liable to be followed by metastasis whereas the second type is unlikely to lead to metastasis. The mechanisms of the two types of invasion may be different.

Vasiliev (1958) suspected that the basic mechanisms of invasion by malignant cells were the same as those in invasion by non-neoplastic cells and depend primarily on the ability of the invading tissue to elicit an appropriate proliferation of connective tissue cells in the invaded tissues. He thought that the main difference between normal and neoplastic invasion might be that malignant neoplastic cells had acquired an intrinsic ability to evoke a suitable proliferation of connective tissue whereas normal cells acquired a similar ability only temporarily at certain stages of development and in response to extrinsic circumstances provided, for example, by their hormonal environment.

Leighton's observations on the invasiveness of neoplastic cells *in vitro* conform well with Vasiliev's general concept. Leighton studied the invasion of normal connective tissue by neoplastic epithelium when the two were cultured close to one another *in vitro* and found that the degree of invasion varied with the extent to which the connective tissue grew. His further observation that in certain places the connective tissue and the neoplastic tissue seemed to be infiltrating one another emphasizes the important consideration that invasion depends on a *mutual* interaction between invading and invaded tissues. Incidentally, Leighton noted that normal fibroblasts often seemed to be more motile than the carcinoma cells (Leighton, 1967).

Both Vasiliev and Leighton drew attention to the affinity between their concept of invasion and Greene's concept of heterotransplantation as an indicator of tumour "autonomy". Briefly stated, the rationale of Greene's heterotransplantation test is that it reveals the capacity of the neoplastic tissue to evoke a stroma reaction in the normal connective tissue of the host and that this capacity is also the basis of invasiveness and metastasizability. As Greene notes, the stroma reaction depends both on the evocative capacity of the neoplasm and on the reactivity of the connective tissue in which it is situated. (Greene, 1952, 1955, 1957a, b.)

The concept of a specific stroma reaction emerged from the early studies of tumour transplantation, as mentioned in the historical introduction to this volume. Its wider applicability to morphological and behavioural phenomena encountered in tumours growing undisturbed in their original hosts has been recognized only fairly recently and has brought about considerable changes in ideas about "tumour-host relationship" (Denoix, 1967b). Our present concern is with invasiveness considered as the result of a mutual interaction between neoplastic and normal connective tissues. The outcome of the interaction can be modified by changes in either of the reactants. The possibility of aggravating the disease by interfering with the tumour-bed is sufficiently illustrated by the phenomenon of spurious malignancy; the danger of aggravating malignant neoplasia in man by injudicious local therapy is also well-recognized, although not, perhaps, sufficiently widely. Palliative treatment is aimed, sometimes with substantial benefit, at producing the converse effect.

The possible outcomes of the interactions between diverse neoplasms and diverse connective tissues are highly varied and graded in degree. It is desirable to determine not only if a particular tumour is invasive but also how invasive it is and also, perhaps, whether or not the invasion is effected by dissociated cells. The difference between a "benign" and a "malignant" tumour may depend in large measure on the nature of the interactions between neoplastic and normal tissues. An overproduction of connective tissue might be expected to impede invasion but this may be an oversimplification; Leighton has pointed out that, despite the abundance and density of their connective tissue stroma, scirrhous carcinomas spread widely and inexorably.

One general conclusion that seems to emerge from this discussion is that the military terminology and value-judgements, which early studies of bacterial infections introduced into medical literature, should be excluded from discussions of invasion and other tumour-host relationships. Even in military parlance, attack and defence or aggression and resistance are apt to acquire different shades of meaning according as to who is engaged in which, and the terms and the ideas they express are at best doubtfully appropriate in application to neoplasia and are probably misleading and not conducive to objective study of neoplastic phenomena.

D. METASTASIS

Metastasis is, properly, a *process* of dissemination whose essential characteristic is discontinuity of extension of the neoplasm. It is the most menacing manifestation of malignant neoplasia in man and, as Willis pointed out, no great improvement in the results of treating metastatic neoplasia can be expected until the present meagre understanding of the mechanisms of metastasis is greatly extended.

The main features of metastasis were known to pathologists in the late nineteenth century and a notable paper by Schmidt (1903) testifies to the acuity of their observations and the general validity of their inferences. In 1934, Willis published the first comprehensive monograph on the subject and his book, later revised (Willis, 1952), remains the most valuable general account. It is usefully supplemented in certain respects by Walther's (1948) detailed analysis of metastasis in man.

The experimental study of metastasis began during the time of pre-occupation with transplantable tumours and has proceeded fitfully over the years but, until recently at least, has proved relatively unrewarding. One reason for this has been the difficulty of finding a satisfactory experimental "model" of metastasis (Wallace, 1961). Metastasis is not nearly so conspicuous a feature of malignant neoplasia in experimental animals as it is in man. Many of the early investigations, inspired perhaps by a too literal interpretation of Bashford's description of transplantation as "artificial metastasis", comprised direct implantation of tumour fragments into various organs to test the "susceptibility" or "resistance" of different tissues to metastatic growths. Bashford's dictum needs qualification. Metastasis is a special form of *autotransplantation*. The early experiments, being necessarily carried out with random-bred animals, introduced irrelevant complications resulting from genetic differences between tumour and host. Moreover, transplantation, as usually practiced, introduced relatively large quantities of tumour tissue into tissues that were inevitably damaged by the act of implantation whereas natural metastasis delivers minute fragments of tumour or cells in small groups or singly to undamaged tissue and most often to the interior of terminal blood or lymph vessels. More recently many investigations have sought to mitigate some of these drawbacks by injecting suspensions of single cells or small clumps of cells into blood vessels or, much less frequently, into lymph vessels.

The value of many experimental investigations has been minimized by a failure to recognize that metastasis is a complex phenomenon depending on at least five processes operating in sequence. Five processes, which require separate consideration in the planning and interpretation of experiments have been named as follows (Foulds, 1957):

(1) *Liberation* of tumour cells from the primary growth, singly, in clusters or in substantial fragments of tissue.

(2) *Transportation* of the liberated cells to a new site by one or other of

several routes, of which the blood and lymph vessels are the most important.

(3) *Deposition* of the transplanted material at the new site.

(4) *Establishment* of the deposited material in its new environment, by which is meant the acquisition of a vascular stroma by a process comparable with the "take" of a transplantable tumour.

(5) *Growth* of the established deposit into a secondary tumour.

Recently L. Weiss (1967) has published the most systematic analysis of the successive steps of metastasis that is yet available under the following headings:

(1) Separation of malignant cells from the primary tumour. Movement and transport of malignant cells.

(2) Entry of malignant cells into vessels. Tumour cells in the peripheral blood.

(3) Attachment of tumour cells to the vascular endthelium.

(4) The penetration of vascular endothelium by tumour cells.

(5) Host-organ/malignant cell interactions.

Aside from differences in phraseology, the two schemes correspond closely. The difference between the two descriptions of step 3 reflects, in part, a change of opinion. It was formerly thought that vessels were "plugged" by tumour emboli, filling the lumen. This certainly happens (Zeidman, 1965a) but current opinion tends to the view that adhesion of tumour cells to the vascular endothelium without occlusion of the lumen is a more important factor in natural metastasis.

It is impracticable, in this book, to consider in detail each step in the sequence of events that leads to overt metastasis. Various aspects of the subject have been well-reviewed by Coman (1953), Zeidman (1957), Wallace (1961), Wood *et al.*, (1961), Fisher and Fisher (1965, 1967), Leighton (1967) and L. Weiss (1967) and by various authors in Denoix (1967a). The book by Weiss is particularly valuable for its critical assessment of the physicochemical phenomena that are, or may be, implicated in metastasis, although Weiss is forced to the conclusion that it is just not possible, at the present time, to give a clear cut picture of the metastic process at the cellular or physicochemical levels. The remainder of this chapter will be devoted to the discussion of a few, rather arbitrarily-selected, general topics.

1. *Experimental Procedures*

One of the chief reasons for the slow progress of the experimental study of metastasis is the lack of wholly-satisfactory experimental "models". Intravascular injections of cell suspensions, as now widely used, by-pass the first crucial step of Liberation and, in my view, it is not justifiable to presume that the experimental procedure closely reproduces the natural process or that the subsequent steps of transportation, deposition, establishment and growth closely approximate to the corresponding steps in natural metastasis unless it can be demonstrated that the experimental procedure yields a characteristic distribution of tumours similar to that resulting from natural metastasis. Unfortunately, the distribution of secondary tumours in small laboratory animals is rarely sufficient characteristic to provide this desirable "control". I am not aware of any transplantable tumours in inbred mice or rats that has the desired qualities with the possible exception of certain lymphoid tumours of inbred mice (Foulds, unpublished observations). As a rule metastatic tumours are restricted to the regional lymph nodes and the lungs and intravenous injections of tumour cells yield tumours only in the lungs.

Except that it was, and seemingly still is, tiresomely difficult to maintain by serial transplantation and that it undesirably introduces the complications of homotransplantation, the transplantable Brown-Pearce rabbit carcinoma has desirable qualities. It has been laboriously maintained by transplantation for some forty-five years. Successful transplants in muscle or testes metastasize most conspicuously to the adrenals, kidney and dyes and intravenous injections of tumour suspensions reproduce this distinctive pattern (Foulds, 1932). It should be noted that "inappropriate" suspensions, which are somewhat the more easily prepared, yield tumours only in the lungs.

More recently, a transplantable carcinoma derived from a Shope rabbit papilloma has proved useful especially when used in parallel with the Brown-Pearce tumour but the satisfactory "model" for the experimental studies has yet to be found.

2. *Neoplastic Cells in the Circulating Blood*

Reports, in 1955, of improved methods of detecting neoplastic cells in human blood seemed to promise new opportunities for obtaining direct evidence about the steps of liberation and transportation in patients bearing malignant tumours (Engell, 1955; Fisher and Turnbull, 1955). The high expectations of that time have not been fulfilled as may be read in discussions by Ritchie and Webster (1961), Wallace (1961) and Wood *et al.* (1961) and in more recent evaluations by Griffiths and Salsbury (1965), Nadel (1965), and Zeidman (1965*b*).

All the technical procedures used for detecting neoplastic cells in human blood are difficult; despite improvements, none is widely agreed to be satisfactory. The criteria for identifying neoplastic cells have become more stringent but the results from different laboratories are often widely discordant.

The neoplastic cells in peripheral blood are few in number, rarely more than 10 per ml. and Ritchie and Webster (1961) suspected that most of them are dead or dying. Nearly all the cells are isolated, clumps being rarely found. Neoplastic cells are found more often and in greater numbers in blood from the veins draining a primary tumour and many of them are in clumps. It seems reasonable to infer that the primary tumour liberates both isolated cells and clumps and that the clumps are filtered out in the lungs, leaving only isolated cells to reach the peripheral circulation. Griffiths and Salsbury (1965) believed that, in all probability, clumps have a far greater clinical significance that isolated cells and in this they were in agreement with the earlier opinion of Willis derived from studies of autopsy material.

Most observers now agree that the prognostic significance of neoplastic cells in the peripheral blood is negligible. Their presence certainly does not signify that early growth of secondary tumours is to be expected. In a general way, neoplastic cells are found with increasing frequency the higher the clinical "grade" and "stage" of the primary tumour but there are numerous exceptions. Neoplastic cells can be found on occasion in patients with early neoplastic disease in whom metastatic tumours do not develop at any later stage. Epidermoid carcinomas seem to liberate cells into the blood stream surprisingly often considering that they rarely metastasize by that route. Although the fate of most of the circulating neoplastic cells has not been determined it is hardly disputable that the great majority of them do *not* give origin to secondary tumours.

Several observers have reported that the manipulation or compression of malignant tumours during surgical or diagnostic procedures cause the first appearance, or an increase in numbers, of neoplastic cells in the emerging venous blood. Even vigorous cleansing of the skin over a tumour has had this effect and so also have rectal or pelvic examinations and the instrumental compression of tumours during such procedures as transurethral resection of growths of the bladder or prostate. Injections of local anasthetics and aspiration biopsies have allegedly had the same consequences. According to Griffiths and Salsbury, cells are often liberated in solitary "showers" and persist in the blood for not more than a few minutes. The failure of some investigators to obtain cordant results may be attributable to sampling of the blood at intervals too widely spaced for detecting "showers".

Goldblatt and Nadel (1965) remained sceptical about the applicability of the foregoing observations to surgical practice but some experiments on animals have provided evidence that manipulation or trauma can cause or aggravate liberation. In a notably early experiment, Jonescu (1930) demonstrated neoplastic cells in the blood of mice after digital compression of transplanted tumours. Many other experimental and clinical observations have led to similar conclusions although usually without proof that the harmful circumstances operated on the process of liberation. There is some evidence that trauma may have a substantial effect on deposition (Black, 1964; Fisher et al., 1967). It is to be hoped that, in the future, investigators will make more serious attempts to localize the point in the sequence of events at which various experimental and clinical procedures operate to modify the end-result of the sequence, the growth of secondary tumours. Liberation of cells from the primary tumour is a necessary condition of metastasis but it is not a sufficient condition for the growth of secondary tumours. Factors operating at other stages of the sequence may be clinically more decisive (Malmgren, 1967).

3. The Fate of Liberated Neoplastic Cells

It is now evident from observations on circulating tumour cells in man and from experiments on animals that large numbers of neoplastic cells can disappear from the blood stream remarkably quickly and that the great majority of them never originate secondary growths. Many, especially the isolated cells, probably die in the blood stream, but probably not all of them (Madden and Malmgren, 1962; Moore and Sandberg, 1965). The persistence of viable cells in various organs can be demonstrated by transplanting fragments of those organs into other mice (Greene, 1965; Greene and Harvey, 1964, 1966; Kim, 1966). The experiments of Greene and Harvey indicated that viable cells might attach to the vascular endothelium but fail to penetrate it or having penetrated might fail to elicit an adequate stroma reaction. Liberated tumour cells may thus fail to originate secondary growths for at least four different reasons (1) failure to survive during transportation; (2) failure of deposition through failure to adhere to the vascular endothelium; (3) failure to penetrate the vascular endothelium; (4) failure to elicit an adequate stroma reaction after penetrating the vascular endothelium.

The separate steps in the metastatic process do not follow automatically, one after the other. The sequence can be halted temporarily or permanently at any one of the steps, even at the last one, so that the established secondary deposit remains "dormant" as described below.

4. Patterns of Distribution of Secondary Tumours

The most controversial problem has been to account for the preferential localization of certain tumours in particular sites and the relative freedom of some tissues from secondary tumours of all kinds. Two different views have been urged. According to one view, supported in particular by Zeidman (1957) with substantial experimental evidence, the distribution of secondary tumours is accountable wholly to the operation of mechanical and hydrodynamic principles. The other view seems to have originated with Paget in

1889 and has been strongly supported by Willis who borrowed from Paget the simile of "seeds" flourishing variously according to the "soils" in which they fall. This is often referred to as the "soil" hypothesis. The predictable outcome of this controversy, as described recently by Fisher and Fisher (1967), is that in all probability the patterns of metastatic spread are dictated by anatomical factors or mechanical factors as well as by intrinsic properties of the tumour cells and of the tissues that they reach. The problems of selective localization of secondary tumours can now be re-stated more clearly but they have not been solved (Weiss, 1967).

E. The Clinical Course of Neoplasia

The clinical course of malignant neoplasia is not always steadily and relentlessly progressive; remissions and exacerbations of growth take place unpredictably and not infrequently. The extreme case of "spontaneous regression of cancer", which will be considered later, is so rare as to be clinically negligible but less rigidly and arbitrarily defined remissions and regressions are not and deserve much more attention.

French authors have emphasized the frequent irregularities in the growth of tumours and in particular the "poussées evolutives" or abrupt "spurts" of growth. Some of the spurts or outbursts are probably attributable to progression to more aggressive types of neoplasia but others are not so easily explicable (Delarue, 1947; Denoix, 1954, 1967b; Dunphy, 1950, 1953; Huguenin, 1946; Smithers, 1964; Smithers *et al.*, 1952).

Untreated malignant tumours kill 75% of the patients within average periods ranging from about fourteen months for cancer of the oesophagus to forty-six months for cancer of the breast but some patients with those types of neoplasia that commonly are rapidly fatal survive for a long time. Occasionally, the course is surprisingly protracted (Shimkin, 1951). In a study of untreated breast cancer, Daland (1927) found that, although the majority of patients died within three years, 22% lived for five years and 5% for ten years. There was no evident correlation between the lengths of survival and the clinical assessments of "malignancy". Nathanson and Welch (1936), in a continuation of Daland's study, found one survival of fifteen years. More recently Rigler (1964) and Bloom (1964) have reported survivals for over ten years of patients with cancer of the lung and of the breast, respectively. Bloom makes the important comment that the survival statistics do not disclose the long, drawn-out, distressing illnesses suffered by untreated patients.

Survivals after incomplete eradications of tumours are similarly varied and unpredictable. Not uncommonly, patients survive for ten or fifteen years before dying with recurrent or metastatic growths and the long survivals are not correlated with any discernible peculiarities of the tumours (Keyes *et al.*,

1954; Morton and Morton, 1953; Shimkin, 1951; Willis, 1952). Paine (1965) quoted in his review a lethal recurrence after thirty-eight years.

Survival statistics provide average figures which have little relevance in clinical practice where each patient's disease differs to a greater or lesser extent from all others (McKenzie, 1956). The results of treatment in individual patients are unpredictable; there are unexpected failures and equally unexpected successes. There are "good" cancers and "bad" cancers (Denoix, 1954); the "good" cancers are curable by any reputable treatment whereas the "bad" cancers respond to none. Denoix and also Macdonald (1951) denied that "curable" cancers differ from "incurable" ones only in being "earlier" or less advanced; they maintained that the two kinds of cancer are not different stages in one process but two distinct processes that are different from the beginning. In his hypothesis of "biological predeteminism", Macdonald proposed that the "biological potential" of a tumour is determined early in the preclinical phase and that the "inherent growth potential" of the tumour and whatever defence mechanisms it may incite in the host together constitute a biological complex that determines the course of the disease. At the time of early clinical detection, the growth pattern is already well established and the clinician has to deal with a disease of biologically-determined potentialities; rigid ideas of prognosis based upon duration and size should be abandoned in favour of attempts to evaluate the biological potential of the tumour in its individual host.

The division of cancers into two groups, "good" and "bad", is an oversimplification; there is no sharp distinction between the "good" and the "bad" which are linked together by tumours with varied degrees of "goodness' and "badness". As Boyd (1966) has recently affirmed neoplasia is "a process of infinite variety". The variety is attributable to the widely diverse assortments of independently variable characters, the possible combinations of which are so numerous that each tumour is an unique entity, differing from all other tumours at least in some minor characteristic (Boyd, 1966; Delarue, 1947; Foulds, 1940; McKenzie, 1956). Nevertheless the "clustering" of tumours into two main groups with less numerous intermediate types may be sufficiently pronounced to justify the separation of "good" and "bad" tumours as an approximation for purposes of discussion and clinical application.

Macdonald's contention that prognosis should be based on an evaluation of the biological potential of the neoplasm, although reasonable enough in theory, is open to the criticism that it is not evident how the evaluation can be made. The evaluation must include an analysis of the assortments of independently variable characters in a particular tumour in a particular patient. Denoix gave evidence obtained from studies of cancer of the human breast that growth rate, local invasion, spread to regional lymph nodes and dissemination by the blood stream are independently variable characters whose associations in a particular tumour determine, or contribute to, its biological potential and, hence, the prognosis. Denoix advanced a provisional classification which, although not wholly satisfactory, gives a practical basis for clinical investigation and discussion. He divided cancers into four main types, which differ from the outset: (1) a localized type is limited to the organ of origin; it is not merely an "early" stage in the development of other types but a particular stable type without tendency to generalization; (2) a regional type exceeds the limits of the organ in which it originated by direct extension without spreading to the lymph nodes; (3) a lymphophile type of tumour which has already invaded the regional lymph nodes when the primary growth is first recognized; (4) a diffused type which is remarkable because, at the time of clinical recognition, the whole body is seeded with tumour cells whose evolutionary potential is revealed at unpredictable times thereafter.

Both Macdonald and Denoix based their arguments mainly on neoplasia of the human breast, and the extent to which they can be extended to other types of neoplasia is uncertain. The word "predeterminism" is unfortunate if understood to mean that the definitive characters of a tumour are always irrevocably fixed at an early stage of its development; on this strict interpretation predeterminism would exclude the commonplace phenomenon of tumour progression. Macdonald's hypothesis is weak and evasive in attributing the individuality of tumours to their "biological potential", an abstraction like "essential malignancy", "growth potential" "growth pattern", and other similar terms which have been used in a like sense. The abstract ideas that these terms are designed to convey need translating into concrete terms by resolving the behaviour of tumours into the several characters and processes that contribute to it. The important and, as I believe, valid inferences drawn by Macdonald and by Denoix from their clinical observations is that the "biological potential" of a neoplasm is established at an early pre-clinical stage of its development. It is proposed here that the basic ideas of Macdonald and Denoix can be better expressed in terms of the early establishment of a programme of neoplastic development. A programme, avoiding the implications of rigidity inherent in predeterminism, allows some elasticity in carrying out the programme and, more important, it refers not to extant characters but to developmental capacities which for the present at least, are not recognizable by overt signs, but only, retrospectively, by subsequent events. Expressed in these terms the important basic ideas receive strong support and extension from experimental investigations notably those concerned with chemical carcinogenesis in rodent skin.

The experiments most pertinent to the present discussion are those of Berenblum and Shubik (1947*a*, *b*, 1949 *a*, *b*) on the initiation and promotion of neoplasia in mouse skin to which reference was made earlier and the extension of these experiments by Shubik and his colleagues (Shubik, 1950, 1961, 1966). The most relevant observations briefly stated are as follows: First, a single low dose of carcinogen followed by repeated applications of croton oil yields a large crop of benign papillomas many of which regress and relatively few of which become malignant. Second, repeated doses of the carcinogen evokes tumours almost none of which regresses but many of which become malignant. Third, a single large dose of carcinogen elicits a response intermediate between the two previous ones; many tumours regress but also many become malignant. The single low dose of carcinogen effects initiation, which is accomplished quickly and which brings about a functional alteration of the skin demonstrated by its capacity for developing tumours mostly benign and regressing ones, when a promoting agent such as croton oil is applied; when croton oil is withheld the capacity or responsiveness persists for as long as forty-three weeks in the absence of any recognizable histological change that lasts as long as the alteration in capacity for response to promotion. Initiation, so far as can be determined, is irreversible. By changing the initiating procedures the course of subsequent neoplastic development can be predictably modified.

According to these observations the biological potential is determined or, as I prefer to say, the programme of neoplastic development is established, extremely quickly at the initiation of neoplasia and thereafter is stable and probably irreversible. The first initiating procedure, using a low dose of carcinogen, is in several ways the most illuminating. It should be noted that in a suitably contrived experiment the programme is not carried through unless and until a promoting stimulus is applied so that neoplastic development is not "predetermined" in the strict sense of the word. It should be noted also that the programme or biological potential is preserved for many months during which the promoting stimulus is withheld; at the end of this time all the cells in the epidermis, which is a clear example of a renewing tissue, must have divided many times. It may be inferred that the programme is vested in some organized structure or system that is stable in the absence of its effective utilization and that is replicable through an indefinite series of cell divisions. A third inference, which will receive more detailed attention later is that the programme can provide instructions for conditional growth, for determinate growth and even for retrogressive neoplasia.

F. Anomalies of Behaviour of Neoplasms in Man

Many anomalies in the behaviour of neoplasms in man are attributable to unexpected but not necessarily rare associations or dissociations of independently variable characters. Many of the Group B lesions described in Chapter 3 are anomalous in one way or another as already indicated. The present section is devoted to certain named varieties of tumour in which the dissociation of characters is especially conspicuous.

Basal-cell carcinoma or rodent ulcer of the skin is described as "locally malignant" because although it is locally invasive and dangerously destructive it rarely metastasizes. By contrast certain other tumours described as "metastasizing benign" tumours disseminate widely without conspicuous local invasion. In truth, these tumours are minimally invasive and the "locally malignant" tumours occasionally metastasize but ordinarily in each type the disproportion between the two cardinal features of malignant neoplasia, local invasion and remote dissemination, is extreme.

Other anomalous tumours are well-recognized and have been described as *occult*, *dormant* and *latent*. These terms apply to three, or more than three, different kinds of lesion but unfortunately there is no unanimity about which name properly belongs to which lesion. Smithers (1960) proposed that *occult* should be applied to those hypothetical neoplastic cells established during the first stage of two-stage experimental chemical carcinogenesis of the rodent skin; Rous described these questionable cells as latent tumour cells and Berenblum preferred to call them dormant so that the confusion of terminology is maximal. Smithers proposed that *dormant* should apply to cells that give origin to late recurrences of primary and metastatic tumours and described as *latent* those carcinomas exemplified most notably by tumours found in the prostrates of old men that are histologically "malignant" but clinically "benign". In a previous account of these lesions (Foulds, 1958a) I used latent and dormant in the senses recommended by Smithers but in accordance with earlier clinical usage applied occult to primary malignant tumours that have escaped clinical recognition despite clinically-recognizable remote metastasis. It seems to me more consistent and desirable to continue to use the terms in which they have been used for a considerable length of time and I propose to adopt the same usage here without arguing whether or not the original attributions of the terms were linguistically the best. This usage leaves the latent, dormant or occult "tumour cells" of the first stage of experimental epidermal carcinogenesis without a special name. Not being convinced that they exist at all, in the sense originally proposed, the deprivation seems to me supportable. Moreover, the terms in their original usage applied to cells, singly or in small groups, that were not microscopically identifiable whereas in clinical usage they apply to tumours within the range of microscopical recognition so that there is probably not much risk of confusion if laboratory workers continue to use the terms in their own ways.

1. *Occult Primary Tumours*

The occult primary tumours are closely allied to the metastasizing benign tumours except for their insignificant size; dissemination is grossly out of proportion to local growth as well as local invasion. The secondary tumours sometimes give a clue to the situation of the occult primary tumour which is often in the thyroid or the lung. Occult primary tumours of the breast yield evident secondary growths in the axillary lymph nodes or, less commonly, in bones causing spontaneous fractures, before the primary tumour can be found even if its presence is suspected and deliberately sought (Willis, 1967).

The reasons for the disproportionate and precocious dissemination from an occult primary tumour are not known but deserve close attention because precocious dissemination is the most serious bar to satisfactory control of some carcinomas, especially carcinomas of the breast. It may be recalled that Denoix distinguished a diffused type of mammary carcinoma, remarkable because the whole body was seeded with tumours cells when the primary tumour was first recognized clinically. It is possible that the occult primary tumours differ from the "diffused" type of carcinoma only in degree.

2. Dormant Tumours

The concept of dormancy was invoked to account for the long-delayed local recurrences of tumours after surgical excisions and for the growth of metastatic tumours a long time after supposedly complete removal or destruction of the primary growth (Hadfield, 1954; Willis, 1952). Delays of 10 or 15 years are not uncommon; sometimes, it is said, they extend to 40 or 50 years. Shimkin *et al.* (1954) argued that if the tumours eventually emerged as the consequence of extremely slow but progressive growth throughout the period of dormancy, there should be a relationship between the duration of the dormant period and the length of survival after emergence, whereas, if host factors precipitate the emergence, the length of life thereafter should be independent of the length of the dormant period. Their own investigations gave no decisive support to either alternative; the earlier observations of Truscott (1947), showing that the rate of growth of emergent mammary tumours was independent of the duration of the dormant period, were consistent with the second. Steady progressive growth, however slow, seems an unlikely explanation for emergence after 40–50 years of dormancy and Hadfield (1954) thought that, when the dormant period lasted six years or longer, it was almost impossible to escape the conclusion that the cells of the dormant tumour had been in a state of temporary mitotic arrest. Hadfield suggested that the cells of dormant tumours might survive for long periods in a "paramorphic state" comparable with that attained during the preservation of cells at low temperatures and that in this state they could withstand separation from their parent tumour until they acquired a new blood supply.

Several possible factors in bringing about the dormant state, such as mechanical or immunological restraints, have been discussed inconclusively by Leighton (1967) who remarks that there is no aspect of neoplasia in man or in animals about which we know less. The eventual emergence from the dormant state is little better understood. Two reasonable possibilities are that the emergence might be "triggered" by some unidentified extrinsic stimulus or that it might be the consequence of progression in the dormant cells. Some observations consistent with the first possibility were made by Fisher and Fisher (1959) who injected small numbers of cells of the Walker carcinoma 256 by way of a large mesenteric vein into rats which they examined five months later for tumours of the liver but found none. Three months after receiving similar injections, other rats were subjected to laparotomy with examination of the liver at seven-day intervals; all of them had a tumour of the liver within a few weeks. This result implies that emergence from dormancy may be precipitated by "stress".

Greene (1952) proposed that emergence from the dormant state was due to progression whereby the dormant cells acquired new powers of invasion and metastasis. Greene believed that tumour cells could be dormant in regional lymph nodes which had been overlooked when the primary growth was excised until, as a result of progression, they acquired new powers of dissemination; thereupon they give origin to widespread secondary (or more properly, tertiary) tumours. According to this interpretation the emergence of widespread secondary tumours many years after surgical removal of a tumour does not necessarily mean that tumour cells had been widely disseminated at the time of operation or earlier.

3. Latent Carcinoma

The original observations of Rich (1935) on the frequency of clinically unsuspected neoplasms found, at autopsy, in the prostates of men past middle age have been amply confirmed (e.g. Edwards *et al.*, 1953; Franks, 1954 *a, b*, 1956; Hirst and Bergman, 1954).

Franks distinguished between *latent* or *unsuspected* cancers and *active* or *clinical* cancer of the prostate. The latent tumours are found especially in old men; they are increasingly common after the age of 80 and are present in most men over 90. They seem to grow very slowly and may be either large or small. Histologically, they may be well or poorly differentiated and are indistinguishable from clinical cancers; they commonly invade blood vessels and lymphatics but distant secondary tumours rarely develop. The difference between latent and clinical carcinoma is a biological one, recognizable only by the behaviour of a tumour in its host. The division between the two sorts of carcinoma is not sharp and different degrees of "latency" probably exist. It is estimated that about 9% of patients with clinically-diagnosed cancer of the prostate survive for five years or longer without treatment. Apparently tumours can grow to clinically-recognizable size and yet retain the characteristics of "latent" tumours.

The available evidence, on balance, suggests that "latent" and "clinical" tumours are qualitatively different from the beginning and not merely consecutive stages in one process of growth and extension. They differ in certain independently variable characters that are expressed in growth rate and in the failure of the "latent" type to yield secondary tumours despite being locally invasive. Latency may depend on the lack of some characters needed for completing all the steps of metastasis or, as Franks (1956) suggested, it may have a hormonal basis or be associated in an unexplained way with aging. Some clinical carcinomas develop in latent ones not, it seems, by mere extension, but as a result of qualitative progression. It is certain that all latent tumours do *not* develop into clinical carcinomas.

In a series of 1300 autopsies on patients over 70 years old, McKeown (1956) found 261 cases of malignant disease of which 20% were "latent" in the sense of being unsuspected during life and not materially responsible for death. The latent tumours included carcinomas of the gastro-intestinal tract and lung as well as carcinoma of the prostate. Franks (1954b) referred to other examples in kidney, thyroid, heart and lung.

II. Retrogressive Neoplasia

Little systematic consideration has been given to retrogressive neoplasia except with reference to the nebulous phenomenon of Spontaneous Regression of Cancer in man. This will be discussed first and the more instructive as well as clinically more important examples of retrogressive neoplasia later.

A. The Spontaneous Regression of Cancer in Man

This contentious matter has been dealt with recently by Smithers (1964), Boyd (1966) and, most exhaustively, by Everson and Cole (1966). From a survey of the world literature covering the years 1900–1965 and from personal communications, Everson and Cole assembled 176 reports of what they considered to be acceptable examples of "spontaneous regression of cancer". They gave summaries of each of the cases but stated that the amount of information available to them was "quite variable" and that some of the reports were "quite brief". Their criteria for acceptance were stated as follows: the partial or complete disappearance of a malignant tumour in the absence of all treatment or in the presence of therapy which is considered inadequate to exert a significant effect on neoplastic disease. They emphasized that they did not require that spontaneous regression need advance to complete disappearance of the tumour or that the regression should be permanent. They did not exclude cases with recurrence at the same site or growth at a different site and insisted that regression was not taken as synonymous with cure.

Boyd, by contrast, stated that in regression the tumour disappears and the patient is restored to health and he distinguished between partial and complete regressions. None of the authors states clearly the criteria for "malignancy" or "cancer" except that Everson and Cole required "histological confirmation" which, it seems, they accepted as incontrovertible. Smithers' paper is entitled Spontaneous Regression of Tumours but he was evidently concerned with malignant tumours.

More than half of the reports accepted by Everson and Cole refer to tumours of four types, namely, adenocarcinoma of the kidney, malignant melanoma, neuroblastoma and choriocarcinoma. Neuroblastoma and choriocarcinoma are dealt with in Chapter 6 primarily as examples of embryonic tumours but with some attention to "spontaneous regression". Their origin in embryonic life may be related to the liability to regression but, as Boyd pointed out, other examples of spontaneous regression are spread over all age groups. The tumours whose spontaneous regression has been reported less frequently include soft tissue sarcomas and carcinomas of the bladder, breast, mammary gland and ovary.

It would be pedantic, in my view, to object strongly to the description of the regressions as "spontaneous" on purely semantic grounds. Nevertheless, as Boyd maintains, the great majority of the patients who enjoyed regressions had been subjected to some kind of "interference", therapeutic or other. The interferences included incomplete removal, palliative irradiation, surgical biopsy and intercurrent febrile reactions. The two most quoted examples of regression of nephroblastoma by maturation were in patients who, at some stage, had received Coley's toxin. None of these procedures is highly regarded as a definitive treatment for malignant neoplasia but it may be prudent to entertain the possibility that if used for palliation they may be responsible in some patients, but probably not in many, for better results than can reasonably be expected. Both Smithers and Boyd accept the regression of some metastatic tumours after removal of their primary growths as established fact. Smithers reports, even more surprisingly, the regression of residual tumour following incomplete removal of malignant melanomas, a procedure ordinarily discountenanced because of the risk of precipitating dissemination. He also notes that the few reported regressions of soft tissue sarcomas seem to have followed a variety of supposedly inadequate treatments. No rational explanation of these regressions is available. Plausible explanations can be advanced for regressions of three kinds of malignant tumours which, on this account, deserve separate mention. The three types are carcinomas of the breast, thyroid gland and urinary bladder.

Smithers, who thinks that the frequency of regression of carcinoma of the breast has been underestimated, tabulated 12 reported examples including three of his own. With two possible exceptions, the regressions occurred at the menopausal age and most of them were specifically reported as having taken place at the time of cessation of the menses. These are reasonably interpreted as regressions of hormone-dependent mammary tumours as a result of withdrawal of the supporting hormonal stimulus at the time of the menopause. The regression of some carcinomas of the thyroid gland following the administration of thyroid extract can be similarly explained; the administration of thyroid extract depresses the secretion, by the pituiatry, of thyrotrophic hormone upon which, it is presumed, the regressed tumour had been dependent.

All or nearly all the reported regressions of carcinoma of the bladder have followed partial or complete diversion of urine from the bladder by transplantation of the ureters and it is inferred that the regressed tumours were conditional tumours dependent for growth on some unidentified stimulating material excreted in the urine. Everson and Cole emphasized that only a small proportion of the patients whose ureters have been transplanted enjoy a regression of their tumours and proposed that the pathogenesis of the tumours that regress must be different from that of the great majority of carcinomas, which do not. Smithers, more specifically, attribute regression to the withdrawal of a chemical carcinogen that is excreted in the urine. Both authors imply that the agent that maintains the growth of a tumour is the same as that which initiated neoplasia, probably many years before. This is not necessarily or even probably, a valid inference. Experimental studies of conditional tumours, to which category Smithers appropriately assigns regressing carcinomas of breast, thyroid and bladder, show that the initiating and promoting or sustaining stimuli are not, as a rule, identical. The studies do not encourage the presumption of Everson and Cole that the pathogenesis of conditional tumours is different from that of unconditional tumours of the same general type.

No one suggests that the kind of spontaneous regression here under consideration, which happens on the average in rather less than three patients in the world per year, taking Eversen and Cole's figures at their face value, is of any significance in clinical practice. The authors quoted believed nevertheless that the fact of it happening at all has important implications which will be discussed later. The question for discussion here is whether or not the figures given by Eversen and Cole can, or should, be taken at their face value since they are based on an essentially arbitrary selection of cases. The use of the word "cancer" is probably nowhere more pernicious than in the present context; it confuses every issue and throws the phenomenon of regression of neoplasia completely out of perspective. In dealing with this subject most authors reveal a rather naïve faith in the infallibility of "histological confirmation" of "cancer" or "malignancy". Accurate diagnosis of vague generalities is not possible. The frequency of histological misdiagnosis of the "cancers" here under consideration is probably far greater than the frequency of "spontaneous regression". This is not a reflection on the professional competence of pathologists but on the innate perversity of neoplasms. The nature of the inquiry has probably selected out a small number of "cancers" with highly unusual associations of independently variable characters. Histological confirmation is further complicated when it is based, as it often is, on the examination of a primary or secondary tumour that has been excised for the purpose and not of the remaining tumour that has been observed to regress. Whether histological confirmation excluded or included too many "cancers" in the series is debatable. Smithers has objected that strict criteria of acceptance have been laid down to exclude as many awkward facts as possible. My own objection is that the criteria have excluded nearly all the instructive facts, which I do not find "awkward". All three authors quoted repeatedly in this chapter exclude the highly instructive spontaneously regressing epitheliomas or keratoacanthomas of human and animal skin for various reasons. Everson and Cole exclude them because of the well-known difficulty of distinguishing between cancer and "certain benign dermatological conditions". Boyd excludes them because he considers that they display a phenomenon of healing rather than of regression, a distinction which seems to me over-subtle. Smithers excludes them for a variety of reasons that are not specified.

B. Spontaneous and Induced Regressions of Neoplasia in Animals and in Man

In his discussion of "spontaneous" regressions of man, Smithers grouped together regressions of tumours of the breast, thyroid gland and urinary bladder as examples of the regression of conditional tumours, whose sustained growth is dependent on some continuing extrinsic stimulus, identified or unidentified. The most important of these in clinical practice are the hormone-dependent tumours which will be considered first.

1. *Hormone-dependent Tumours*

The clinical demonstration that carcinomas of the prostate and of the breast could be induced to regress in a substantial proportion of patients by some kind of disturbance of their endocrine systems was of revolutionary importance in at least two respects: first, the tumours are indisputably malignant "cancers" in any and every reasonable usage of the term; they are almost inevitably fatal in the absence of radical therapy, the few reported "spontaneous regressions" being probably due, as Smithers maintained, to endocrine disturbances; second, the sustaining stimuli required for progressive growth are known, in a general way, if not precisely specified. Bielschowsky (1955) discussed the regression of hormone-dependent tumours in a valuable review and the phenomenon of hormone-dependence will be discussed in detail in Volume 2 of this book. The pregnancy-responsive mammary plaques of mice described in Chapter 3 of this volume illustrate and illuminate some of the important characteristics of hormone-dependent tumours.

Little can be added here to Bielschowsky's account of the mechanism of regression. It is widely presumed that regression is attributable to withdrawal of a required sustaining stimulus that operates directly on the neoplastic parenchyma although some have attributed it to a stimulation of natural defensive processes (Emerson et al., 1953) and recently Leighton (1967) has questioned whether the primary effect of therapy might not be on the stroma instead of on the parenchyma. My own limited observations on the regression of mammary plaques in mice revealed simultaneous changes in stroma and in parenchyma and seemed to be the more consistent with a primary effect on the parenchyma cells. The regressed plaques recurred almost without exception when the pregnancy stimulus was restored, however long the regression had persisted and sometimes they recurred in the absence of pregnancy. Comparable observations have been reported more recently on rats. The most depressing feature of endocrine therapy for mammary carcinoma in women is the near-certainty of ultimate uncontrollable relapse after a longer or shorter period of satisfactory remissions. It seems to be an almost invariable rule that endocrine control of mammary neoplasia, even to the point of complete clinical regression, whether in mice, rats or women, leaves behind a region of *residual neoplasia*, which persists indefinitely and from which fatal recurrence can stem. The relapses after temporary remissions despite continued therapy in women correspond with the "spontaneous" recurrence of plaques during intermissions of breeding in mice and imply, similarly, that progression from the hormone-dependent state had taken place during the quiescent phase. Several steps of progression may precede the loss of responsiveness to all practicable endocrine procedures and consequently several cycles of remission and relapse, brought about by various endocrine derangements, precede the ultimate fatal recurrence.

The nature of the residual neoplasia is not known. The site of a long-regressed plaque was often recognizable by the naked eye at autopsy, if only by a slight discolouration in the mammary tissue. Little was to be seen microscopically except slight fibrosis and a few shrunken irregular tubules yet the capacity to react to pregnancy by forming a plaque, was apparently always and indefinitely maintained. Recently, Leighton (1967) has advanced the interesting suggestion that the residual neoplasia may be attributable to the persistence of neoplastic stem-cells, whose progeny differ somewhat from their parents as a consequence of repeated cell division, or some form of "differentiation" or which for some other reason are more vulnerable to adverse circumstances to which they succumb whereas the stem-cells survive. The phenomenon is of great clinical as well as theoretical importance and deserves thorough investigation. It is evident that *something* persists indefinitely and that in this sense the regression is not complete. The programme of neoplastic development persists in the residual neoplasia but an extrinsic stimulus is needed to activate it; in this respect residual neoplasia resembles incipient neoplasia.

2. Epidermal Neoplasia

The regressing neoplasms of the human skin include the rare "self-healing carcinomas" of the Ferguson-Smith type, the common keratoacanthomas, which formerly were often diagnosed as epidermoid carcinomas and the Group B lesions seen in xeroderma pigmentosum and in the severe regional neoplasias induced by prolonged occupational exposures to chemical carcinogens or sunlight. Detailed consideration of these lesions is deferred to Volume 2. Attention here will be devoted mainly to the regression of comparable lesions induced experimentally in rabbit and mouse epidermis by chemical carcinogens.

Reference has been made already to the observations made on the early stages of chemical carcinogenesis in mouse skin showing that initiation quickly established a region of incipient neoplasia that is apparently permanent and irreversible. So far as can be determined, incipient neoplasia is *not* subject to regression. The observations indicated further that the initiating procedure determines the quality of the neoplastic capacity or programme of development of the incipient neoplasia and that certain procedures establish a programme that specifies the regression of most of the early-emerging lesions. These regressions, determined by intrinsic characters of the neoplastic cells, may reasonably be called "spontaneous". The regression of the self-healing carcinomas and adenocanthomas of human skin is probably "spontaneous" in this sense.

When the initiating procedure is suitably contrived, the early-emerging Group B lesions are numerous and varied. Rous and Kidd (1939, 1941) studied the B lesions induced by coal-tar on rabbits' ears, where the fates of individual lesions can be more easily traced than on loose mouse skin, and gave detailed descriptions of the early warts and of their behaviour. In several respects, their observations are closely relevant to the present discussion.

Rous and Kidd distinguished three main kinds of wart, namely, common papillomas, frill horns and carcinomatoids. Many of the warts other than frill-horns regressed when applications of tar were withheld and recurred when they were resumed; they were conditional tumours but differed from hormone-dependent tumours in that recurrence could be induced also by non-specific stimulation provided by wound healing at the site of regression or by applications of turpentine. The cycles of growth and regression could be repeated several times. Histological studies indicated that the normal-looking epidermis that covered the scars left by regressed warts was neoplastic epithelium derived from the warts. Rous and Kidd inferred that the deviations from normal that result in tar papillomas and carcinomatoids are so inconsiderable that the cells, when not extrinsically stimulated, may conform with the laws of normal organization and resume the appearance and habit of life of normal epidermal cells but, Rous and Kidd went on to say, "this does not mean that they have become normal, only that their neoplastic potentialities are in abeyance." Although the epithelium looked normal, it was not functionally normal as plainly demonstrated by its capacity for quickly producing a wart in response to a non-specific stimulus such as wound-healing or applications of turpentine. The epithelium, in short, was in a state of residual neoplasia comparable with that succeeding the regression of hormone-dependent tumours; the programme of neoplastic development endured intact in the residual neoplasia but it was not put into operation without an extrinsic stimulus. The stimulus needed to reactivate the residual neoplasia of hormone-dependent tumours is a specific hormonal one whereas non-specific injurious stimuli suffice to reactivate the residual neoplasia of warts. Two different mechanisms of reactivation may be implicated; it seems likely that the reactivating hormones act directly on the neoplastic cells but the reactivation of wart residual neoplasia by wound-healing or turpentine is more credibly due to a direct action on connective tissue and only an indirect action on neoplastic cells.

Only a small proportion of the warts behaved in the way described; sooner or later most of them disappeared permanently, some of them despite continued exposure to tar. Rous and Kidd were satisfied from their histological studies of these regressing tumours that the neoplastic epithelium at the site of regression was replaced by normal epidermal cells. The careful histological studies of Rous and Kidd give no support to the view that regression of neoplasia whether spontaneous or induced, entails a reversion or conversion of neoplastic cells into normal cells. The normal appearance is restored either by actual replacement of neoplastic epithelium by normal epithelium or by inactivation of an intact programme of neoplastic development in a region of persistent residual neoplasia.

3. *Defective Neoplastic Cells*

In another section of their paper, Rous and Kidd (1941) discussed, under heading *Liabilities Entailed by the Neoplastic Change*, some insufficiently-heeded limitations of the neoplastic state. They pointed out that the abnormalities responsible for, or accompanying the neoplastic state, range all the way from those in substhreshold neoplastic states, corresponding with what is here called "incipient neoplasia", to those in clinical "cancers". Even in clinical "cancers" the neoplastic cells are often "sick cells" that die young and the tumours grow only because the cells proliferate more quickly than they die. Neoplasia is often attended by a lesser or greater loss of functional capacities and Rous and Kidd said that it seemed certain from their observations on tar warts that the disabilities resulting from such losses might be lethal to the tumour cells *ab initio*. This applies, in particular to the carcinomatoids; their emergence is conditional upon the operation of certain extrinsic stimuli but the continuation of those stimuli does not ensure the persistence of the tumours, most of which eventually regress and do not recur. The general conclusion of Rous and Kidd expressed in their own words is that "It is plain that the neoplastic state does not necessarily connote independence of behaviour or success in tumour formation. On the contrary it may render cells unable to survive or endow them with powers which they can exert only under favouring conditions".

Willis (1967), highly experienced in the pathology of neoplasia in man, reached a similar conclusion. It is conceivable, in his view, that tumours have a finite life span which, under special circumstances, may be shorter than that of their hosts. He supposed that the finite life span is determined, on the one hand, by the intrinsic proliferative energy with which a tumour is endowed at its inception and, on the other hand by the adequacy with which its peculiar nutritional requirements are met.

Rous and Kidd, as well as Willis, imply that the eventual regression of tumours may be inherent in the neoplastic capacity or programme of development impressed on cells at the time of the initiation of neoplasia. It is probable that this capacity can be acquired also at later stages of neoplastic development as a result of random disturbances of the karyotype or by progression. During prolonged culture *in vitro*, malignant neoplastic cells are apt to lose the ability to yield tumours when implanted into appropriate animals and, since transplantation provides the only reliable criterion, they seem to lose their "malignancy" (Sanford, 1965). It is not difficult to imagine that a further step might lead to the loss of viability.

In summary, the implication of this discussion is that some tumours regress on account of an intrinsic defectiveness of their cells; the cells have neoplastic characteristics but lack the basic maintenance mechanisms for keeping alive through more than a finite number of cell generations. The defectiveness may be inherent in the tumour cells from the beginning or be brought about later by progression or by random changes in karyotype.

C. Varieties of Retrogressive Neoplasia

The "spontaneous regression of cancer"—a phrase in which three of the four words, namely "spontaneous" "regression" and "cancer" are loaded with ambiguities—is scarcely amenable to rational discussion and would not have been discussed here except for the dubiously valid inferences that have been drawn from this ill-defined phenomenon about the "reversibility" of neoplasia and about "host-resistance" to it. It is evident from the preceding sections of this chapter that a comprehensive but by no means exhaustive study of retrogressive neoplastic changes discloses a wide range of regressive phenomena which differ in their mechanisms and in their significance and which are not susceptible to tidy classification.

For convenience of discussion three categories of regression will be distinguished although, as will become apparent, they are not sharply separable.

1. Regression of Conditional Tumours

The hormone-dependent carcinomas of the human breast and prostate are not distinguishable by any available pathological or clinical procedure from the majority of those carcinomas that are not hormone-dependent and that for generations have accepted without question as lethal, "cancers"; the only way to identify them reliably, is by observing their response to derangements of the endocrine system. (The possibility of making a clinically-useful distinction by means of biochemical studies of excreted hormones will be discussed in Volume 2.)

They can be discussed, as few other tumours in man or in animals can, without encountering the semantic morass of disputation about their "malignancy". Their regression, like that of other conditional tumours, leaves behind a persistent state of residual neoplasia from which the carcinoma can, and almost certainly will, recur. It should not be presumed that no other authentic hormone-dependent carcinomas exist, only that their authenticity has not been decisively established. In his review, Bielschowsky (1955) described other carcinomas in man that with varied degrees of probability might be hormone-dependent. He mentions the existence of uncommon neoplastic lesions in which regression is the rule and instances papillomatosis of the larynx in children, which is usually self-limiting by regression at the time of puberty although the rate of recurrence is high and the tumours are potentially malignant.

It is noteworthy that it is widely believed, although not certainly proved, that the extrinsic stimulus that is needed to sustain the growth of the dependent tumour is either identical with or of the same general nature as the stimulus that was needed to provoke its first emergence. This view may need revision in the light of the experiments on epidermal neoplasia indicating that a tumour evoked by a chemical carcinogen may be sustained by turpentine or by wound-healing. This may apply to the small proportion of carcinomas of the human bladder that regress after diversion of the urine. It may be remarked, also, the some epidermal tumours of experimental animals require an extrinsic stimulus to provoke their emergence but then persist and grow without it. On the other hand, some other conditional epidermal tumours fail to grow progressively despite continued operation of the extrinsic stimulus and these may be as appropriately assigned to the next category.

2. Regression Specified by the Programme of Development

These regressions reasonably may be described as spontaneous since they are determined by the intrinsic characters of the neoplastic cells and not by an extrinsic stimulus. They are exemplified by the numerous and varied "benign" or "anomalous" Group B lesions of rodent epidermis. Similar lesions occur in severe regional neoplasias of the human skin and keratoacanthoma. The initiating procedure seems to determine the details of the programme and extrinsic factors may be needed to activate it but extrinsic factors apparently do not affect the end-result of its operation. None of the neoplasias in this category is unequivocally "malignant"; the keratoacanthomas are invasive but their "malignancy" is of the "spurious" or "factitious" type.

It is shown in Chapter 6 that many tumours, including many plainly "malignant" tumours follow a programme of differentiation similar to that of their normal parent tissues. Epidermoid carcinomas for example may keratinize abundantly but they do not, on that account regress. Some benign tumours, on the other hand, seem to differentiate themselves to death as though impelled by a positive feedback mechanism; epidermal warts may keratinize completely and drop off. Neoplasms originating during embryonic life sometimes carry a programme of differentiation closely resembling that of their embryonic parent tissue and in consequence may differentiate or "mature" in a similar way. This kind of programme-specified "maturation", most evident in neuroblastomas and teratomas may lead, exceptionally, to regression.

3. Regression Due to Acquired Defectiveness of Neoplastic Cells

This category is designed to include lethal defects of cells that are not prescribed in the programme established at the inception of neoplasia but produced much later either by progression leading to progressive intracellular disorganization or to random changes in karytope. These kinds of defectiveness are probably responsible, at least in part, for a high death rate of individual cells; whether or not they ever lead to regression of whole tumours is questionable but worth considering. The occasional regression of rapidly growing choriocarcinomas and their relatively good response to chemotherapy might be attributable to acquired defectiveness.

D. THE SIGNIFICANCE OF RETROGRESSIVE NEOPLASIA

Retrogressive neoplasia has been cited as evidence for the reversibility of the neoplastic state (Smithers), for a natural resistance of tissues to cancer growth (Boyd) and for a "biological control" of cancer (Everson and Cole). These matters will receive limited attention here in advance of more comprehensive discussion elsewhere.

1. Reversibility of Neoplasia

Smithers, writing in support of reversibility about the regression of hormone-dependent tumours, said that cells that have already been called "cancer cells" cease to be so and whole masses of them degenerate and die; the site of this regression heals and it may be difficult to see where they have been. The "cancer cells" stop being "cancer cells" by becoming dead cells; as evidence of "reversibility", this is massively unconvincing. The reversion to a normal appearance is not accompanied by a reversion to normal behaviour; as already shown, the dead cells are removed but leave behind a region of residual neoplasia with a permanent capacity for neoplastic development that is not present in normal cells of the same lineage. Rous and Kidd reached the same conclusion from their studies of conditional epidermal neoplasms. They described also an alternative process of healing of the defect left by rejection of a degenerated neoplasm, namely, by immigration of normal epithelium from the periphery. This is almost certainly true after regressions of the second category. The evidence for reversal of neoplasia in regressions of both the first and second categories is nil. The regressions of the third category provide convincing evidence against reversal; these regressions are attributable to progressive internal disorganization of cells and increasing alienation from the "normal" state.

2. Tissue Resistance to Neoplasia

In the discussion of invasion in Section C, which is closely relevant to the matter now under consideration, I proposed that concepts of aggression and resistance should be replaced by a concept of host-tumour relationships and, more particularly, by a concept of varied mutual interactions between the neoplastic parenchyma and the non-neoplastic host connective tissues and vasculature. On this view, the capsule of a benign tumour is not interpreted as a defensive barrier against extension but as a product of interaction between neoplastic and host connective tissue and blood vessels. The "resistance" is a negative quality, the failure to establish a relationship conducive to invasion and the primary "cause" of this is the failure of the benign tumour cells to exert an appropriate evocative stimulus on the host connective tissue and blood vessels. Ordinarily, the stimulus provided by the tumour parenchyma is the dominant factor but the quality of the responding tissue can greatly affect the outcome. Whether or not it can do so to the extent of bringing about regression of the tumour is not clear. On the other hand, there is no doubt that interference with the host tissues can aggravate neoplasia to a dangerous degree. Wallace (1959) and others have described the disastrous consequences of surgical intervention that entails the spilling of cells of a carcinoma of the urinary bladder into a wound made by opening the bladder or into the extravesical tissues; judging by clinical criteria the "malignancy" of the tumour is greatly increased in the new sites. Two factors may be implicated in this increase; first, any raw or damaged surface provides a favourable site for the implantation, establishment and growth of tumour fragments and, second, the connective and vascular tissues react to the tumour by providing the conditions favouring invasion.

This aggravation of "malignancy" recalls the "spurious" or "factitious malignancy" of benign Shope papilloma tissue that has been implanted in various internal sites as described in this chapter. Transplantation of the "malignant" tumours showed that it still behaved as a benign papilloma when implanted in other sites. The spurious malignancy was not attributable to changes in the tumour parenchyma but to a change in the tumour-host relationship that was especially favourable for invasive growth. The same considerations probably apply to the artificially disseminated cells of the vesical tumours of man. The enhanced invasiveness is no less dangerous for being "spurious".

CHAPTER 6

The Histological Analysis of Neoplasms

The earliest experimental studies of neoplasia in animals relied on an already-established pathology of neoplasms of man. The study of the morbid anatomy and histology of human neoplasms by a succession of notable pathologists has provided a mass of important and accurate observations that laboratory investigators cannot ignore. These methods remain indispensable for diagnosis and prognosis in medical practice, as well as for the recognition, naming and classification of sporadic and experimentally-induced neoplasms of animals. The descriptive and diagnostic histology of neoplasms is well covered in standard text-books of pathology and in several specialized treatises of which the books of Willis (1967) and of Evans (1966) and the "Atlas" of the Armed Forces Institute of Pathology may be mentioned. No attempt to cover the same ground will be made in this chapter whose intent is essentially different. Briefly, the aim here is to consider to what extent histological studies, together with experimental procedures, can illuminate fundamental problems in the biology of neoplasia. Attention will be directed to resemblances and differences between neoplastic cells and tissues and their normal precursors in, for example, their capacities for differentiation and morphogenesis, and to the interactions of neoplastic cells with one another and with non-neoplastic cells and tissues. In a review published a long time ago (Foulds, 1940), I tried to show how experimental procedures, in particular tumour transplantation, could supplement the histological observations and enlarge their interpretation so as to yield limited but important insight into the biology of neoplasia. Some of the same material will be summarized in this chapter, and in some places re-interpreted, but much more attention will be given to the developmental aspects of neoplasia, to the phenomenon of tumour progression and to the independent variability of characters, as well as to the newer concepts of developmental biology, in particular to contemporary ideas about "differentiation".

Some difficulties are almost inevitably met with in applying the concepts and terminology of the pathology of human tumours to purposes other than those for which they were originally designed and for which they are still ordinarily used. It may be well to draw attention, at the outset, to some of these difficulties and limitations. In the first place, it may be noted that the facility and relative accuracy which many pathologists acquire in the histological diagnosis and prognosis of tumours in man does not, strictly speaking, depend solely on their own personal histological observations; it depends also on other clinical information at their disposal and, in considerable measure, on the cumulative experience of several generations of predecessors who have sought to correlate pathological observations with clinical behaviour and in so doing have established a fairly reliable estimate of the statistical probability that a particular histological picture will correspond with a particular mode of clinical behaviour. In experimental research, a reliance on similar histological criteria without a comparable background of clinical observation and experience can be misleading; as experienced pathologists well know, the relationship between histological structure and clinical behaviour is not simple and straightforward.

The pathology of neoplasia in man is based primarily on relatively late stages of neoplastic development. The earliest stages are sub-clinical or even sub-microscopic and most clinically-recognizable "tumours" are late or terminal manifestations of a process of neoplastic development extending far backwards in time to inconspicuous beginnings. Clinically "early" is biologically "late". In the study of neoplasia, as in navigation, it is imprudent to ignore the submerged portion of the iceberg. The pathology of neoplasia in man has provided no adequate concepts or terminology for dealing with the early stages. It is seldom possible to trace the whole history of a tumour from beginning to end even in experimental animals but, by piecing together the fragmentary information available, it is sometimes possible to form provisional generalizations about the developmental history of histological structures. Haaland (1911) recognized that traces of the whole history of development through successive stages may be discernible in a gross tumour. The gross tumour itself may be only a transitory stage in the development of a different kind of tumour.

A cardinal difficulty in tissue histology, based of necessity mainly on the examination of "dead meat", is to introduce the time dimension, which is crucial in the study of developmental processes. Presuming good fixation, histological sections give a satisfactory representation of the spatial relationships of cells and tissues at a particular point in time. For most histological, as contrasted with cytological, purposes it is not essential, although it is desirable, that the representation be exact provided that it be consistent. The "still" pictures are the "frames" that the histologist must try to animate into a moving picture. The arrangement of the individual frames in the correct sequence is a difficult and hazardous task, especially when the frames are obtained from many tumours derived from numerous patients. The distinction between collateral and consecutive events, to which reference was made earlier (Chapter 3) needs especial care. Identification of the parent cells of a tumour, as attempted in histogenetic classifications is not the only problem in early neoplastic development. The point of origin of a tumour is often inferred, more or less plausibly, but it is scarcely ever seen and the validity of inferences about histogenesis is rarely subjected to crucial tests.

In the following discussion, neoplasia of the epidermis and of the mammary gland will again be drawn on freely for illustrative purposes for reasons of convenience and of abundance of both clinical and experimental observations. Some observations that are accurate in spite of being old will be discussed and re-interpreted. Certain unusual or even "freak" tumours whose biological interest is out of all proportion to their frequency or clinical importance will be discussed in detail. After more than 100 years of "cellular pathology" and 60 years of experimental cancer research, it is opportune to pay increasing attention to the exceptions that prove (in the original sense of "test") the rule and to minor discrepancies of observation such as those that led to the discovery of the rare gases of the atmosphere and to the acceptance of Einstein's theory of relativity.

I. The Components of Neoplasms

Tumours are composed of a parenchyma, a stroma and, in malignant tumours, varied amounts of invaded tissues. The parenchyma is the proper neoplastic component; it survives transplantation, whereas the stroma dies and is replaced by a new stroma formed by a *specific stroma reaction* of the host tissues elicited by the parenchyma which determines its quantity and quality. Although the parenchyma is dominant, the stroma and invaded tissues contribute substantially to certain complexities of histological structure.

Throughout this chapter and some later ones, the discussion will repeatedly turn to the subject of *differentiation*. It is desirable, therefore, to digress at this point to deal briefly with contemporary ideas about differentiation and the mechanisms that are involved in it. The evidence for these ideas will be dealt with in Part III.

Differentiation has to do with *differences* between cells and tissues. In histopathology, it applies almost exclusively to the specializations of structure whereby the differences between the various kinds of cell become visibly manifest. For the most part, the term will be used in this sense in the present chapter but occasionally it will be stretched to include biochemical specializations that are not necessarily revealed by visible signs. This usage is sanctioned by many biologists who, nevertheless, recognize other admissible usages. In general, pathologists interpret the term more narrowly than do biologists.

Contemporary ideas about differentiation are based on the newer concepts of genetic materials and genetic actions derived from "molecular biology". One basic, if subtle, change in the concept of genetic material deserves emphasis; it is no longer conceived as something concerned only or even mainly with Mendelian inheritance but something that is continuously implicated in the daily lives of cells. In particular, the genetic material is involved in the synthesis of specific proteins. The genetic material proper, the deoxyribonucleic acid (DNA) of chromosomes bears the *code* specifying the sequence of amino acids in the completed proteins. Different segments of the DNA are coded for different specific proteins. Differentiation, as now supposed, depends on the pattern of protein synthesis in particular cells and this in turn depends on the *selective utilization* of the genetic material or genome. The essential features of this hypothesis are conveniently expressed by the use of three new terms (Foulds, 1963): (1) the *total genome* applies to the genetic material carrying the whole of the "genetic information" present in the fertilized ovum. Two characteristics of the total genome are of particular relevance to the present discussion: first, the whole of it is not in effective use at any one time and second, the total genome is not, or at least is not necessarily, diminished in differentiated cells.

(2) The *effective genome* is that portion of the total genome that is in effective use in a particular cell at a particular time and place. Different portions of the total genome are in effective use in different kinds of cells or in one kind of cell at different stages of development. The type of differentiation manifested by a cell depends on the particular *effective genome* that is engaged.

(3) The terms "total genome" and "effective genome" are almost self-explanatory and are scarcely controversial, whereas the third of the suggested terms, the *facultative genome*, calls for more explanation. The preceding paragraph implies that more potential effective genomes are *available* in the total genome than are actually used at any one time. The term *facultative genome* was coined to apply to the collection of genetic patterns (or potential effective genomes) that are *available* for effective use under various normal and abnormal conditions. The facultative genome provides for a selection or choice of effective genomes; it provides a genetic basis for embryonic *competence* or *capacity* and for pathological *metaplasia*. Cells which normally produce mucin, for example, but which under pathological conditions produce keratin are credited with a facultative genome that carries the potential effective genomes for both mucin and for keratin synthesis.

The cells that have a capacity for producing either mucin or keratin may, in fact, produce neither; the two potential effective genomes are *available* but neither is brought into effective use. It is desirable to consider briefly what constitutes "effective use". The first step in effective utilization or, in other terms, in the "expression" of genetic information, is the *selection* of one of the potential effective genomes available in the facultative genome. The mechanism of selection is scarcely understood at all. In general, the selection of one effective genome automatically suppresses the utilization of any alternative one so that, by a process of *mutual exclusion*, a particular cell may produce *either* mucin *or* keratin but not both at the same time. Some possible exceptions to this general rule will be mentioned elsewhere. The next step is *transcription* of the genetic code of the DNA constituting the selected effective genome into RNA, more specifically into *messenger* or informational RNA (mRNA). The next step of *translation* consists of the synthesis of a polypeptide having a sequence of amino acids specified by the sequence of nucleotides in the mRNA. Transcription and translation have been extensively studied but, as yet, much less attention has been given to the many subsequent steps leading to genetic "expression". Broadly speaking, polypeptides must be converted into proteins, especially enzymatic proteins which, in turn, catalyse a longer or shorter chain of biochemical processes that culminates in the formation of the specialized differentiative product, mucin or keratin, for example. According to this interpretation the decisive event in differentiation is the synthesis of a specific protein or group of proteins. The statement that "DNA makes RNA and RNA makes protein" is an oversimplification but, for the purposes of this chapter, it is, perhaps, an adequate summary of current views about the effective utilization of selected portions of the genome.

Most of the observations to be dealt with in the following pages were first reported in orthodox pathological terms and for the most part will be similarly described here, but also, where practicable, re-interpretation in the newer terms will be attempted. One of the chief aims of a detailed analysis of special cases of differentiation in neoplasia is to seek information about the total, effective and facultative genomes. Do one or more of these genomes differ from those in non-neoplastic cells and if they do, how do they differ and to what extent do the differences help to explain important characteristics of neoplasia?

A. The Parenchyma

The presumption, inherent in histogenetic classifications of tumours, that tumours inherit differentiative properties from their parent tissues is an oversimplification and subject to important qualifications. This section is devoted mainly to a discussion of the validity of three more specific propositions. The three propositions are as follows:

(1) The *type* of differentiation in a tumour corresponds with one or another of the types of differentiation that its parent cells can manifest either normally or as a consequence of experimental interference or disease. The range of

types of differentiation possible amongst tumours derived from a single parent tissue corresponds, in general, with the types of metaplasia or divergent differentiation of which the parent tissue is capable. Otherwise expressed, the range is circumscribed by the facultative genome of undifferentiated cells of the parent tissue. The type of differentiation is not necessarily "determined" by the normal effective genome of the parent tissue cells as some have implied (e.g. Kleinsmith and Pierce, 1964).

(2) The *degree* of differentiation in a tumour may be equal to or greater or less than that of its normal parent tissue.

(3) The type and degree of differentiation are stable and heritable characteristics of neoplasms, being maintained through an indefinite series of transplantations, with the reservation that they are liable to reversible change by *modulation* and to irreversible change by *progression* or, perhaps, by *mutation*.

1. Type of Differentiation

Although the tumours derived from a single parent tissue, amongst them can probably manifest all the types of differentiation of which that tissue is capable in health or disease, a particular tumour, with exceptions to which reference will be made later, is usually restricted rather narrowly to one of the possible types and the type may correspond with one that the parent tissue manifests only under the abnormal conditions of experiment or disease. The choice of a differentiative pathway entails the *selection* of an effective genome from amongst the potential effective genomes carried in the facultative genome of undifferentiated *stem cells*, which are often situated in distinct basal layers of epithelia or in germinative foci. The selection usually corresponds with the selection in normal differentiation, as will be shown later, in two important ways. First, the selection of one effective genome inhibits the utilization of all possible alternative effective genomes by the poorly understood process of mutual exclusion. Second, the selection is usually *biased* in favour of the "normal" type of differentiation of the parent tissue; the selection of an unexpected, "abnormal" type of differentiation probably implies the operation of some extrinsic direction or constraint. The selection of an effective genome is probably made at the inception of neoplasia; thereafter it seems to be stable and replicable even in the absence of phenotypic "expression". It should be emphasized that the facultative genomes of the undifferentiated cells of most normal tissues and the capacities of the tissues for divergent differentiation are far more versatile than is commonly supposed.

It is particularly instructive to draw concrete illustrative examples of the phenomena outlined in the preceding paragraph from epidermal neoplasia and mammary neoplasia because of the close embryonic relationship of the parent tissues. The rudiments of the mammary glands are formed by downgrowths of epidermal cells and the glands are widely believed to be homologous with sweat glands. The special pathology of those two forms of neoplasia will be considered in detail in Volume 2 where the summary interpretations advanced in the following paragraphs will be submitted to more rigorous critical examination.

The epidermis is continually being renewed from the unspecialized cells of its basal layer. These cells multiply throughout life; on balance, half of their progeny become differentiated cells and eventually die, whereas the other half remain in the basal layer and live to divide again. The basal cells illustrate especially well the concept of the facultative genome. In the first place, it is clearly stable and replicable and is preserved in the basal layer throughout life; in the second place, it provides for multiple types of divergent differentiation. It is now recognized that under appropriate conditions, notably in regeneration, the basal cells can produce not only the stratified keratinizing squamous epithelium of the surface epidermis but also any of the epidermal adnexa including hair follicles, sebaceous glands and sweat glands (Montagna, 1962). Neoplasms derived from basal epidermal cells can also differentiate in these several ways, but there is a strong bias towards the keratinizing squamous epithelial type of differentiation. It should be noted that the distinction between basal cell carcinoma and epidermoid carcinoma is not a histogenetic one. Probably *all* types of tumours of the epidermis originate from cells of the basal layer.

The "normal" function of the mammary epithelium is to secrete milk but capacities for other types of differentiation have been demonstrated, a capacity for epidermoid differentiation being especially firmly established. It is found in chronic cystic disease of the breast in women (e.g. Pasternack and Wirth, 1936) and in cysts distant from the nipple in mice and it is easily evoked experimentally in animals (White, 1910). The experiments show decisively that the epidermoid differentiation is not restricted to the region of the nipple or primary ducts or to "embryonic rudiments" or "rests" but can be manifested in any part of the mammary tree. Powell White commented with admirable clarity ". . . when we find squamous epithelium in a mammary gland or a tumour arising from it, it is not necessary to suppose an aberrant germ of epithelium at its point of origin. The squamous epithelium can originate from the mammary epithelium itself." The usual form of metaplasia in the mammary gland is keratinizing stratified squamous epithelium, but Wieser (1934) described also sebaceous metaplasia in mice.

The most controversial type of differentiation in the breast is the *myoepithelial* or *myothelial* type. The existence of myoepithelial cells has been disputed but it seems to me that several highly competent observers have adequately identified them as cells between and outside the basal mammary epithelial cells but inside the basement membrane and probably having a contractile function, useful in aiding the excretion of milk (Hamperl, 1940; Kuzma, 1943; Linzell, 1952; Richardson, 1949). Unfortunately, the term has been applied to abnormal cells and tissues that possess neither myoid nor epithelial qualities on the presumption that they are derived from myothelial cells. The presumption is probably wrong. Myothelial cells are differentiated cells and their capacity for proliferation and divergent differentiation has not been adequately demonstrated. The abnormal cells are more probably formed by differentiation of unspecialized stem cells towards a connective tissue instead of an epithelial type of tissue, as Hamperl agrees in principle. The basal cells can differentiate, normally, into either myothelial or secretory epithelial cells. The application to abnormal cells and tissues of a terminology based on histogenetic inferences of dubious validity has probably done more to confuse discussions of the role of "myothelial" cells in neoplasia than have discrepancies of observation.

Several observers have described "myothelial" proliferation in the human breast especially in senile involution and fibrosis, chronic cystic disease and adenosis. Hamperl (1940), in particular, gave detailed accounts of non-neoplastic "myothelial" proliferations and credited the cells with a capacity for producing reticulum and collagen fibres, cartilage and bone, as well as typical smooth muscle fibres. The proposition that cells of epithelial origin can produce connective tissues is uncongenial to those pathologists who hold to a rigid interpretation of "the specificity of the germ layers" long after embryologists have abandoned it. Nevertheless, Nicholson (1950) recalling that all three germ layers contribute to the embryonic mesenchyme found no inherent improbability in the post-embryonic differentiation from epithelium.

The expectation that mammary tumours amongst them will show all the types of differentiation of which non-neoplastic mammary epithelium is capable is fulfilled or, perhaps, even exceeded. The capacity of primary and transplanted mouse mammary tumours for exercising the normal function of mammary epithelium to secrete milk in response to pregnancy in the host or to administered oestrogen was described in Chapter 3. Hilf and his colleagues made similar observations on a spontaneous mammary carcinoma in a Fischer rat and on the transplanted tumours derived from it. The administration of oestrogen evoked a copious lactational secretion in the primary tumour and the capacity for that response to oestrogen was retained in the transplanted descendants. It was shown, moreover, that the milk-like material contained lactose, fatty acids and proteins and that the proteins had electrophoretic properties similar to those of the casein and whey of normal rat milk (Hilf et al., 1965; Hilf, 1967). Many of the mammary tumours induced in rats by feeding with dimethylbenzanthracene were, in the experience of Daniel and Pritchard (1964, 1967), milk-secreting growths. It is noteworthy that these "milk-secreting adenomas", as the authors designate them, were found in non-pregnant rats to which no hormones had been administered. Recently, Archer and Orlando (1968) have reported confirmatory observations. Daniel and Pritchard suggested that the carcinogen had brought about conditions that stimulated the pituitary gland to secrete the hormones that are normally associated with lactation. Milky secretion in tumours of the human breast seems to be restricted almost completely to unequivocally "benign" growths.

Epidermoid differentiation is frequent and extensive in mammary fibroadenoma and carcinoma in rats, especially in transplanted tumours as also in the "mixed" mammary tumours of dogs. It is frequent also in the commonplace adenocarcinomas of the mouse breast but often only in a minor degree in the form of whorling and stratification with central keratinization. Major degrees of keratinizing squamous epithelial differentiation are found in two named varieties of mouse mammary tumours, the fairly common *adenocanthoma*, formerly called adenocancroid, and the rare *molluscoid* tumour distinguished by a remarkably organoid structure, which will be described later. Structures identifiable with more or less probability, as hair-shafts, sweat glands and sebaceous glands have been described less often in mammary tumours of mice and rats. The descriptions of sebaceous cell differentiation are the most convincing and some rare mammary tumours of mice are predominantly of this type. (Bielschowsky *et al.*, 1956; Bonser, 1958) Epidermoid differentiation seems to be less frequent in the mammary tumour of women than in those of experimental animals but it has been described in mammary carcinoma and in some rare "mixed" tumours comparable with the mixed tumours of dogs (Pasternack and Wirth, 1936; Tudhope, 1939).

A classical paper by J. A. Murray (1908b) describing a transplantable "adenocancroid" of the mouse breast provides much instructive information about differentiation in tumours. The primary tumour contained solid epithelial alveoli, keratinized alveoli and adenomatous areas in intimate association. Some acini were lined in part by cubical epithelium and in part by stratified squamous epithelium, the two forms of epithelium together forming a continuous layer. Keratinization was prominent in some of the early generations of transplanted tumours but diminished in later ones as the growth rate increased. According to a later report by Bashford (1911), the tumour was maintained through eighty-two serial passages as a pure alveolar carcinoma with no squamous differentiation save in one tumour of long-standing. Murray had no doubt that the two types of differentiation in his tumour were inherent in cells of one kind. He described them as different "growth forms" of cells of one type; in a newer terminology they were *divergent differentiations* or modulations of cells of one type.

Murray's tumour changed during serial transplantation from adenocancroid to adenocarcinoma. A transplantable rat mammary carcinoma described by Lewin (1908) changed, under observation, in the opposite direction. The primary growth was a typical adenocarcinoma. During transplantation, the tumour became complex and it was maintained through many serial transplantations as a mixture of adenocarcinoma, keratinizing carcinoma, solid carcinoma and sarcoma-like tissue. Lewin interpreted his observations in the same sense as Murray. The two tumours differed from commonplace mammary carcinomas in their capacity for divergent differentiation and not in their histogenesis. They supply exceptions to the general rule that the type of differentiation in a tumour is rather narrowly restricted to *one* of the types of which its parent tissue is capable.

The question whether the "myothelial" type of differentiation is possible in mammary neoplasms remains for discussion. The first clear references to myoepithelial proliferation as an important element in neoplasia are those of Peyron and his colleagues, most of whose observations were made on the "mixed" mammary tumours of dogs. Peyron distinguished external and internal layer of epithelium in mammary ducts and acini and believed that the inner layer plays a predominant or even an exclusive role in mammary neoplasia in women whereas in dogs, the external zone, which he calls myoepithelial, is predominant. Peyron and his colleagues described various modifications of the myoepithelial component of the mixed mammary tumours of dogs and noted, in particular, an evolution towards connective tissue. They reported resemblances to leiomyosarcoma and stellate myosarcoma, syncytia in which collagen fibres develop and progressive incorporation of myoepithelial cells into the stroma and they believed that the osteocartilaginous and myxomatous portions of the mixed tumours were derived from the myoepithelial cells (Peyron, 1924a, b; Peyron et al., 1926; Petit and Peyron, 1927). Several more recent authors have substantially corroborated Peyron's observations on the mixed mammary tumours of dogs and have assigned a major role to myothelial proliferation in producing the "mixed" structure. Hamperl reinforced Peyron's assertion that myoepithelial cells formed collagen fibres by giving detailed descriptions of the formation of reticulum fibres in myoepithelial cells. Hamperl maintained that the differentiation of neoplastic myoepithelial cells could advance so far as to yield pure myxoma, fibroma and chondroma or sarcomas of various kinds (Allen, 1940; Biggs, 1947; Hamperl, 1940; Mulligan, 1949).

Varied degrees of myoepithelial proliferation have been described in mammary fibroadenoma and carcinoma in rats (Archer and Orlando, 1968) and it is plausibly responsible for the high liability of the relatively uncommon sporadic mammary carcinomas and their transplanted descendents to become complicated by various connective tissue components. The commonest sporadic mammary tumour in rats is fibroadenoma. Most pathologists have accepted the opinion of Nicholson and others that fibroadenoma of the human breast consists of two neoplastic components, the one of epithelial, and the other of connective, tissue origin. Extensive observation and experimentation on the mammary fibroadenomas of rats support the view that both components are neoplastic but throw doubt on their different origins. Both primary and secondary tumours are prone to wide variations in structure and are notably responsive to hormones. Oestrogens stimulate epithelial proliferation so that the fibroadenoma becomes an adenoma, secreting adenoma or cystadenoma all of which are variations of one basic type. Transplantation of a wholly glandular tumour yields, once again, a fibroadenoma. Deficiency of oestrogen or relative excess of androgen favours growth of the connective tissue component and change to a fibromatous structure. There is at least one report of the reappearance of the glandular component after several serial passages as a pure fibroma (Guérin, 1954; Oberling *et al.*, 1933, 1935, 1937, 1938; Roussy *et al.*, 1943; Shay *et al.*, 1952). It was not rigorously proved that the transplanted portions of the "pure" glandular or connective tissue tumours were completely free from the other component; this proof would be extremely difficult without cell cloning but it is reasonable to infer from the experiments that the glandular and the connective tissue components are divergent differentiations of one basic cell type.

To my knowledge, the only authors who have explicitly derived all types of mammary tumours from one kind of parent cell are Shay *et al.* (1952) who induced mammary tumours in rats by gastric instillation of methylcholanthrene. They described a wide variety of tumours so induced, including fibroadenoma, carcinoma and spindle-celled tumours, and believed that all were derived from a single neoplastic component that could provide both epithelial and fibrous variants; all the tumours were "mixed" in the sense of containing both derivatives in varied proportions. Shay and his colleagues identified the common parent cell of all the variants as myoepithelial. Archer and Orlando (1968) have recently described peripheral layers of cells, which they think are myothelial, around tumour nests in chemically-induced mammary tumours of rats.

The relatively uncommon sporadic mammary carcinomas of rats and their transplanted descendents are highly prone to become complicated by various connective-tissue-like tissues as exemplified by the "sarcoma-like tissue" in Lewin's tumour already mentioned in a different connection. Similar complications have been described in tumours of the human breast (Biggs, 1947; Gaudier *et al.*, 1931; Günther, 1937; Hamperl, 1940). Some at least of the structures described by these authors are probably commonplace but not being called "myoepithelial" by most pathologists, their true frequency is difficult to estimate. This seems to apply also to mammary tumours of mice in which, so far as I can recall, "myoepithelial" differentiation has not been specifically described, although something comparable almost certainly takes place. The over growth of epithelium in mammary tumours of mice, as described first by J. A. Murray (1908a) and more recently by Bonser (1945, 1949, 1954) and Foulds (1956) may extend either inwards to block the tubule or outwards to form a halo of irregular epithelial cells around a patent tubule. On occasion this outward or centrifugal growth has features similar to those that in other species have been called "myoepithelial". Tumours in which this type of growth predominates correspond with a subdivision of mouse mammary carcinoma named by Apolant (1906) *fissured carcinoma* (Figs 48–50). It is rare for a tumour to be composed entirely of this type of growth but, in my experience, its presence in lesser amounts is not uncommon and its contribution to the structure of the commonplace mammary tumours may be substantial. Dunn's Type C mammary tumour (Dunn, 1945, 1958) in which cystic spaces are surrounded by mantles of elongated cells resembling smooth muscle cells is a probable example of centrifugal growth, which merits the description "myoepithelial" as well or better than some of the other growths to which it has been applied. I have seen small mammary growths in which cysts were lined by a single layer of epithelial cells which had long tongue-like projections of "myoid" appearance extending into the stroma. More commonly centrifugal growths lack distinctive epithelial characteristics and tend towards the elongated spindle form but without notable resemblance to smooth muscle cells.

One other phenomenon deserves attention here because from time to time it has been cited as evidence that carcinoma cells harbour an infective principle that can be transmitted to normal connective tissue cells and transform them into sarcoma cells. Ehrlich and Apolant (1905) first described a "sarcomatous transformation of the stroma during serial transplantation of mammary tumours of mice and independent confirmation was soon forthcoming from several laboratories. Haaland (1908) reported the most detailed and comprehensive study of the phenomenon. He described transitional stages between epithelial cells and "sarcoma" cells and inferred from his observations that a stimulus emanting from the carcinoma cells effected a true conversion of connective tissue cells of the stroma into sarcoma cells. Haaland's interpretation has been widely accepted by laboratory workers but less readily by most pathologists who are experienced in the histopathology of neoplasia in man; in their opinion the "sarcoma" cells were more probably spindle-shaped carcinoma cells. After a careful re-examination of Haaland's descriptions and illustrations, I reached the opinion that they provided strong but by no means decisive evidence of transformation of the stroma. In particular, and in general agreement with Peyron *et al.* (1926), I interpret the "halos" described by Haaland as intermediate stages in the "transformation" of the stroma as centrifugal proliferations of epithelial tumour cells. I have often seen similar "halos" in mouse mammary tumours under circumstances where their epithelial origin seems hardly questionable.

Bashford (1911) described observations on two transplantable mammary tumours that are pertinent to this discussion. One tumour (no. 91) began as a haemorrhagic adenocarcinoma but, during serial transplantation, acquired areas resembling sarcoma. Further transplantation of portions composed wholly of sarcoma cells yielded tumours that eventually became wholly glandular. The second tumour (no. 129) began as a carcinosarcoma and continued as one through repeated serial transplantations. Transplantation of artificially separated regions containing only spindle-shaped, sarcoma-like cells yielded tumours of the original carcinosarcomatous type. It is arguable that, in the absence of cell cloning, the transplantation experiments are not completely decisive but the reasonable inference from them is that both tumours were throughout epithelial tumours whose neoplastic epithelium had a strong tendency to modulate towards the "sarcomatous" type. The phenomenon described as "sarcomatous transformation of the stroma" is explicable by the circumstance that the variation of carcinoma cells towards the sarcomatous type is not merely a reversible *modulation* as it was in Bashford's tumours but an irreversible *progression*, after which modulation back to the carcinomatous form is difficult and infrequent but not necessarily wholly impossible.

The foregoing discussion substantiates the proposition that the range of types of differentiation possible in tumours corresponds with the range of divergent differentiations that are possible in non-neoplastic tissues of the same provenance in various normal and abnormal conditions. It is noteworthy that the ranges of divergent differentiation possible in mammary epithelial tissues and in epidermal tissues respectively are remarkably similar but not identical. To the best of my knowledge, milky secretion has not been demonstrated in epidermal tissue under any natural or contrived circumstances that have been studied. Mammary epithelial cells seem to be capable of all or nearly all the types of divergent differentiations of which epidermal cells are capable, although it may be permissible to question the identification of sweat gland differentiation in mammary tumours. It is noteworthy that the epidermal types of differentiation are seen only in abnormal mammary tissues. The reasonable inference from the observations is that the facultative genomes of undifferentiated epidermal cells and mammary epithelial cells respectively overlap extensively but are not identical with one another. It should be noted that a great variety of epithelia, perhaps all of them, are capable of epidermoid "metaplasia" in response to disease or experimental interference and, inferentially, should be able to give origin, on occasion, to epidermoid neoplasms.

I cannot recall that histological analysis has provided any decisive evidence that neoplasms can manifest types of differentiative behaviour of which their parent tissues are not demonstrably capable under normal or abnormal circumstances but biochemical investigations have indicated that they may be able to do so. Most of these investigations have been concerned with endocrine functions. It has been shown that tumours in certain endocrine organs can secrete hormones that are characteristic of other endocrine organs. Dorfman and his colleagues, for example, reported that a tumour situated in the ovary provided chemical and biochemical evidence that the tumour originated in adrenal tissue (Dorfman, 1960). In a discussion of Dorfman's paper, Huseby proposed that each of the steroid-producing tissues probably had at least a majority of the enzymes present in the others and that at the inception of neoplasia certain enzymes disappeared in a rather random way so that a synthesis of corticoid hormones, for example, might become predominant. Huseby maintained, therefore, that there was no need to presume that corticoid-producing tumours in ovaries or testes developed from "rests" of adrenal tissue in those organs. Dominguez *et al.* (1960), who found a 21-hydroxylase characteristic of the adrenal gland in interstitial cell tumours of the testis believed that their own observations supported the hypothesis that the differences between the biosynthetic systems in the various cells that form steroid hormones were quantitative rather than qualitative.

It has become evident during the past few years that neoplasia in man can produce clinical syndromes referable to abnormal endocrine function. Lipsett (1965) noted that hepatomas could be associated with hypoglyaemia, polycythaemia and hypercalcaemia and that tumours of endocrine glands could secrete the hormone proper to the gland and in addition could secrete an unrelated hormonal substance. Even more remarkably, anaplastic carcinomas of the human lung can secrete hormones with the physiological actions of adrenocorticotrophic hormone or anti-diuretic hormone. In one of the most recently available reports, Lipscomb *et al.* (1968) describe a clinical syndrome attributable to the "inappropriate secretion" of antiduretic hormone in a patient with an oat-cell carcinoma of the lung, from which they isolated and characterized a hormone-like antidruretic material. Lipsett drew attention to the biological significance of the "inappropriate secretion" as did Lipscomb *et al.*, who agreed further with Lipsett in suggesting that the neoplastic cells may have lost a "repressor" mechanism that normally inhibits secretion of the "inappropriate" hormone. Some remote effects of "non-endocrine cancer" on the blood and on the skin have been summarized, in a recent discussion, by Malpas *et al.* (1968) and by Alexander (1968) respectively.

The general phenomenon of "inappropriate secretion" now being widely recognized, it is natural and proper that special cases are beginning to be subjected to a more severe critical analysis. Azzopardi and Williams (1968), who were concerned in particular with tumours associated with Cushings' syndrome, drew attention to the need for greater care and precision in the identification of the normal cells from which the tumours are presumed to originate. They accept as beyond question the association of Cushings' syndrome with oat-cell carcinoma of the bronchus and the presumed "non-endocrine" quality of the parent tissue with the reservation that, in their opinion, the cell of origin has not been satisfactorily identified. They believe that other types of tumour less commonly associated with Cushings' syndrome the parent tissues are not properly described as "non-endocrine". Azzopardi and Williams think also that the "repressor" hypothesis, to which reference was made in the preceding paragraph should be reconsidered.

Goodall (1969), in a comprehensive review of "para-endocrine cancer syndromes", has drawn attention to other important observations. He applies the term to the occurrence of clinical symptoms and signs usually pathognomic of over-secretion of the hormone of some particular endocrine tissue but associated with and due to a tumour, not of the relevant endocrine organ but of some other tissue. Although most of the reported cases implicate tumours of supposedly non-endocrine tissues, the definition also covers tumours of endocrine tissues which secrete "inappropriate" hormones. Goodall emphasizes with specific examples that the material secreted by the tumour may be indistinguishable from the normal hormone by all available tests, it may be very similar to that hormone with only minor differences in chemical or immunological properties or its activity may be attributable to some other substance lacking the properties considered characteristic of the normal hormone. The para-endocrine syndromes, as Goodal points out, may be of considerable significance in clinical practice.

2. *Degree of Differentiation*

The prevalent view that neoplasms in general and malignant neoplasms in particular are characterized by the loss of specialized histological and cytological structures typical of their parent tissues and of the specialized biological functions which those histological and cytological features reflect, needs considerable qualification. Askanazy (1936) reviewed the functional activities of neoplasms of man with special reference to "functions" visibly manifested by histological and cytological "differentiation" such as the formation of keratin, mucin, cartilage and bone. His general conclusion was that the degree of these functions in neoplasms might be equal to or greater or less than that in the normal parent tissues and that an inverse relationship between the degree of "function" and the degree of "malignancy" was not consistently demonstrable. Keratinization of epithelial "pearls" in metastatic epidermoid carcinomas is a commonplace example of specialized structure and function in malignant tumours.

Various degrees of differentiation are demonstrable similarly in primary and transplanted tumours of animals. Secretion of bile, for example, has been observed in a primary mouse hepatoma and in its transplanted descendants (Strong and Smith, 1936). Milky secretion in both primary and transplanted mammary carcinomas of mice and rats has been described already. Physiological function matched by near-normal structural differentiation is apparent in many tumours of man and of laboratory animals. This characteristic of endocrine tumours is referred to in a later discussion of trophoblastic tumours. As judged clinically in man, the hormone secretion may be normal or near-normal in quantity and quality. There is on record at least one patient whose needs for thyroid hormone following thyroidectomy for carcinoma of the thyroid were satisfied by the secretion of a persisting metastatic tumour in the lung; when the secondary tumour was removed, the patient developed symptoms of hypothyroidism (Willis, 1952). More generally, the degree of functional activity may differ widely in individual tumours of one general kind and so also may the quality of the product. Secretory activity may be especially conspicuous in endocrine tumours perhaps because the continued functioning of normal or neoplastic endocrine tissues, contrasted with that of exocrine glands, does not depend on an orderly relationship between secreting cells and excretory ducts (Foulds, 1940). The lack of functional activity and corresponding structural differentiation in some other kinds of tumours may be attributable not so much to a primary failure of *differentiation* as to a secondary consequence of structural *disorganization*.

The early investigators of transplantable tumours found considerable differences between the various tumours derived from a single tissue, but each transplantable strain derived from them had its own characteristic level of differentiation and function which it retained through many serial transplantations. The *degree* of differentiation was a stable and transmissible characteristic of transplantable tumour strains. Haaland (1911) could find no strict correlation between structure and clinical behaviour in transplanted mammary tumours; tumours with similar structures might behave quite differently although, when the rate of growth varied, the tumours of slower growth were usually the more completely differentiated. Haddow (1938) reached similar conclusions from a study of primary mammary tumours of mice. Murray (1908a) and Bashford (1911) discussed the loss of differentiation that they sometimes saw in transplanted tumours and decided that it was neither progressive nor irreversible but attributable mainly to rapid growth. Bashford remarked that the differentiation seen in a primary growth or in the early generations of transplanted tumours might be lacking from some tumours of later generations when they were examined soon after implantation but might be present in older growths. He believed that metastatic tumours in man are often undifferentiated because there has not been enough time for them to differentiate. In a general way, rapid cell proliferation discourages differentiation but the available evidence does not justify the inference that there is a close and obligatory negative correlation between proliferation and differentiation in neoplasms. Biologists have given a good deal of attention to the supposed inverse relationship between the two in normal tissues and recently have shown increasing scepticism about its generality. This complex matter will be discussed further in another place (Chapter 13). Many tumours, especially fast-growing malignant ones, are undifferentiated or poorly differentiated but the outcome of the foregoing discussion is to indicate that lack of differentiation is not a necessary condition or consequence of either benign or malignant neoplasia. The undifferentiated state is often described as *anaplasia*. Von Hansemann (1907), who coined the term, insisted that it signified *two* properties of neoplastic cells, namely a greater independence (grössere selbständige Existenzfahigkeit) and de-differentiation or loss of specificity; it is not, therefore, an accurate synonym for "undifferentiated". The proposition that the cells are de-differentiated implies that they or their normal or neoplastic parent cells were at one time differentiated, which has not been demonstrated. It is more likely that they are the offspring of undifferentiated "stem-cells". It may be doubted, also, if "independence" and lack of differentiation are closely correlated properties. The possibility of an imperfect correlation between growth rate and degree of differentiation should be admitted but a correlation between degree of differentiation and the abstract quality "malignancy" does not exist.

B. THE STROMA

One of the earliest outcomes of the study of tumour transplantation was the concept of the stroma as a manifestation of a tumour-host interaction. Histological examination of tumour grafts during the early stages of their growth in new hosts showed that whereas parenchyma cells survived the the transfer and proliferated actively, especially at the periphery of the grafts, the transferred stroma degenerated progressively and was replaced by the ingrowth of blood vessels and connective tissue cells from the ambient "normal" tissues of the host (Bashford *et al.*, 1905). Bashford and his colleagues, finding that the new stroma was constant and qualitatively and quantitatively characteristic for each particular transplantable tumour described the reaction of the host as a *specific stroma reaction*. The stroma reaction was essential for the "take" of a graft. As expressed by Ehrlich (1907),
. . . the transplanted cells exert a direct chemotactic influence on the fibroblasts of the host. This influence is essential for the growth of the tumour cells; if it is lacking, no tumour can develop. Russell (1908) gave further evidence that the failure of implants to grow in resistant animals was due to

a failure of the stroma reaction and other pertinent investigations were reviewed by Woglom (1929).

Algire and his colleagues reinvestigated the stroma reaction by implanting transplantable tumours tissue and, for comparison, normal connective tissues in transparent chambers introduced into skin flaps in mice. They inferred from their observations that tumours cells are distinguished from normal cells by an outstanding capacity to elicit an early and sustained ingrowth of new capillaries from the host's tissues. They suggested that this capacity might constitute a primary difference between neoplastic and normal cells and be responsible for the rapid growth of tumour transplants (Algire et al., 1945). More recently, Day (1964) also has expressed general agreement with the concept of a specific stroma reaction and, like Algire, extends it to apply to the growth of established primary tumours as well as to the "take" of tumour implants in new hosts. Day recalled the distinction made during the early studies by Gierke (1908) between fibroplastic and angioplastic reactions. Gierke pointed out that a delicate balance between the two was necessary for successful growth; too much connective tissue could interfere with nourishment and lead to central necrosis whereas overabundance of blood vessels together with insufficient supporting connective tissue could also interfere with nutrition of the tumour cells by leading to oedema and haemorrhage. The fibroplastic reaction was as important as the angioplastic one in ensuring effective vascularization of the neoplastic parenchyma. Consideration of the possibility that *invasiveness* of neoplasms also is related to fibroplastic and angioplastic reactions of normal tissue to neoplastic parenchyma is referred to elsewhere.

Day reviewed some of the more recent studies of the vascularization of tumours but found no satisfactory answer to the question as to how vascularization of undisturbed primary tumours is established. He suggested that two mechanisms may be implicated. He described the vascularization mentioned in the preceding paragraphs as a *generated blood supply*, established by ingrowth of new capillaries. He described also an *appropriated blood supply* established by an increase in length and diameter of prexisting vessels without an increase in their number. It seems that the vessels of an *appropriated* blood supply retain their ability to contract in response to stimuli, whereas those of a *generated* blood supply may be devoid of active, contractile elements.

In certain special cases, a specific stroma reaction may contribute prominently to the histological structure of a tumour. Cartilage and bone, for example, are often present in the stroma of certain kinds of epithelial tumours but not in most other kinds. Of the tumours that invade the human skeleton, where conditions for osteogenesis are favourable, only particular types of tumour, of which carcinoma of the prostate is a notable example, elicit secondary bone formation. In a classical paper on the subject, Axhausen (1909) deduced from a review of the evidence that some unidentified product of prostatic carcinoma cells acts on connective tissue cells of the stroma so as to incite them to osteogenesis. A transplantable chicken carcinoma, mentioned elsewhere in another context (p. 177) provided evidence in the same sense. Stromal connective tissue cells produced osteoid tissue instead of fibrous tissue when they penetrated a hyaline matrix produced by the carcinoma cell (Foulds, 1937). These observations recall those of Huggins (1931) which showed that rabbit urinary epithelium, but not other kinds of epithelium subjected to test, regularly induced osteogenesis when implanted under the sheath of the rectus muscle of rabbits. The same kind of epithelium did not induce osteogenesis in certain other situations. The osteogenesis in the rectus sheath was a *dependent* or *directed differentiation* of connective tissue cells not biased towards that activity and it was elecited by a stimulus provided by a particular type of epithelium. Evidently the capacity of connective tissue cells of similar appearance to react to the stimulus differed according to their location.

C. Interactions of Neoplastic and Non-Neoplastic Tissues

The tissues invaded by tumours are usually passive although it is noteworthy that some tissues stubbornly resist invasion and survive to modify the structure of the invading tumour. Infiltrating tumours of the human lung often use persistent alveolar walls as a preformed stroma (Willis, 1952). A transmissible chicken tumour (MH2), thought to be an endothelioma or reticulosarcoma, sometimes displaced the epithelium of organs that it invaded and, especially in pancreas and kidney, MH2 cells applied themselves to the surviving basement membranes and by so doing produced an unfamiliar appearance similar to that of a carcinoma (Foulds, 1934a).

Invaded tissues complicate the structure of tumours more severely, although uncommonly, by undergoing what von Hansemann (1902) called *collateral hyperplasia*. Borst (1902) and Ewing (1928) described examples. According to Ewing, either gland cells or stroma cells of invaded tissues may undergo a collateral hyperplasia that is hardly distinguishable from neoplasia. Ewing instances acini of neoplastic appearance in adrenal glands invaded by carcinoma of the kidney and extensive hyperplasia of liver cells around hepatomas. Comparable hyperplasia of liver cells is not seen around simple inflammatory and degenerative lesions in the liver and, although overnutrition is probably the main cause and although inflammation may contribute to the process, the peculiar structure suggests that other influences are at work. Ewing, therefore, described the collateral hyperplasia as "rather specific" and quoted the opinion of Beneke to the effect that tumour cells exert a peculiar trophic action in addition to producing nutritive and regenerative effects.

Collateral hyperplasia has received little attention from later writers although it can profoundly alter the structure as I found many years ago in the course of histological studies of transmissible tumours of domestic fowls (Foulds, 1934a). The most conspicuous examples were found in thymus glands containing metastatic growths of transplanted MH2 tumours. About 25% of the metastatic growths contained histologically atypical structures. The simplest divergence from the typical MH2 tumour pattern was the presence within tumour nodules of hyperplastic thymic epithelium, which, here and there, formed concentric bodies resembling mammalian Hassall corpuscles. The most conspicuous changes were present in a greatly enlarged thymus which contained some regions with the usual structure of the MH2 tumour and others with the appearance of squamous carcinoma containing epithelial pearls. Other metastatic tumours were complicated to intermediate degrees by squamous epithelial hyperplasia which was conspicuously present, also, in one metastatic thymic growth of a transmissible fibro-sarcoma MH1.

The attribution of the atypical regions in the metastatic tumours in the thymus to collateral hyperplasia of thymic epithelial cells entails a presumption that thymic epithelium has the capacity for epidermoid metaplasia. This was plainly demonstrated by auto-transplantation of normal thymus tissue into muscle. The first change in the implants was degeneration, especially of the more vulnerable thymocytes. Epithelial reticulum cells survived and soon began to proliferate. Some grafts became re-populated with thymocytes and approximated in structure to normal thymus gland. In other grafts, the surviving epithelial cells underwent epidermoid differentiation; tonofibrils first appeared in them and later keratinization began. If their origin had not been known, some of the grafts could have been pardonably mistaken for metastatic epidermoid carcinomas with keratinizing pearls. (Foulds, 1934b).

Autotransplantation and tumour invasion alike led to a degree and type of epithelial differentiation not found in the intact normal gland at any age. Under such circumstances, destruction of thymic tissue was followed by hyperplasia of thymic epithelial cells in an abnormal environment, which in some way evoked an abnormal type and degree of epithelial differentiation. Possibly the ingrowth of capillaries and connective tissue allowed the epithelial cells to assume a polarity which was absent from cells of the normal gland, where connective tissue is restricted to sheaths around blood vessels. Ludford (1925) showed that where epithelial cells of tar cancers of mice were in contact with stroma, the distribution of their mitochondria and golgi bodies was polarized; elsewhere polarization was absent. A similar polarization of thymic epithelial cells might be responsible for their epidermoid differention (Foulds, 1934b). Fell (1964) has recently advanced a similar explanation of metaplasia of chick epidermis.

Recent reports of probable examples of collateral hyperplasia in nephroblastomas are discussed below. Other possible examples were encountered, uncommonly, during studies of pregnancy-responsive mammary plaques in BR mice. Some plaques were atypical in being composed of tubules which for the most part were parallel with one another and with the skin surface and which showed no conspicuous branching. The most notable of these plaques were not, like typical plaques, wholly subcutaneous but invaded the dermis. Some tubules near the superficial surface turned at right angles to the bulk of their fellows and extended towards the epidermis, immediately below which, many of them terminated in expansions lined by hyperplastic epithelium. The overlying epidermis was epilated and thickened. Histologically it contained many empty hair follicles into which superabundant sebaceous glands opened. Hyperplastic epidermis lined the mouths of the follicles and covered the intervening skin surface with a fully stratified squamous epithelium. In some places the hyperplastic epithelium grew downwards in solid cords, with or without traces of hairs or to form hollow tubes. Most remarkably, some of the downgrowths fused with sub-epidermal tubules derived from the underlying plaque so as to form a sinus lined in its more superficial portion by squamous epithelium of epidermal origin and more deeply by gland-like neoplastic mammary epithelium derived from the plaque. The two types of epithelium together formed a continuous lining for the sinus and no clear-cut line of fusion between them was discernible; the two types of epithelium merged imperceptibly. (Figs 42–44.)

The interpretation of these extraordinary histological findings is not easy and must take into account the clinical history of the lesions. The greater portion of the plaques regressed during intermissions of breeding but the epidermal changes, the sinuses and hyperplastic sub-epidermal extensions of plaque tubules persisted; during intermissions of breeding they neither advanced nor regressed. Many measurements made during the intermissions were recorded at the time as being of doubtful accuracy; some of them seemed incompatible with the measurements made at autopsy. It became

FIG. 42. Thin superficial regressed plaque from a non-pregnant mouse with focal progression to carcinoma at the right. Epidermal changes extend far beyond the carcinoma and correspond with the extent of a plaque measured during a preceding pregnancy. × 13·5

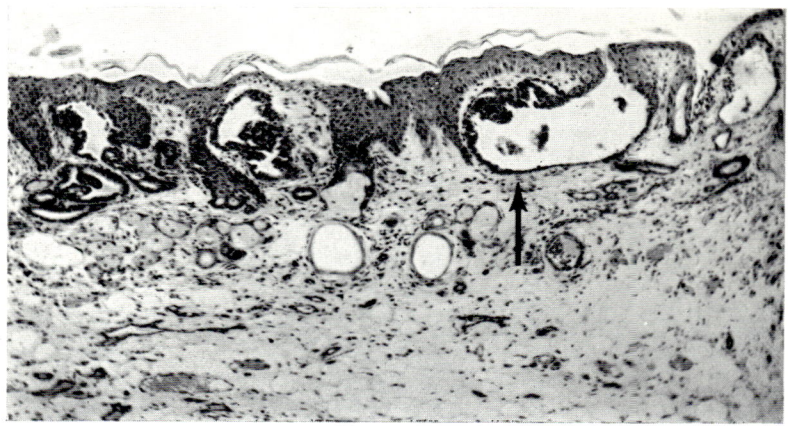

FIG. 43. A portion of the lesion illustrated in Fig. 42 showing thickening of the epidermis, subepidermal tubules and a sinus (arrow). X 60

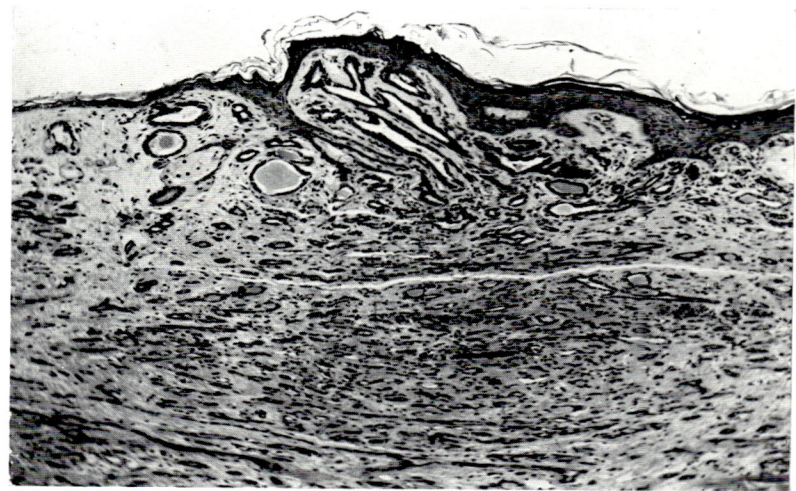

FIG. 44. Another superficial regressed plaque. The tubules mostly run parallel to the skin surface but some of the more superficial ones bend towards the epidermis, which is thickened. X 60

evident that the "doubtful" measurements made during life applied to the epidermal and sub-epidermal changes, whereas the autopsy measurements applied to the underlying plaques or to carcinomas that had developed by progression within them. The epidermal changes were co-extensive with the maximum extension of the plaques during pregnancy and unrelated to the size of carcinomas that developed in the plaques. No comparable epidermal lesions of mouse skin were seen in the absence of an underlying plaque. The most conspicuous hyperplasias of epidermis and of subepidermal plaque tubules were often closely apposed as if they had induced growth of and towards one another, the culmination of the process being fusion of the two dissimilar tissues to form a sinus (Foulds, 1956). Similar lesions were described and illustrated by Bonser (1958) in mice with chemically induced mammary tumours. She also noted that subepidermal mammary tubules might fuse with the epidermis and even open onto its surface. In the course of a discussion, Heston (1965) mentioned having seen "malignant papillomas" of the skin overlying regressing plaques but the relationship of these lesions to those discussed here is less evident.

The phenomena under discussion are of considerable biological interest. They are possibly related to *collateral neoplasia* but provide, also, suggestive evidence of a mutual interaction between neoplastic and non-neoplastic tissues. The neoplastic tubules, it seems, could both supply and react to

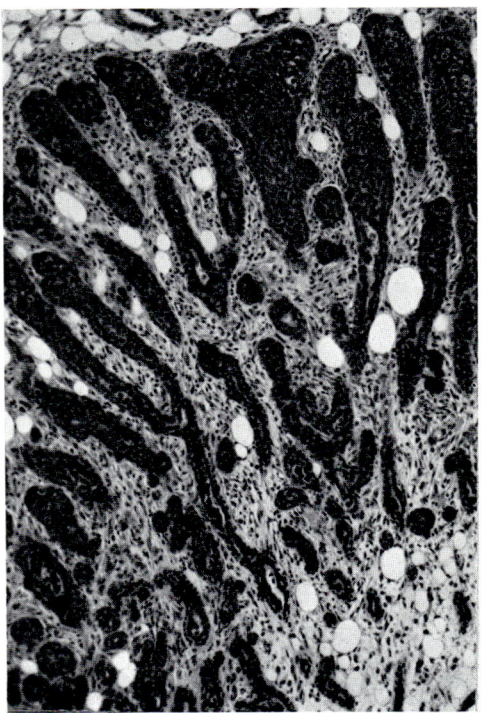

FIG. 45. Intracanalicular carcinoma with organoid structure at its periphery. X 60

stimuli that can be described, not unreasonably, as inductive or morphogenetic. The gradual merging of the two epithelia to form a sinus also implies a mutual adjustment or morphogenetic interaction between them.

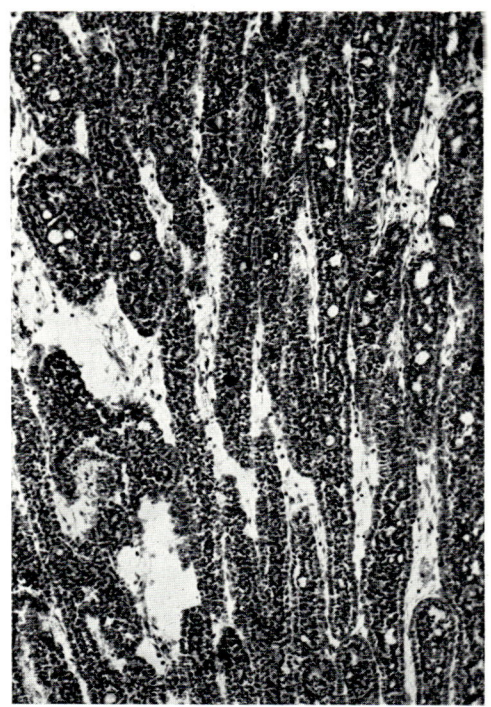

FIG. 46. Intracanalicular growth with organoid features. X 60

II. The Structural Organization of Tumours

Tumours in general are not composed of random conglomerations of cells but of organized *tissues* with characteristic histological patterns, which pathologists use to name and classify them. Tumours, in general, are not aptly described as "cell populations" and do not behave as such. Some degree of structural organization is rarely absent from tumours of man (Willis, 1967) but it varies greatly in degree being maximal in certain "organoid" tumours and minimal in the most "anaplastic" ones. The ascites tumours stand at the extreme "anaplastic" end of the range and are almost alone in being amenable to study as cell populations. For certain purposes and in judicious hands they are invaluable but when used less judiciously as "models" for wide-ranging speculations about the nature and properties of

The Cancer Cell, to the mythology of which Smithers so strongly and justifiably objects, they can be perniciously misleading. The ascites tumours are almost without a counterpart in clinical medicine (Leighton, 1967).

More than half a century ago, Albrecht and other German pathologists maintained that the structure of tumours in general is basically organoid. Nicholson (1950) and Böhmig (1937) have given references to the extensive early German literature on the subject. Several workers in W. Fischer's laboratory described regular organoid growths of various types of primary and metastatic carcinomas of man, which they studied with the use of serial sections and wax-plate reconstructions. Bauman (1935) maintained that a primary adenocarcinoma of the large intestine developed like normal glandular tissue and manifested a regularity of growth and structure that contrasted strongly with the disorderly proliferation widely believed to be characteristic of malignant neoplasia. In adenocarcinomas of the stomach and of the intestine studied by Franck (1935), the parenchyma grew and branched like normal glands and complexity of structure resulted from fusion of the glands

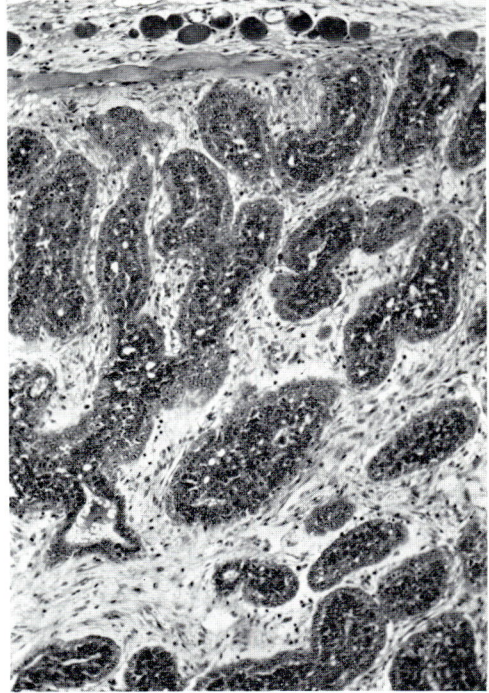

FIG. 47. Intracanalicular growth. The tubules are outlined by a regular layer of pale-staining epithelium within which are irregular masses of darker-staining cells. X 60

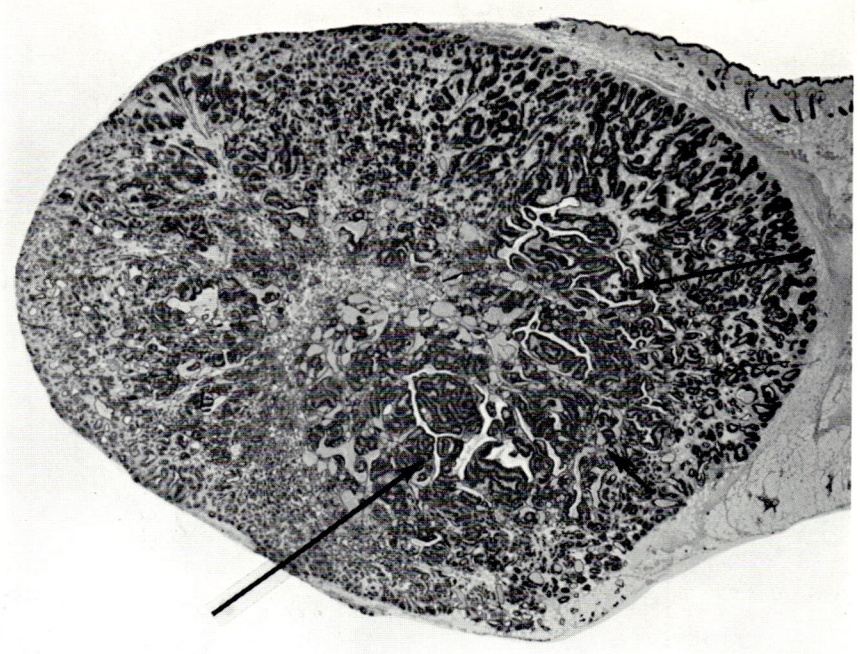

FIG 48. The arrows indicate regions of fissured carcinoma within a simple radiate organoid tumour.
X 15

and subsequent proliferation to form a network. Havemann's (1936) observations pertained to four masses of adenocarcinoma widely separated in the mammary fat. Each seemed to be a confused mass of solid and glandular epithelial strands but wax-plate reconstructions disclosed the orderly pattern of a tree with trunk, branches and twigs. Rapp (1936) described organoid growth in the hepatic metastatic tumours of two adenocarcinomas of the stomach and similar observations were made by Dabelstein (1937) on primary and metastatic duct carcinomas of the breast. From these varied investigations, Böhmig drew the broad conclusion that the tumours developed in an orderly way, in obedience to definite laws, like their parent tissues but they grew more quickly and were soon too big to be contained in the pre-existing tissue spaces; the basic regularity of structure was then obscured by mechanical distortion.

Nicholson (1950) analysed some examples of organized growth in detail and came to the conclusion that the majority of tumours, perhaps all, are "organoid" in Albrecht's sense; they are built on the same plan as the tissues in which they originate, their cells are grouped to form tissues and organs in a way that is sometimes remarkable and their structure, growth and development are intelligible only if tumours are viewed as aborted organs. Although hesitant to accept tumours as "aborted organs", a histological study of mammary neoplasia in about 700 BR mice led me to broad agreement with most of his conclusions (Foulds, 1956).

Haaland (1905) first emphasized that the most characteristic structure of mammary adenocarcinoma in mice is its tubular structure. The unit of structure of the mammary plaques, whence a majority of the carcinomas in BR mice developed, is a branching tubule. Typically, plaques are composed of symmetrically-arranged systems of branching tubules radiating from a central core or medulla. During pregnancy, the tubules start in the medulla with a small bore and single-layered epithelial wall, extend outwards in a straight course, apart from branchings, with an increasing diameter and epithelial proliferation and terminate at the periphery of the enlarging plaque in pyriform expansions comparable with the "end-bulbs" of a normally-developing mammary gland.

Organoid structure in gross mammary carcinomas is, in essentials, the manifestation in late stages of neoplastic development, of habits of growth

FIG. 49. Tumour composed of separated units of fissured carcinoma. X 13·5

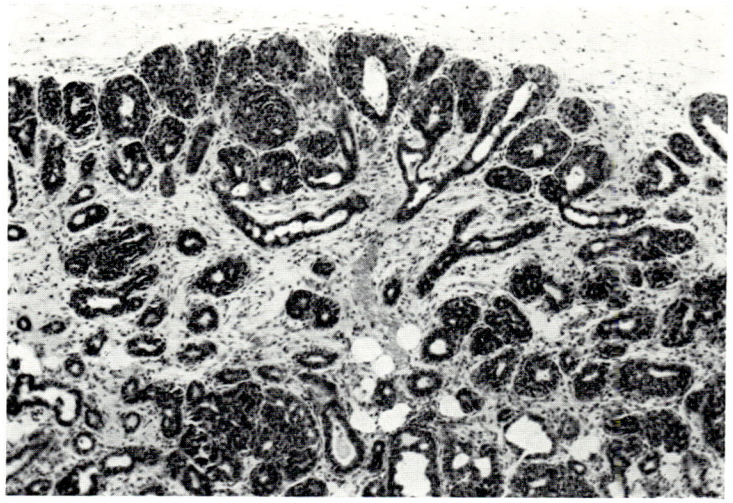

FIG. 50. Higher magnification of the edge of the tumour illustrated in the previous figure. X 60

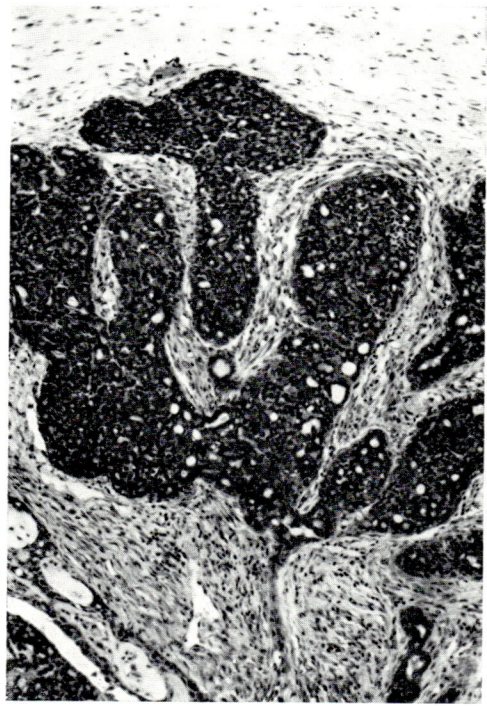

FIG. 51. Organoid structure at the edge of an intracanalicular carcinoma. X 60

FIG. 52. Irregular growth of tubules with branching and prominent end-bulbs at the edge of a tumour that was not, as a whole, conspicuously organoid. × 60

that are evident in mammary plaques and, still earlier, in the development of a normal mammary gland. Histologically the mammary tumours of mice are extremely varied and defy neat classification (Dunn, 1958). The diversities are accountable to variations in the type and degree of epithelial differentiation, to which reference was made earlier and also to variations in the architecture of the tubules and in the degree and direction of epithelial proliferation. The tubules may be long or short, straight, curved or tightly coiled; their bores may be narrow, irregular or wide or, not uncommonly, dilated into cysts with or without invagination or evagination of their walls. Epithelial proliferation which just keeps pace with the growth of the tubules in length provides a uniformly single-layered lining. Sometimes the tubules are uniformly double layered. A greater degree of epithelial proliferation, if directed inwards or centripetally leads to an *intratubular* or *intracanalicular* type of carcinoma; the growth encroaches on, and finally, eliminates the lumen of the tubule which is converted into a solid cylinder with a sharp outline (Figs 45–47). When the proliferation is predominantly outwards or centrifugal the lumen of the tubule remains conspicuous and often lined by a single layer of regular epithelial cells surrounded by a more or less massive halo of irregular proliferating cells. Some of these tumours have been called *fissured* or *macroglandular* tumours (Figs 48–50).

The possible combinations of the foregoing characters, most of which are independently variable, are extremely numerous and, together with the consequences of oedema, haemorrhage, necrosis and mechanical distortion, they sufficiently account for the great diversities of histological structure, even within individual tumours, for the masking of the basic tubular architecture in many of them and for the disorderly arrangement of tissues in others. Nevertheless, J. A. Murray (1908a) noted "a tendency to organoid arrangement" and radially disposed branching tubules or cords are discernable fairly often at the growing edges of tumours which, elsewhere, are not evidently organoid (Figs 51–52). Exceptionally, the orderly arrangement of tubular units to produce an unmistakably "organoid" appearance dominates the architecture of whole tumours. Some examples of these plainly organoid growths, which in spite of being uncommon are theoretically important, will

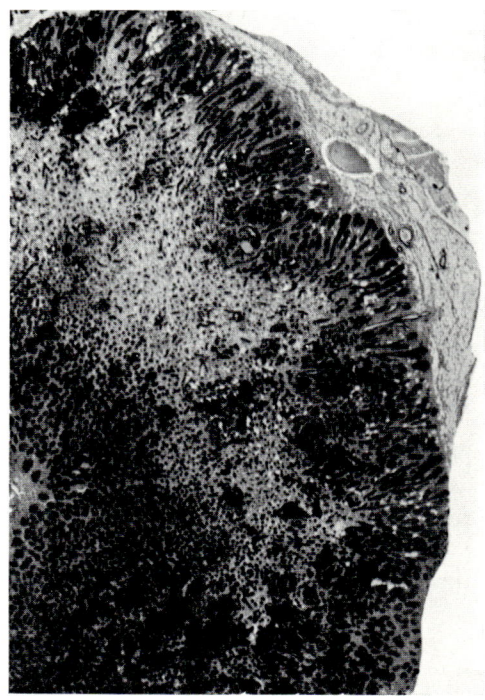

FIG. 53. Compound organoid tumour Y × 30; biopsy specimen from the primary growth. There is a peripheral zone of branching hyperplastic tubules, a more central region of narrow tubules and, to the left and lower margins of the field, denser areas of irregular growth. × 15

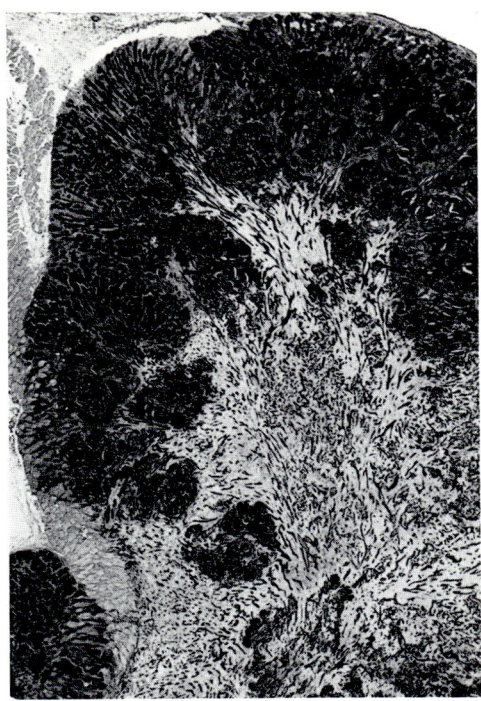

FIG. 54. A Y×30 tumour of the third transplanted generation, which, in form, resembled a giant plaque. Radiate growth is most conspicuous at the left-hand border; at the upper, right-hand portion, growth is less regular and there is an extensive central region of long fine tubules. X 15

be considered in the following paragraphs based on a more detailed report published previously (Foulds, 1956).

The most conspicuously organoid tumours are made up, like mammary plaques, of straight branching tubules or cords with terminal expansions at their growing tips. In the simplest ones, conveniently designated *simple radiate* tumours, the tubules or cords radiate more or less symmetrically from a single circumscribed region. These are uncommon tumours. The more frequent *compound organoid* tumours are built up of similar units growing from multiple centres and radiate structure is less conspicuous or absent. Less distinctively "organoid" structure is recognizable in intratubular and microtubular carcinomas and in fissured and macroglandular tumours.

The organoid tumours of my own series were found during a period when tumours, as well as plaques, were being measured routinely at frequent intervals. When the remarkable histological structure was eventually recognized it was possible to refer to the recorded measurements with the unexpected result that the organoid tumours proved to be amongst the fastest growing of all mammary tumours in the BR mice. Exceptionally rapid growth might be a useful clue in seeking other examples in living mice; organoid tumours are not otherwise recognizable clinically. The organoid tumours are invasive but apparently do not metastasize. The only one tested was transplantable.

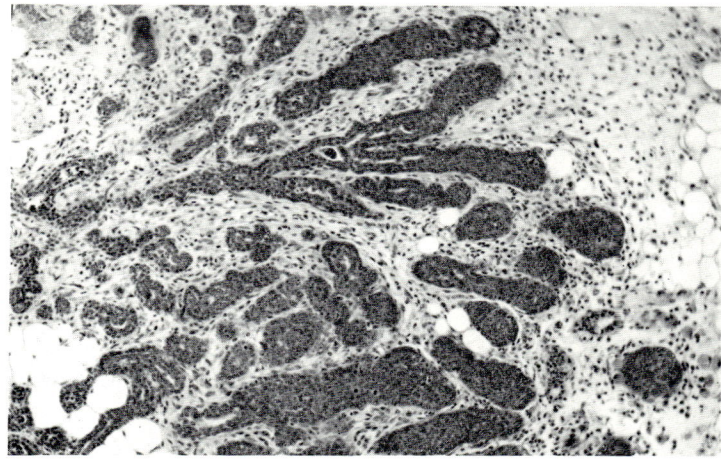

FIG. 55. Edge of recurrent primary growth of Y×30 showing dichotomous branching of tubules and invasive end-bulbs. X 60

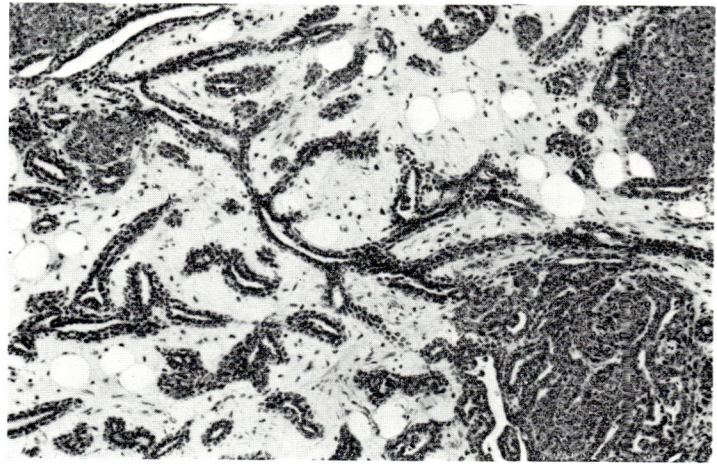

FIG. 56. Fine branching tubules in a loose delicate stroma together with nodules of less regular growth from the central portion of Fig. 54.

The most instructive and important example of a compound organoid tumour, which was also transplantable, was found by my former colleague, Dr. B. D. Pullinger, who supplied me with records and histological material from the primary and earliest transplanted tumours as well as tumour-bearing mice from which serial transplantation was continued. Unfortunately, transplantation had to be discontinued after the fifth passage but up to that time the transplanted tumours grew rapidly in all the innoculated mice and, more remarkably, they retained their organoid structure to the end (Figs 53, 54).

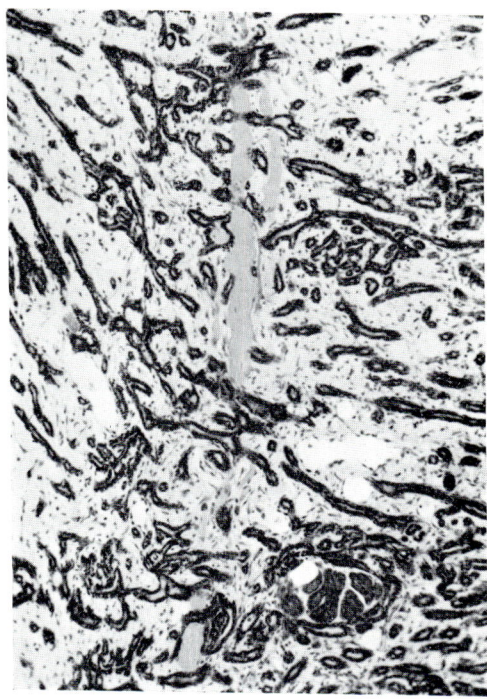

FIG. 57. A section through the central portion of the tumour illustrated in Fig. 54 showing narrow branching tubules, some of which transect at right angles a band of muscle traversing the centre of the field vertically. X 60

All tumours had essentially similar histological structures and all contained some areas of radiate organoid structure. Tubules of narrow bore extended for considerable distances, branched often and expanded peripherally into solid cords (Figs 55, 56, 57). After further branching and thickening the cords terminated in end-bulbs, which were usually solid. Mitotic figures were abundant, especially in the end-bulbs which, evidently, were the growing points of the tumour and which were invasive although they retained

FIG. 58. Molluscoid tumour. The growth radiates from a central core. Progressive keratinization extends outwards and almost reaches the periphery at some places on the upper and lower borders.
× 13·5

a smooth outline and shed no free, infiltrating cells. The advancing cords did not, in general, follow lines of least resistance but seemed to push forward through the tissues in a straight line, irrespective of obstacles. The more centrally-placed fine tubules presumably mark the path along which the growing points had advanced and some of them indicate that growth had penetrated muscle at right angles to the fibres (Fig. 57). Evidence of metastasis was not found in the mouse with the primary tumours, in the mice bearing transplanted tumours or in any of the mice bearing other primary organoid tumours.

It is noteworthy as another example of the diverse associations of independently variable characters that organoid architecture may be associated with varied types of differentiation and specialized function as will be next illustrated.

One variant of the simple radiate tumour has received a special name, the *molluscoid* mammary tumour, described first by Haaland (1905, 1911) and later, in more detail, by Teutschlaender (1926). According to Haaland's original descriptions and illustrations, there are two essential criteria of a molluscoid tumour, namely progressive keratinization and radiate structure. In the central portion of the tumour, keratinization is complete; at the periphery there is a rim of growing tumour, which is rarely wide and is sometimes extremely thin (Figs 58, 59). The wholly keratinized cylinders

FIG. 59. Edge of a larger molluscoid tumour measuring $2 \cdot 8 \times 2 \cdot 5 \times 1 \cdot 6$ cm. X 15

of the central region extend radially and taper peripherally to form a core with a mantle of healthy epithelial cells. The epithelial mantle is often slightly constricted where keratinization stops but then it broadens again and terminates as a rule, in a terminal expansion or end-bulb, which is most often rounded or else, as Teutschlaender describes, bifurcate (Fig. 60). Mitotic figures are abundant in the end bulbs which are the growing tips of the tumour. My own observations confirm those of Haaland and Teutschlaender on the invasiveness of the tumours. The radiate organoid structure is maintained at the growing edge and invasion is not attributable to growth along lines of least resistance. The invading tips have sharp outline and do not shed isolated infiltrating cells. There is no convincing record of metastasis. Progressive keratinization evidently keeps pace with progressive growth at the periphery and to this extent the tumours are self-limiting. The few available clinical records indicate that the rate of growth is moderate, exceeding that of most adenoacanthomas but falling short of that of non-keratinizing simple radiate tumours. Histologically, the molluscoid tumour differs from the simple radiate organoid tumour in one independently variable character, namely the keratinizing type of differentiation of neoplastic mammary epithelium (Foulds, 1956).

An uncommon organoid variant of the mammary plaque differs from the usual type in secreting a milky fluid in parallel with the secretion of the normal mammary gland during pregnancy and lactation and in persisting during lactation instead of regressing immediately after parturition as ordinary plaques do (Figs 63, 64). So far it has not been recognized during life and

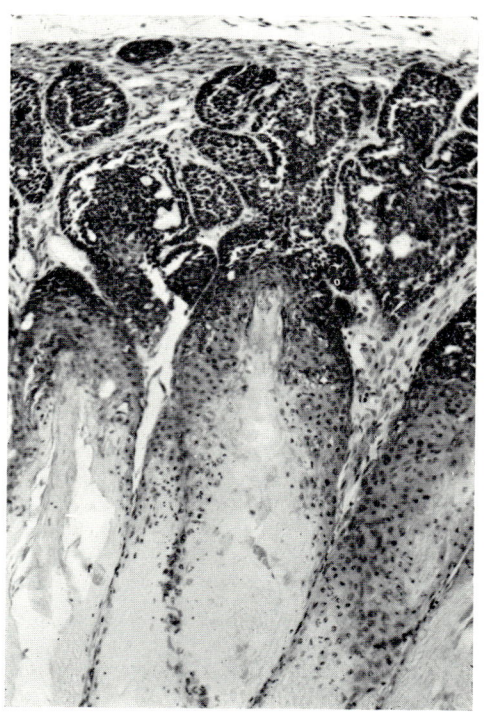

FIG. 60. Edge of the molluscoid tumour shown in Fig. 58 with keratinized tubules and end-bulbs without keratin. X 60

few examples have been described. Judging from this small material, secretion begins in the centrally-placed tubules, which become distended and, when pronounced, extends outwards until most of the tubules contain secreted material. The tubules are variously, but not maximally, distended or mutually compressed. At this stage the growths correspond with the cystadenoma simplex described by Apolant (1906). After further secretion, they match Apolants "cystadenoma simplex with scanty stroma". Their ultimate fate is uncertain owing to the inability to recognize any of them during life but tumours found by chance in three mice, which had littered from three to ten days previously and had not nursed their young, are probably regressing phases and fit into another of Apolant's categories, namely "Papillares Cystadenoma mit relativ viel Stroma und geschichteten Inhaltmassen" (Fig. 63). Two other lesions found accidentally are possibly regressed cystadenomas and the counterparts of regressed plaques (Fig. 64). The inference from these observations is that cystadenoma is the result of self-limiting secretion at an early stage of neoplastic development in response

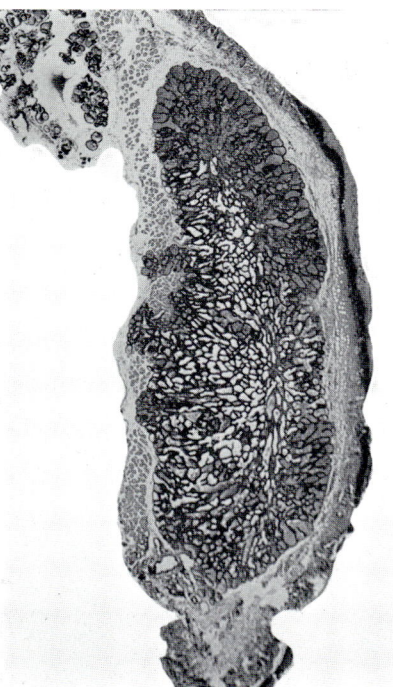

FIG. 61. Cystadenoma with apparent invasion of muscle at the left edge. The mouse had been nursing for three days. X 15

174 NEOPLASTIC DEVELOPMENT

to hormonal stimulation during pregnancy and lactation Physiologically, cystadenoma is the smallest recognizable deviation from normal mammary gland. Structurally, it is distinctly organoid but differs from normal gland in lacking a communication with an excretory duct. The basic fault, therefore, seems to be the failure of a portion of near-normal mammary gland to achieve or retain morphological integration into the normal mammary tree (Foulds, 1956). It may be noted in passing that, in the absence of a knowledge of the developmental history, different histological labels may be attached to different phases in the life of a single neoplasm.

A. The Significance of Organoid Tumours

The preservation of organizing capacity throughout the growth of a tumour to a large size and through repeated serial transplantations must depend on some stable replicable character of the neoplastic cells which, it

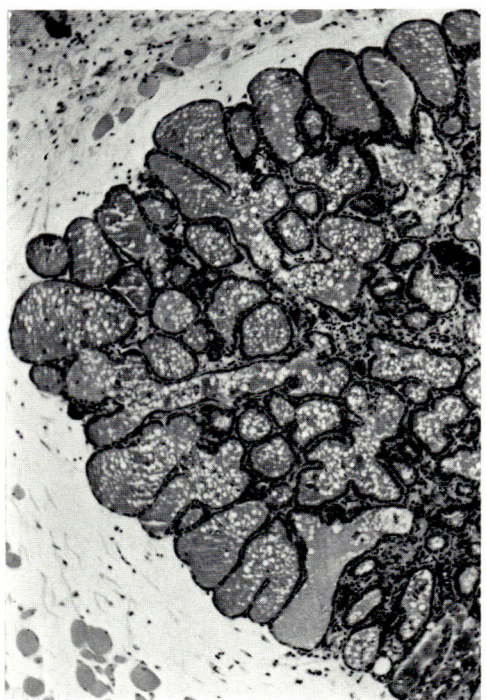

FIG. 62. Same tumour as in Fig. 63 showing the radial arrangement of branching tubules distended with vacuolated secretion. × 60

seems, have a built-in "morphogenetic memory". The organoid characteristic can associate with a variety of other independently variable characters to produce, on the one hand, an unexpected, out-of-place tumour like the mammary molluscoid tumour or, on the other hand, a near-normal tissue like mammary cystadenoma. All the organoid tumours, including the cystadenoma share one characteristic that distinguishes them from normal tissues: they are not morphologically integrated into the architecture of the organ or tissue in which they grow.

The concurrence of conspicuously organoid structure with rapid, invasive growth and transplantability refutes many facile definitions and theories of "cancer" that are based on presumptions of uncontrolled, chaotic growth or cell "autonomy". Albrecht and Nicholson were right to draw the lesson from these tumours that the problem of neoplasia is not simply a problem of growth; it is also a problem of morphological *pattern*.

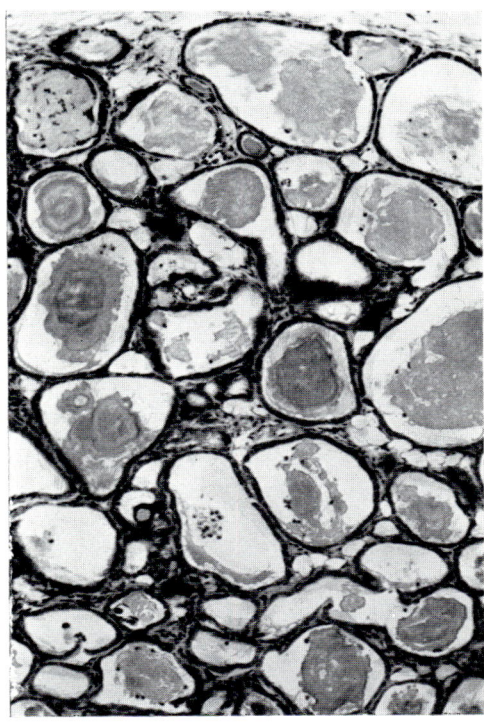

FIG. 63. Probable regressing cystadenoma. Inspissated secretion with concentric markings is present in spaces lined by flattened inert epithelium. The mouse had littered ten days previously.
X 60

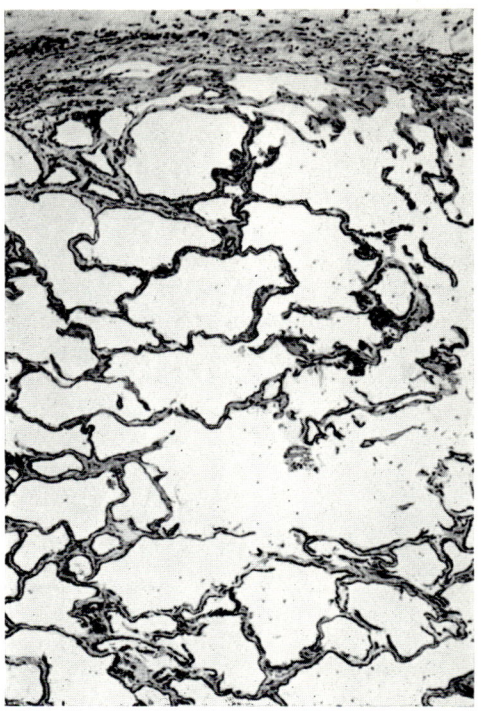

FIG. 64. Possible regressed cystadenoma from a mouse that had not been pregnant for two months. X 60

III. Complex Tumours

Certain complex tumours contain a variety of tissues whose presence is not easily or even plausibly explicable on a histogenetic basis by the origin of the tumours from a single type of cell or tissue. Most argument has been devoted to "mixed" tumours of which the "mixed" salivary gland tumours of man and the "mixed" mammary tumours of dogs are the most familiar. The complexities include varied, unexpected types of epithelial differentiation and the presence of unexpected "heterotopic" tissues such as cartilage and bone. Huguenin (1929) asserted uncompromisingly that all the mixed tumours were purely epithelial in origin and he was probably nearly right, most of the complexities being attributable to divergent differentiations of the epithelial parenchyma, secondary dependent differentiations of the stroma or collateral hyperplasia and metaplasia of invaded tissues. In the course of serial transplantation in animals, it is sometimes possible to analyse a complex tumour experimentally as exemplified by the analysis of complex metatastic tumours of the thymus gland described in a previous section or, still better, to trace how a primarily unexceptional tumour becomes complex. Within my own experience, a transplantable carcinoma of the domestic fowl acquired under observation a remarkable resemblance to a mixed salivary gland tumour of man and illustrated how the complexities could come about (Foulds, 1937).

The original tumour was a carcinoma simplex originating apparently in the ovary. Transplantation was successful but at first extremely difficult; evidently the grafts were in uncongenial environments to which they responded by profound modifications in structure. The carcinoma cells produced an intercellular matrix having the staining properties of the ground substance of cartilage. The accumulation of this material had several effects. Occasionally it collected in vesicles and the tumour then resembled a colloid carcinoma. Elsewhere it surrounded individual cells which took on an oval or spindle shape so that in some regions the tumour looked like a spindle-celled sarcoma. The end result of the imprisonment of carcinoma cells in the avascular matrix was to transform whole lobules of the tumour into a tissue resembling cartilage. Moreover, true cartilage formed in the stroma of the tumour. The transplanted tumours also contained osteoid tissue in the stroma and, more remarkably, between columns of epithelial cells. The hyaline matrix usually broke up lobules of carcinoma cells into columns with a radial arrangement. At first the matrix contained no connective tissue cells which were restricted to the periphery of the lobules. Later, connective tissue cells grew into the matrix but instead of producing fibrous tissue they formed osteoid tissue and bone. Cartilage and bone were constant and prominent features of several successive generations of transplanted tumours. They were not themselves transplantable but developed anew in each host in part by a dependent differentiation of connective tissue cells evoked by the abnormal environment provided by the extra-cellular matrix produced by the carcinoma cells and in part, it seemed, by the formation of cartilage by carcinoma cells.

One substrain of the tumour, after producing cartilage and bone in several successive generations, began to grow more quickly and then ceased to do so and reverted to a wholly carcinomatous structure approximating to that of the primary tumour. It is evident that the tumour began as a carcinoma of epithelial origin and continued as one throughout the series of transplantations. The main difficulty in accepting a purely epithelial origin for the mixed tumours like the salivary gland tumours of man and the mixed mammary tumours of dogs is to account for the constancy of the connective tissue alterations. This difficulty is removed when it is recognized that the connective tissue changes are dependent differentions evoked by specific activities of the epithelial carcinoma cells. The chondrogenesis and osteogenesis may be interpreted as special cases of specific stroma reaction. Tudhope (1939) accepted this interpretation and showed that it was applicable to a complex tumour of the human breast. It is applicable also to the mixed mammary tumours of dogs.

*Complex Tumours**

A. *Tumours with One Neoplastic Component*

  I. *Tumours with complex parenchyma*

   (i) Changes in form or disposition of cells caused by their environment, e.g. by conditions imposed by invaded tissues.

   (ii) Due to progression within an established tumour.

   (iii) Variations in the degree of differentiation of parenchyma cells.

   (iv) Divergent differentiations of parenchyma cells.

  II. *Tumours with complications in stroma or invaded tissues*

   (i) Accidental modifications, e.g. osteogenesis in necrotic tissues.

   (ii) Dependent differentiations of the stroma.

   (iii) Collateral hyperplasia or dependent differentiation of invaded tissues.

B. *Tumours with Multiple Neoplastic Components*

   (i) Due to multifocal progression in a region of incipient neoplasia.

   (ii) Independent primary tumours due to intermingling of tumours primarily unconnected. (Kollisionstumoren of Meyer)

   (iii) Composite tumours. Simultaneous neoplasia in two closely associated tissues, e.g. parenchyma and stroma (Kompositionstumoren of Meyer)

   (iv) Dependent tumours: Neoplasia of stroma or invaded tissues in a primarily simple tumour.

C. *Embryonic Tumours and Teratomas*

Complex tumours are of several kinds which are tabulated above. By far the most numerous and important of them are the tumours with a single neoplastic component complicated by various modifications of the parenchyma, the stroma or invaded tissues. Complexities in the parenchyma are summarized under four headings. The first subdivision comprises all the reversible modulations of structure including those due to mechanical factors and the secretory responses to hormones. Tumours of the second subdivision show more than one of the possible divergent differentiations of the parent tissue and are exceptions to the general rule that differentiation in tumours is usually rather strictly canalized in one of the possible paths. The divergent

* Slightly modified from Foulds (1940), where some of the earlier literature is discussed more fully.

differentiation of a single parenchyma into both glandular and keratinizing cells in some mammary tumours of mice is a good example. Varied degrees of differentiation in different portions of a single tumour are less likely to cause serious confusion but the persistence of traces of earlier stages of progression may sometimes do so. The secondary complications in the stroma or invaded tissues include the accidental modifications exemplified by calcification or osteogenesis in necrotic tissues which are usually not difficult to recognize. The complexities due to dependent differentiation in the stroma exemplified in the transplantable fowl sarcoma and those due to collateral hyperplasia of invaded tissue, as illustrated in the metastatic tumours of the thymus described in an earlier section, are more important and may be difficult to interpret.

If the foregoing analysis is correct, few tumours have more than one neoplastic component. Multifocal progression within a single region of incipient neoplasia may yield histologically diverse tumours, which if close together, may complicate one anothers' structure but the constituent tumours being of the same general kind, the composite growth might be better described as heterogeneous than as "complex" or "mixed". The intermingling of truly independent and unrelated tumours, being governed wholly by chance, is probably rare. Composite tumours, resulting from neoplasia in two closely related tissues such as the parenchyma and stroma of an organ, are most plausibly exemplified by fibroadenoma of the breast and by some carcinosarcomas, which, nevertheless can be differently interpreted. Consequential tumours, including "sarcomatous transformation" of the stroma are even more questionable and I doubt their existence except, perhaps, as extreme rarities. Uncertainty about the nature of certain composite tumours, especially some carcinosarcomas, is admitted. The embryonic tumours and teratomas will be dealt with in more detail in the following pages.

IV. Embryonic Tumours

Pathologists have long been prone to explain away unexpected and disconcerting properties of tumours by recourse to "embryological" interpretations which credit them with an origin in embryonic tissues or with having accomplished a "reversion to embryonic type". As a rule these interpretations are neither plausible nor necessary (Ewing, 1936). Truly "embryonic" or "embryonal" tumours are rare but disproportionately instructive. They include neuroepithelioma and medulloblastoma of the brain, retinoblastoma of the eye, neuroblastoma of the sympathetic nervous system, nephroblastoma of the kidney, hepatoblastoma of the liver and rhabdomyosarcomas of the pelvic organs and some other regions (Willis, 1958, 1967.)

Willis describes an embryonic tumour as one that arises during embryonic, foetal or early post-natal development from a particular organ rudiment or tissue whilst this is still immature. They are, at the same time, both neoplasms and malformations. They are neoplastic from the beginning and are formed during the very process of formation of the affected tissue; a nephroblastoma, for example, is a neoplastic part of the renal blastema itself.

Some embryonic tumours, including some that have already metastasized, have been found in human foetuses; many are recognized during infancy or early childhood at ages compatible with an origin during pre-natal or early post-natal life. The later age-incidence of certain embryonic tumours is explicable by the late maturation of their parent tissues; developmental stage is more important than chronological age.

Some specific examples of human embryonic tumours of particular biological interest will be discussed in the following pages; further details of their pathological features may be found in the books by Willis and in standard text books of pathology. Relevant experimental investigations will then be considered. The related teratomas are dealt with separately.

A. Retinoblastoma

This rare tumour has the characteristic age-incidence of embryonal tumours. It is found almost exclusively during infancy and childhood and many tumours have been detected at birth or soon after. Although it often seems to occur randomly in the population there are well-substantiated reports of a high incidence in certain families, implying a strong genetic factor in its origin. As an example, ten out of sixteen children in one Australian family were afflicted. The clinical course is often rapid and fatal as a consequence of intra-cranial extension along the optic nerve. Nevertheless, spontaneous regression, though rare, is well attested (Smithers, 1964), and enucleation of the eye is curative in about half of the patients who have no visible extension of the tumour outside the eye, even when microscopic deposits are present in the optic nerve.

The disease often originates in multiple foci and, especially in the familial examples, it is often bilateral. Histologically, the tumours consist of randomly-arranged round cells without distinctive features, but some portions may be structurally differentiated to form cell rosettes and fibrils. Willis considers these structures to be identical with the rod and cone layer and external limiting membrane of the normal retina and believes that the bipolar cells present in certain fibrillar areas are homologous with the bipolar cells of the inner nuclear layer. In sum, he concludes that the tumours are essentially embryonic and represent immature developing retinal tissue (Willis, 1958, 1967).

B. Neuroblastoma and Ganglioneuroma

Neuroblastoma is the second commonest form of malignant neoplasia in children. Ordinarily, it presents itself clinically during infancy or early childhood, but sometimes it is not detected before adolescence or early adult life. The neoplasia originates most often in ganglia of the sympathetic nervous system or in the adrenal glands. As a rule no familial or hereditary predisposition is demonstrable, but Everson and Cole (1966) mentioned neuroblastoma in each of two homozygous twins; the tumour in one regressed spontaneously but the ultimate fate of the growth in the other twin was not stated.

Both neuroblastomas and ganglioneuromas originate from embryonic neuroblasts. Willis insists that the two types of growth are not sharply distinct; many tumours have a transitional or mixed structure and the two names indicate merely the poorly differentiated and the highly differentiated members of one class. His further contention that the tumours differ amongst themselves only in rate of growth "and hence malignancy" is, in my opinion, a serious oversimplification. Some neuroblastomas are highly malignant and may metastasize widely whilst the primary growth is still small; histologically these tumours show little or no recognizable differentiation of neuroblasts and were described formerly as round-celled sarcomas. The less rapidly growing ones show various stages in the differentiation of neuroblasts; some of the cells are grouped together in rosettes and form processes and young nerve fibres. The fully differentiated benign ganglioneuromas contain cells closely resembling normal nerve cells and many bundles of normal-looking non-medullated nerve fibres accompanied by Schwann cells. More commonly the tumour cells differ to a greater or less extent from normal nerve cells and the fibres are less perfectly organized into bundles. Many tumours contain abundant immature proliferating cells and are assigned by Willis to a group of tumours of transitional or mixed structure, intermediate between neuroblastoma and ganglioneuromas and sometimes called ganglioneuroblastoma.

Neuroblastoma comes second in order of frequency in the collection of cases accepted by Everson and Cole (1966) as authentic examples of "spontaneous regression of cancer"; these accounted for 29 of the total of 176 cases. The relative frequency of regression should not be allowed to obscure its absolute rarity. Everson and Cole include in their list five examples of what other authors have distinguished as *maturation*. Apparently neuroblastoma cells can differentiate or mature in much the same way as their normal progenitors, the embryonic neuroblasts, and in so doing they produce well differentiated ganglion cells and non-medullated nerve fibres; in this way a neuroblastoma matures into a ganglioneuroma. Fox *et al.* (1959)

remarked that well-documented cases of spontaneous maturation are even rarer than those of spontaneous regression. They themselves re-studied two of the most satisfactory examples; both were included in Everson and Cole's list of twenty-nine spontaneous regressions. Both patients had received Coley's toxin at an early stage. One patient (P.W.) was reported on by Phillips in 1953. A diagnosis of malignant neuroblastoma was made at the age of seven months. When examined nearly twenty years later, the patient had multiple subcutaneous ganglioneuromas and a retroperitoneal tumour composed of immature ganglion cells, together with some tissue resembling neuroblastoma and regions of necrosis, fibrosis and calcification. The retroperitoneal tumour was removed except for a portion adherent to the inferior vena cava and the patient was alive and clinically well a year later. The other patient (W.W.) was first reported on by Cushing and Wolbach in 1927. The original diagnosis was *paravertebral malignant sympathicoblastoma*. About ten years later a tumour was removed from the vertebral canal and described as a ganglioneuroma. The patient came to the notice of Fox, Davidson and Thomas nearly forty-six years after the original diagnosis of neuroblastoma and was then apparently free from residual or recurrent tumours. After reviewing some of the early material, which had been preserved, Fox and his colleagues accepted the evidence for maturation of neuroblastoma into ganglioneuroma. Another good example has been reported more recently by Visfeldt (1963).

Observations by Goldstein, *et al.* (1964) on neuroblastoma cells cultivated *in vitro* disclosed a comparable maturation. Cultures of cells from thirteen neuroblastomas formed more mature neural elements in the primary cultures within five months; the morphological changes included an increase in nuclear size, the formation of a single large nucleolus, the accumulation of Nissl RNA, the production of neurofibrils and the hypertrophy of axons. One of the neuroblastomas appeared more mature than the others at the outset, and its cells were transformed into mature ganglion cells within twenty days after explantation *in vitro*.

The evidence for maturation is substantial but a change in the opposite direction is more common; it effects acceleration of growth, structural dedifferentiation and enhancement of malignancy. The description by Willis of actively growing neuroblastoma overtaking and invading a relatively well-differentiated ganglioneuroma strongly implies that neuroblastoma can originate by progression in a ganglioneuroma.

The well-differentiated nerve cells in ganglioneuromas do not multiply. The embryonic neuroblasts and the immature nerve cells are the proliferative elements; they are present in every growing tumour, and transitional forms link them with the fully developed ganglion cells. Willis maintains that every ganglioneuroma that is growing by cell proliferation is really a transitional or mixed tumour, that is, a ganglioneuroblastoma. Furthermore, every ganglioneuroblastoma was once a neuroblastoma. This is probably true as a histological description, but the descriptive terminology can be misleading and liable to obscure the most important biological problems at issue; in particular it is apt to encourage the false presumption that things that *look* alike *are* alike in every way. I am not aware of any justification for supposing that an early "neuroblastoma" that ends up as a fully differentiated inert ganglioneuroma is equivalent in all its properties to an equally early "neuroblastoma" that eventually becomes an invasive, metastasizing and lethal neuroblastoma. The two kinds of "neuroblastoma" may be histologically indistinguishable at the beginning, but they differ in their developmental fates and, by inference, in their developmental capacities. Developmental capacity cannot be detected microscopically; cells that look alike may differ enormously in prospective developmental fates.

It is pertinent to this discussion to anticipate here a fuller discussion in Chapter 13 by noting that the sequence of events in the life history of cells during normal development is multiplication, differentiation, maturation, senescence and death. This is the *programme* of normal development; the programmes of neoplastic development differ from it to widely varied extents. The programme of benign ganglioneuroma resembles the normal programme but differs from it in at least two ways: first, the timing of the programme is out-of-step with that of the normal parent tissue and, second, although differentiation advances a long way along the normal pathway, morphogenesis is rudimentary. The maturing, regressing neuroblastoma is an extremely rare tumour. Its programme of development is similar to that of ganglioneuroma but it apparently specifies a further advance towards senesence and death. Plainly the timing of the programme is crucially important; the maturation phase must begin before the proliferative phase has done irreparable harm. It is not clear to what extent the absence of a maturation can be presumed in the more numerous cases of spontaneous regression in which it has not been recognized or reported but, whether or not a phase of maturation is interposed, it is reasonable to suppose that the ultimate regression is specified in the programme of development.

Benson *et al.* (1962) stated that spontaneous regression had never been reported in a neuroblastoma that was first detected in a patient more than two years old. The information collected by Everson and Cole was consistent with this statement. Benson and his colleagues suggested that although cell maturation increases with increasing age, the tumours in the younger children might grow so quickly that they are liable to outgrow their blood supply with the result that total necrosis of the tumours takes place. This is not a convincing explanation of spontaneous regression. Although an inadequate blood supply can cause extensive necrosis, it is unlikely to cause *total* necrosis; at some places at the periphery the blood supply will be adequate to allow the survival of regions of residual neoplasia from which recurrence of growth is almost inevitable. In some, at least, of the cases reviewed by Everson and Cole, regression was very gradual and inconsistent with an acute vascular crisis, which if it did occur, would be more likely to kill the patient by toxaemia than to cure the neoplastic disease.

C. Nephroblastoma (Wilms Tumour)

Nephroblastoma, the commonest malignant tumour in children, is detected most often during infancy and early childhood. A high proportion of the tumours are undoubtedly present at birth and some have been found in the foetus. No hereditary factor in the aetiology of nephroblastoma is mentioned in most descriptions but Miller et al. (1964) drew attention to an accumulation of case reports during the preceding decade linking nephroblastoma with an undue frequency of developmental anomalies including hemihypertrophy, hypospadias, cryptorchidism and horse-shoe kidney. They referred also to the occurrence of nephroblastoma in siblings in at least eight families, in one of which three, or possibly four, siblings were affected. These observations suggest that heritable genetic factors may be implicated in at least some cases of the disease.

The prognosis of nephroblastoma is relatively good. The survival rate of all children subjected to nephrectomy, irrespective of age, is about 47%; if nephrectomy is carried out during infancy, the survival rate is about 80%. Survival for two years without recurrence is usually accepted as a "cure" although, occasionally, recurrence may become apparent up to eight years after nephrectomy (Lucké and Schlumberger, 1957). By contrast, Everson and Cole do not mention spontaneous regression. Pathologists have been interested chiefly in the structural heterogeneity of the tumours and especially in the frequent presence of tissues not normally found in the kidneys.

It is generally believed that the tumours originate from the metanephrogenic blastema, which, as described by Willis, is a stem-cell tissue serving as a prescursor of both nephrons and interstitial tissues of the kidney. During normal development nephrons develop in the blastema only in response to a stimulus provided by the ingrowth of collecting tubules from the renal pelvis. Nicholson (1931) provided evidence to show that this applies also to the development of nephrons in nephroblastomas. As a rule the tumours consist mainly of embryonic renal tissue resembling the metanephrogenic blastema, together with varied amounts of heterotopic tissues which is usually explained by the capacity of the metanephrogenic blastema for divergent differentiation. Under abnormal circumstances the blastema can produce various mesodermal tissues, including smooth muscle, striated muscle, cartilage and, possibly, bone. The presence of any one or more of these heterotopic tissues in a nephroblastoma is reasonably attributable to divergent differentiation of multipotent neoplastic stem cells and is consistent with an origin of tumours from the metanephrogenic blastema as also with the general proposition advanced earlier that a tumour may manifest any of the types of differentiation of which its parent tissue is capable under normal or abnormal circumstances. The occasional presence of keratinizing squamous epithelium cannot be accounted for in the same way. Willis (1958) suggested that it might be derived from epithelium of the renal pelvis that had been incorporated in the growing tumour. Hou and Azzopardi (1967) found supporting evidence for this suggestion in one of thirty-four nephroblastomas examined by them. The tumour contained keratinizing squamous epithelium and, also, mucin-secreting epithelium and argentaffin cells, which had not been found previously in nephroblastomas. Histological studies indicated that all three of these tissues originated from tubular derivatives of the renal pelvis which represented ramifications of collecting tubules. The three unexpected heterotopic tissues were, therefore, attributed to metaplasia of the collecting tubule epithelium. Willis had previously noted the difficulty of deciding whether these derivatives of collecting tubules were integral parts of the tumour or were non-neoplastic tissues. Hou and Azzopardi suggested that the metanephrogenic blastema and the collecting tubules exerted a reciprocal effect on one another, and that the blastema could probably incite proliferation and metaplasia of the epithelium of the collecting tubules. They remarked that it was hard to prove that the collecting tubules were *not* neoplastic but thought that it might be significant that they had never been demonstrated in metastatic tumours. The reaction of the collecting tubules seems, indeed, to be a good example of *collateral hyperplasia*.

Hou and Holman (1961) had previously adduced evidence for a reciprocal action, namely the incitement of differentiation of the metanephrogenic blastema by contiguous derivatives of collecting tubules. Their evidence was obtained from an extraordinary growth having several features that are hard to interpret. The patient was an infant who survived only thirteen hours after premature birth in the thirty-second week of gestation. Both kidneys were enormously and symmetrically enlarged, but retained the usual kidney shape and each was connected by an apparently normal ureter to an apparently normal bladder. The pelvis in each was normally situated and had seven to nine calyces which divided into small branches. The branches tapered sharply and disappeared into the substance of the kidney. Grossly, the kidneys were devoid of the usual differentiation into cortex and medulla. Histologically their structure was extremely complex. Tissue resembling metanephric blastema containing immature glomeruli and tubules was abundant, and, apart from the absence of heterotopic tissues, resembled the structure of nephroblastomas so closely that Hou and Holman thought that their tumour could be justifiably diagnosed as nephroblastoma. Nevertheless, the growths were exceptional, and apparently unique, in three respects: (1) The overgrowths of the kidneys were bilaterally symmetrical, (2) the tumours contained glomeruli and various tubules in all stages of differentiation which resembled those of a developing normal kidney to an unusual degree, and (3) both kidneys were altered diffusely, uniformly and completely so that no normal kidney tissue was discernible; the tumours represented, and actually were, the entired developing kidneys themselves. The authors considered that the histological evidence was inconsistent with the replacement of an originally normal kidney by invasive tumour growth or with the confluence of tumours of widespread multifocal origin.

The tumours contained two kinds of tubules namely secretory tubules connected with glomeruli and straight tubules extending from the renal pelvis, and were comparable with the collecting tubules in a developing normal kidney. Hou and Holman inferred from their histological observations that in some regions collecting tubules penetrated into condensations of blastema cells and induced in them the differentiation of near-normal glomeruli and secretory tubules. An analysis of serial sections showed that both kinds of tubule ended blindly; they did not establish connections with one another. The authors thought that a similar relationship might obtain also in localized nephroblastomas and that glomeruli and secretory tubules might not differentiate from the blastema unless induced to do so by collecting tubules, as Nicholson (1931) had proposed earlier. On the other hand, blastema cells might stimulate neighbouring collecting tubules to proliferate.

Hou and Holman found it difficult to decide whether the bilateral tumours were two malformed kidneys with overgrowth of both metanephric blastema and collecting tubule epithelium or genuine neoplasms of the blastema cells originating during normal development and accompanied by extravagant proliferation of collecting tubules. In view of the extensive proliferation of the immature blastema cells they thought that the growths must be considered neoplastic and that they were embryonal tumours as well as malformations. Taking into account the many resemblances to nephroblastoma they decided that the most appropriate designation would be nephroblastomatosis.

Observations on the embryogenesis of the mouse kidney *in vitro* and *in vivo* will be dealt with in Chapter 12. Briefly they indicate that under certain circumstances glomeruli and tubules can differentiate from the blastema *in vitro* without the stimulus of the ureteric bud or its branches, but that the apposition of ureteric bud and metanephrogenic mesenchyme is necessary for the kidney to develop normally *in vivo*. In certain mutant mice where this apposition is not achieved, no kidneys develop. In the tumours now under consideration, apposition of the two components is effected and the basic developmental fault seems to be their failure to establish continuity with one another. It is difficult to evade the conclusion reached by Hou and Holman that one or both components are neoplastic. The most likely interpretation seems to be that total neoplastic conversion of the metanephric blastema has taken place in both kidneys simultaneously, and that the collecting tubules derived from the ureteric bud have undergone collateral hyperplasia. The relative normality of the proximal branches of the ureteric bud may account for the retention of the normal kidney shape and the relatively normal renal pelvis and ureters. The derivatives of the ureteric bud, in short, seem to have exerted their normal morphogenetic function.

Despite many uncertainties of interpretation, which are inevitable when experimental analysis of histological observations is impracticable, the two tumours studied by Hou and his colleagues supply impressive evidence for mutual interactions between neoplastic and non-neoplastic tissues. Their observations imply that neoplastic cells are competent to respond to inducing and regulatory stimuli provided by non-neoplastic cells, and further that they can provide similar inducing stimuli to which non-neoplastic tissues can respond. The contention that the bilateral renal overgrowths are embryonal tumours as well as malformations raises the important question of the relationship between neoplasia and errors of development to which further reference will be made later.

Willis has drawn attention to the substantial degree of *maturation* of the components of nephroblastomas into well-differentiated structures but could find only one report of this advancing as far, as in some neuroblastomas, to yield a completely mature benign tumour. Some apparently genuine nephroblastomas reported in adults contained well-differentiated tissues. Attention may be drawn also to the *organoid* structures present in many nephroblastomas.

D. THE TROPHOBLAST AND ITS NEOPLASTIC DERIVATIVES

The justification for introducing neoplasms of the trophoblast under the heading of embryonal tumours depends on the presumption that the trophoblast is derived from the embryo and not from the maternal uterine tissues. If this be correct, the trophoblast and the maternal tissues are genetically different, and the apparent lack of a homograft reaction in the uterus to the genetically different trophoblast calls for a more persuasive explanation than it has yet received. An embryonic origin, although generally accepted, cannot be held to be proved beyond question, and Park and Lees (1950) suggested that the origin was worth reinvestigating. More recently, Gordon (1960), finding that the sex chromatin in the trophoblastic syncytium of hydatidiform moles was of female type, suggested that the syncytium was derived from maternal ovarian granulosa cells that surround the ovum when it leaves the ovary and that its function is to protect the foetus from the immunological processes of the mother. Willis (1967) expressed misgivings about the validity of conclusions drawn from observations on sex chromatin, but, whatever be the histogenesis of normal trophoblast, it is not disputed that the tumours to be dealt with in this section are referable to peculiarities of the parent tissue which, therefore, need prior attention. Much of the material on which the following account is based is discussed in greater detail in the report of a symposium on the trophoblast and its tumours edited by Ober (1959a), who himself contributes an instructive and interesting historical introduction to the subject (1959b).

During the implantation of the ovum, the trophoblast erodes the maternal endometrium. This "invasiveness" is an intrinsic characteristic of active trophoblast cells but its mechanism is still obscure. It is often presumed that invasion is effected by means of proteolytic enzymes produced by the trophoblast, but none has been demonstrated. Experiments on rodents have shown that in order to penetrate the endometrium and survive thereafter, the trophoblast must have "matured" up to or beyond the corresponding stage of endometrial "maturation". If endometrial development is one day ahead of that of the trophoblast, implantation is inhibited. The observations suggest that the endometrium is not passive during its invasion but, on the contrary, actively "receives" or "accepts" the trophoblast. The syncytium of the trophoblast seems to be chiefly responsible for the erosion, whereas the cytotrophoblast penetrates into the maternal blood vessels (Noyes, 1959). It is noteworthy that the invasion is not indeterminately progressive; it halts when the trophoblast has achieved its physiological objective by reaching the maternal blood vessels and bringing the maternal and foetal blood circulations into their proper physiological relationship. When its physiological purpose is ended at parturition, it regresses although remnants are said to persist in the endometrium until the third week *post partum* (Böving, 1959; Hertig and Mansell, 1956; Orsini, 1959).

It has been estimated that emboli of trophoblast are present in the lungs of about 50% of the women who die from various causes during or after the termination of a normal pregnancy. There is some doubt as to whether embolism occurs throughout pregnancy or only or mainly in its closing stages. Park considers that the most important causative factor is the act of separation of the placenta or "placental commotion" and that the effect of eclampsia, pre-eclampsia and toxaemia of pregnancy, which have been impugned as causative agents, depends almost certainly on the attendant systemic convulsions and probably increased degrees of uterine contraction. The circumstances of dying, from whatever cause, may themselves influence dissemination of trophoblastic emboli. There remains sufficient evidence that dissemination of trophoblastic cells and their deposition in the lungs do occur during normal pregnancy, and that, after surviving for a while, they die and autolyse. Park estimates that their maximal survival time is of the order of three days (Bardawil and Toy, 1959; Park, 1958, 1959b; Ober, 1959b).

A further peculiarity of the normal trophoblast, shared by its neoplastic derivatives, is that it secretes chorionic gonadotrophin; secretion of this hormone is an important diagnostic sign of trophoblastic neoplasms.

1. Neoplasms of the Trophoblast

Neoplasms of the trophoblast are usually divided into three groups namely benign *hydatidiform mole*, *malignant choriocarcinoma* and an intermediate form, invasive mole or *chorioadenoma destruens* which, although clinically "benign" as a rule, is described as being locally invasive and, rarely, capable of metastasis. *Chorioadenoma destruens* is often reported as *choriocarcinoma*. The classification is basically arbitrary and is no more than a way of describing microscopic appearances, which correlate extremely poorly with clinical behaviour. It is difficult, if not impossible, to predict the outcome of trophoblastic neoplasia from histological structure (Hertig and Mansell, 1956; Park, 1958, 1959a; Park and Lees, 1950). In the experience of Hertz et al. (1964) the clinical course of any individual patient with neoplasia of the trophoblast is unpredictable and the histological diagnoses of lesions in a single individual may range from hydatidiform mole to metastatic choriocarcinoma. Park and Lees maintain that the facts concerning the behaviour of individual tumours are "the most casual and accidental aggregation of unrelated facts in the whole of medicine", which implies that neoplasia of the trophoblast is characterized by an exceptional degree of dissociation of independently variable characters. The unravelling of the resulting complexity is hampered, as Ober pointed out, by the rarity of the tumours; few observers have had the opportunity to acquire wide personal experience of them. Some of the problems recognized over half a century ago are still unsolved.

2. Hydatidiform Mole

Hydatidiform mole has been defined as a conceptus devoid of an intact foetus in which all or many of the chorionic villi show (1) gross nodular swelling culminating in the formation of cysts, (2) disintegration of blood vessels and (3) variable proliferation of trophoblast (Edmonds, 1959). The histogenesis of hydatidiform mole is obscure. Hertig and Mansell (1956) and Edmonds believed that mole is the late stage of a process that is manifested first by focal degenerations and thickenings of the trophoblast, next by "transitional moles" and finally, provided that abortion is sufficiently long delayed, by typical hydatidiform mole. Edmonds describes these lesions as "successive stages in the development of villar abnormality". It is not clear wherein the primary abnormality lies, whether in the trophoblast itself, in a developmental malformation or in the foetus with consequent secondary disturbances in the blood supply of the villi leading to an accumulation of fluid and proliferation of the trophoblast. Hertig and Mansell maintained that hydatidiform mole is "in no sense a true tumour". Park, in a discussion of Edmonds' paper, disagreed. He maintained that the two essential features of moles are oversecretion of fluid into the villi and overgrowth of trophoblast, the former being secondary to the latter, and held that there is no *a priori* reason why hydatidiform mole should not be attributed to benign neoplasia of the trophoblast.

Hydatidiform mole, on occasion, sheds trophoblastic cells into the blood stream, whence they are deposited in the lungs; the pulmonary emboli, it seems, always degenerate. Hydatidiform mole is clinically benign. It is reported in one in about 2000 to 2500 pregnancies but many early moles, which might give deeper insight into their histogenesis, are probably aborted and escape examination. Of those that are diagnosed about 0·5–1·0 % are followed by choriocarcinoma. As a rule, *chorioadenoma destruens* is equally benign clinically and seems to entail no greater risk of subsequent choriocarcinoma than does a hydatidiform mole.

3. *Choriocarcinoma*

The small but definite risk that a hydatidiform mole will be followed by a choriocarcinoma has led some authors to describe moles as being "precancerous". Roughly 50% of choriocarcinomas are preceded by moles; the antecedents of the other 50% are not known. Park and Lees could find no graded transitional stages between mole and choriocarcinoma from which a direct connection might be inferred; they were inclined to think that the mole was "an irrelevant by-product or side-line" in the neoplastic process that culminated in choriocarcinoma. An independent origin of mole and carcinoma at two different sites in one region of incipient neoplasia would be entirely compatible with the general principles of neoplastic development enunciated earlier, but my impression is that decisive evidence about the connection between mole and choriocarcinoma is not now available and will not easily be obtained.

Typical choriocarcinoma is unusual in several ways. The clinical course is always acute compared with that of most other malignant neoplasms; if untreated, it is nearly always fatal within a year, or not at all and, on the average, patients live only about four months after diagnosis. The neoplasm grows quickly and metastasizes widely. It is remarkable how often it is impossible to find any primary tumour in the uterus, or to find more than a minute area of apparently normal trophoblast. Hou and Pang were unable to find a tumour in nine out of twenty autopsies on patients in whom metastasis had been recognized. Apparently local growth and remote metastasis were inversely related; when the primary growth was extensive and infiltrated the myometrium, the distant metastatic tumours were less widely distributed and less massive than they were when the primary tumour was small and localized (Hertig and Mansell, 1956; Hou and Pang, 1956; Park and Lees, 1950). Perhaps in some patients, the primary tumour is expelled with the conceptus after dissemination has taken place.

The high degree of clinical "malignancy" manifested by choriocarcinoma is not reflected in its histological or cytological characters. Mitotic figures are surprisingly hard to find and the microscopic diagnosis of choriocarcinoma is extremely unreliable (Park, 1959*a*).

Choriocarcinoma secretes the characteristic hormone of its parent tissue, chorionic gonadotrophin, more prominently and consistently than any other functioning malignant tumour. In several respects it resembles the tumours of other endocrine tissues, and Park and Lees (1950) proposed that all of them should be included in a single category designated *endocrinoma*. The common features include the following: all the tumours originate in endocrine tissues; they are formed of mature cells individually similar to their normal parent cells and together form a structure more or less similar to that of the parent normal tissue; they produce the hormones appropriate to their parent tissues and the hormones exert their normal physiological effects; when they behave like "cancer", as they sometimes do, they invade and metastasize but retain the physiological and morphological characteristics of the normal parent tissue or of its benign neoplastic derivatives so that these characteristics cannot be reliably used to predict clinical behaviour.

Choriocarcinoma shares third place in order of frequency in the series of spontaneously regressing tumours collected by Everson and Cole. The tumours in this series are said to have been "histologically confirmed". In view of what has already been said about the histological diagnosis of choriocarcinoma, the histological "confirmation" must be accepted with substantial reserve especially as it rarely applies to the lesion that actually regressed but to another one that was excised. Nearly all the observations apply to presumed metastatic tumours, most often in the lungs or sometimes in the vagina. The recognition of the pulmonary tumours depends almost wholly on radiography of the chest. Ober (1959*c*) pointed out that the X-ray evidence may be misleading; the shadows on the radiographs may be caused by a tissue reaction surrounding emboli of non-neoplastic trophoblast and not by neoplastic trophoblast invading lung tissue. Ober was unconvinced, also, of the relevance of the so-called vaginal metastases as evidence of malignancy. He thought that these lesions were attributable to permeation of venous channels by trophoblast and that true infiltration of maternal tissues was rare. Some authors have suspected that most reported examples of spontaneous regression of choriocarcinoma are explicable by misdiagnosis. Whether or not this be true, spontaneous regression is of negligible importance in clinical practice. The outstandingly important advance in the management of trophoblastic tumours has been the demonstration by Hertz and his colleagues of the therapeutic value of the folic acid antagonist methotrexate and some related compounds. Lasting remissions have been obtained in about 75% patients with a diagnosis of *chorioadenoma destruens* and in about 60% of those with one of choriocarcinoma.

E. Some Characteristics of Embryonic Tumours

Embryonic tumours commonly attract attention during infancy or early childhood but some may not be apparent until adolescence. Some tumours undoubtedly originate during embryonic life and many others probably do so but the common feature is that all originate in immature tissues; the immaturity of the tissue has a more decisive effect than the chronological age on the embryonic qualities of the tumours. A genetic or hereditary factor is important in the origin of some of them but apparently not in all.

The embryonic tumours are not common; their substantial biological interest stems from their retaining, in varied degrees, some features of their parent normal tissues. In particular they have similar capacities for divergent differentiation and for some degree of morphogenesis. The embryonic tumours provide a large proportion of the plausible reported examples of "the spontaneous regression of cancer". Of more practical importance, the prognosis is often more favourable and the response to treatment more satisfactory than could be inferred from microscopic examination of the tumours. This is particularly true of choriocarcinoma; the correlation between histological structure and clinical behaviour is extremely poor. The dissociation of independently variable characters is especially conspicuous in these embryonic tumours.

V. Experimental Studies of Embryonal Tumours

A. The Chemical Induction of Neoplasia in Transplanted Embryonic Tissues

Rous and Smith tested the neoplastic potentialities of embryonic mouse epidermis, gastric epithelium and lung by transplanting those tissues, together with methylcholanthrene, into adult mice. The tissues were obtained during the second half of gestation, tissues from younger embryos did not survive the technical procedures and failed to grow in adult mice. Tumours developed quickly in the successfully transplanted tissues. The epidermal tumours differed in certain details from those usually seen after the exposure of adult mouse skin to chemical carcinogens. The majority of the emergent tumours were carcinomas, not papillomas. Some of the carcinomas originated in papillomas but most of them were carcinomas from the beginning and although the apparently "malignant" carcinomas grew quickly to a considerable size, they usually regressed eventually and even the most malignant-looking of them failed to metastasize. The discrepancy between the histological and clinical signs of "malignancy" should be noted.

None of the tumours had "embryonic" histological characteristics but ample time had been available for them to have matured before they were examined. Furthermore, tumours did not develop as a rule until a little after the time when the donor embryos would have been born if undisturbed so that it was possible, that the tissue was not strictly "embryonic" when the carcinogen initiated neoplasia (Rous and Smith, 1945; Smith, 1947; Smith and Rous, 1945). Contemporaneously, Greene (1945) also reported the chemical induction of tumours in transplanted embryonic tissues.

B. Exposure of Embryonic Tissues to Urethane *in Utero*

Larsen (1947) showed that a single intraperitoneal injection of urethane into late-pregnant mice of the A strain evoked a high incidence of pulmonary adenomas in the offspring by the time they reached the age of six months. Untreated strain A mice have a high natural incidence of pulmonary adenomas and the effect of urethane is to increase the percentage of mice developing adenomas, to increase the number of adenomas per mouse and to lower the age at which adenomas first appear. The experiment provides, indeed, a model demonstration of the phenomenon described as *enhancement*.

Smith and Rous (1948) repeated Larsen's experiment with some modifications. They used strain C mice, whose natural incidence of pulmonary adenomas, although substantial, is much lower than that of strain A mice, they varied the number and the timing of injections of urethane and they examined the lungs of the offspring mice soon after birth as well as sixty to seventy days later. Repetition of the injections did not increase the yield of adenomas; indeed a single dose seemed to be the more effective but the numbers of mice available for comparison were too small to justify a firm conclusion about this. Urethane being a volatile compound, quickly excreted, the effective exposure of the lung tissue must have been of short duration. M. Klein (1954) found later that the enhancing effect was evident in offspring that had been removed by Caesarean section five minutes after the administration of urethane to their mothers. Smith and Rous found that the mothers, also developed pulmonary adenomas in excess of those to be expected in uninjected controls but could establish no correlation between the numbers of adenomas in the mothers and the numbers in their offspring.

Microscopic search revealed adenomas occasionally in mice examined three days after birth and fairly often in mice ten or fifteen days old. Evidently the adenomas had developed quickly. Mitotic figures were correspondingly abundant, although they are rare in the adenomas of adult mice, and the adenomatous pattern was often less well defined and the individual cells less well-differentiated and less basophilic than in the tumours of adults. These differences were not permanent; when mice were kept until they were sixty to seventy days old, their adenomas were relatively quiescent and had few mitotic figures and good differentiation comparable with that of a normal lung. Smith and Rous inferred from their observations that when adenomatous change takes place in a very young animal, a stimulation of cell division resulting from the neoplastic change is super-imposed on the proliferative activity natural to embryonic cells so as to bring about rapid proliferation. This relationship does not endure; as the animal matures and its normal pulmonary cells almost stop multiplying, the adenoma cells stop to almost the same extent. The tumour cells, it was believed, remain susceptible to the normal ageing influences and, when no longer active, differentiate until they resemble, more or less closely, the normal pulmonary epithelium. The relevance of these observations to the examples already mentioned of "maturation" and "differentiation" of embryonal tumours in children is evident.

Larsen reported that urethane injected into pregnant strain A mice during the final 24 hr of gestation evoked at least five times as many pulmonary adenomas in the offspring as comparable injections made at an earlier time and increased the proportion of mice developing adenomas to 100%. M. Klein (1952) confirmed the greater effectiveness of administration during the final 24 hr of gestation in an experiment with mice that were nominally of strain A but suspected of having been outcrossed at some time with strain C. Smith and Rous, by contrast, obtained much the same result by giving urethane on any one of the last few days of pregnancy. They used a different strain of mice (Strain C) and somewhat different procedures. More recent evidence about the importance of the specific timing of exposure to carcinogenic stimuli will be given below.

The early experiments with urethane were focused on the enhancement of the development of pulmonary adenomas which, at that time, was thought to be the only neoplastic response to urethane. This is now known to be erroneous. The preponderance of pulmonary adenomas in the early experiments was due, in considerable measure, to the use of inbred mice with a relatively high natural incidence of these neoplasms and, in part, to inadequate doses of urethane given over too short a period of time and too short periods of observation. Under appropriate conditions urethane can initiate neoplasia of the epidermis of mice when administered remotely and can elicit neoplasia in a wide variety of tissues when given at various times during pre-natal and post-natal life. Not all of this carcinogenic action can be attributed to enhancement.

The scope of experimentation has been widened in two ways; carcinogens other than urethane have been shown to induce neoplasia *in utero* and observations have been extended to the early post-natal period. There are wide differences in the times at which various tissues attain comparable stages of maturity and, consequently, in the times at which they acquire and then lose the ability to give origin to embryonal tumours. The main proliferative or "embryonic" phase is completed in some tissues well before birth whereas in other tissues, in the mammary glands, in the epiphyses of bones and tooth tissue, for example, developmental maturity is not reached until puberty or adolescence. Each tissue may be expected to have a characteristic time-pattern of responsiveness to carcinogenic stimuli; the maximal responsiveness of some tissues may be anticipated during pre-natal life and of others in early or later post-natal life. The demonstrated value of newborn animals for the detection of tumour-inducing viruses, notably the Gross virus and the polyoma virus prompted an investigation of their response to carcinogenic agent and it was at once apparent that the reactivity of some tissues was high (Pietra *et al*., 1961; Pietra *et al*., 1959). These findings were soon confirmed and extended by various carcinogens and a variety of experimental animals (De Benedictis *et al*., 1962; Chieco-Bianchi *et al*., 1963; Kelly and O'Gara, 1961; Roe *et al*., 1961; Toth *et al*., 1961).

The more recent experiments have shown that the question "Are newborns more responsive than adult animals to the carcinogenic effects of chemicals?" is an oversimplification of the problem (Toth, 1968). The tissues of newborn animals do not consistently respond more strongly to carcinogenic stimuli than do adults. The epidermis of newborn mice is less responsive to the action of chemical carcinogens than is that of adult mice; this is explicable by the absence of hair-follicles from newborn mice. In general, pre-natal or early post-natal administration of carcinogens does not increase the incidence of tumours in tissues that mature late; the effect on mammary neoplasia, for example is, if any, a depressing one. The results differ widely in animals of different species and of different strains within the same species and many, although not all of them, can be ascribed to *enhancement* of a genetically-determined liability to "spontaneous" neoplasia in particular tissues and organs. Chronological age, of itself is probably not the important factor in determining sensitivity to carcinogenic stimuli. Sensitivity seems to depend more on the stage of development or differentiation of the exposed tissue and different tissues attain the same critical stage of sensitivity at highly various chronological times.

Some of the results will be summarized here under two headings: (a) Experiments with urethane and (b) Experiments with carcinogenic polycyclic hydrocarbons and nitrosamines. Further information and references may be found in reviews by Magee and Barnes (1967), Roe et al. (1967), Shubik (1966, 1967b) and most recently Toth (1968) dealing with the hydrocarbons and nitrosamines and in various papers by Vesselinovitch and his colleagues dealing with urethane.

C. Exposure of Tissues to Carcinogens during later Pre-natal or Early Post-natal Life

1. Experiments with Urethane

When administered to mothers during pregnancy or to newborn, infant or young adult mice, urethane increases the incidence of pulmonary adenomas to a degree that is proportional to the natural, "spontaneous" incidence of those tumours in the particular strain of mouse being used. If given early in pregnancy, urethane has no demonstrable effect on the incidence of pulmonary adenomas in the progeny. Also, if the natural incidence in a particular strain is low, the effect of pre-natal exposure to urethane is inconspicuous or undetectable. In some of the strains in which pre-natal exposure to urethane has little effect on pulmonary adenomas, the conspicuous result is an increase on the incidence of leukaemia or hepatoma. The effect on these forms of neoplasia diminishes with increasing age at the time of first administration of urethane.

When given to young adult mice of appropriate strains, urethane strongly enhances the development of mammary tumours in later life but if given during the pre-natal or early post-natal periods it has no effect except, possibly, a mildly depressing one on mammary neoplasia. This is not surprising in view of the late development of the mammary gland to morphological and functional maturity. Adenomas of the Harderian gland develop in small numbers in untreated mice of certain inbred strains, the highest incidence that has been observed being about 1%. Urethane enhances the development of these tumours to an extent that seems to be strongly conditioned by genetic factors. Incidences of 29% have been reported by Deringer and of about 10% by Tannenbaum, following the administration of urethane to adult mice of different strains. Vesselinovitch and Mihailovich made the interesting observation, worth further investigation, that the responsiveness to urethane given during pre-natal and early post-natal life rose to a maximum, maintained for a short period about the third week after birth, and then declined (Della Porta *et al.*, 1957; Deringer, 1965; Klein, 1966; Tannenbaum, 1961; Tannenbaum and Silverstone, 1958; Vesselinovitch and Mihailovich, 1967, *a, b*; Vesselinovitch *et al.*, 1967).

Most recently Vesselinovitch and Mihailovich (1968, *a, b*) have compared the action of urethane in new-born rats with that on new-born mice and found several differences in the response of the liver. On the basis of the species differences in the response to urethane, the authors suggested that the "intrinsic predisposition" of the organ might be more important than the specific nature of the inciting agent in determining the type of the resulting neoplasia. Tumours were found less often in eleven other organs. The results indicated that urethane could incite neoplasia in the majority of tissues of rats provided that it was administered systemically during the first day of life. It seems that the stage of intra-uterine development is important in determining whether or not neoplasia will be incited in a particular tissue *in utero* (Vesselinovitch *et al.*, 1967).

2. *Experiments with Carcinogenic Polycyclic Hydrocarbons and Nitrosamines*

In general, the experiments have demonstrated that certain tissues of newborn mice are highly sensitive to the carcinogenic action of a variety of chemical agents. This section is concerned primarily with the age-dependent qualitative variations in the response to carcinogens.

In their earliest experiments, Shubik and his colleagues administered 9·10 dimethyl-1, 2-benzanthracene subcutaneously to newborn mice and found, unexpectedly, that this procedure increased the incidence of lymphomas but rarely induced sarcomas at the site of injection. Contrarily, in adult mice sarcomas developed abundantly at the site of injection but the incidence of tumours at distant sites was little affected. Other observers confirmed the findings in mice but not in experiments with newborn rats, hamsters and rabbits, which reacted like adult mice; the carcinogen acted predominantly at the site of injection and had little effect on incidence of tumours at distant sites. Evidently age was not the only factor implicated in determining the qualitative response of new born animals to carcinogens; a species-specific factor, inferentially a genetic factor, was also implicated (Roe *et al.*, 1967).

More recent experiments have shown that the chemical nature of the carcinogen may have an important effect on the qualitative response of newborn mice. In experiments with newborn AKR mice, 3, 4-benzpyrene (BP) accelerated the energence of lymphomas, many of which were of the stem cell variety and considerably increased the incidence of lung adenomas; dimethylnitrosamine (DMN) by contrast, had no effect on the incidence of lymphomas and only a moderate effect on the incidence of lung adenomas but it evoked a high incidence of hepatomas and hepatocarcinomas (Shubik, 1966, 1967b).

Another experiment, which again demonstrated the importance of species differences in response, revealed an apparently organ-specific depression of neoplasia. In the Swiss strain of mice which develop both lymphomas and mammary tumours when a year old without treatment, the drug isonicotinic acid hydrazide (INH) had little effect on the incidence of lymphomas but depressed mammary neoplasia; some mammary tumours emerged but persisted only briefly and then regressed. In C_3H mice, the same compound increased the incidence of lung tumours but profoundly inhibited mammary neoplasia (Toth and Shubik, 1966, a, b). It may be recalled that ante-natal administration of urethane effected a similar but apparently less conspicuous depression of mammary neoplasia probably correlated in some way with the late maturation of the mammary gland.

Until recently, few observations on the results of transplacental exposure of foetal tissues have been carried out with carcinogens other than urethane but it is now evident that carcinogenic nitrosamines and polycyclic hydrocarbons can be used conveniently. By administering diethylnitrosamine (DENA) to pregnant mice, Mohr and Althoff (1965) significantly increased the incidence of pulmonary adenomas in the progeny. Mohr *et al.* (1966) then carried out a similar experiment with hamsters. The characteristic and predominant action of DENA on adult hamsters is to evoke papillomas of the trachea and squamous-cell carcinomas of the lung. The offspring of female hamsters who had received small doses of DENA during late pregnancy developed what were presumed to be early stages in the same neoplastic responses. Hyperplasia and squamous metaplasia of the tracheal epithelium were found in the progeny with a frequency that increased with lapse of time until, after twenty-five weeks, only about 20% of the animals had healthy tracheas and 40% had papillomas of the trachea. Only epithelial hyperplasia and, occasionally, metaplasia were found in the lung tissue; no squamous cell carcinoma of the lung developed, presumably because the dose of DENA was too small.

Mohr and his colleagues reported good evidence that the carcinogen reached the foetuses through the placenta and not through their mothers' milk. This had usually been taken for granted in experiments with urethane, although De Benedictis *et al.* (1962) and Chieco-Bianchi *et al.* (1963) reported that urethane was transferred to the offspring in mothers milk, and it is useful to have a direct demonstration of transplacental transmission of DENA.

The experiment with mice almost certainly exemplifiies the phenomenon of enhancement and this may be the mechanism also in the hamsters. A more recent experiment reported by Druckrey *et al.* (1967), who used a different nitroso compound, ethylnitrourea, seems to provide a good example of *induction*. The compound was administered to late-pregnant rats with the result that nearly all the progeny died with malignant tumours of the brain, spinal cord or peripheral nervous system. The effective dose of carcinogen was low and would have been subthreshold in adult rats. Druckrey *et al.* estimated that the foetal tissues were at least ten times as sensitive as adult tissues to the carcinogenic action of the nitroso compound, probably as they supposed, because of a greater reactivity of the nucleic acids as targets for the primary carcinogenic action. The identification of nucleic acids as the primary targets of carcinogenic action is plausible but controversial. It is sufficient here to emphasize, in more general terms, the importance of a "sensitivity" or "reactivity" of unspecified targets that changes with time. There is, at least, a formal analogy with embryonic competence, which similarly and characteristically changes with the passage of time.

D. Effect of Carcinogens Administered during Early Pregnancy

Most investigators of pre-natal carcinogenesis started their experiments during the second half of pregnancy, having found that to begin earlier, especially during the first week, entailed severe "wastage" of pregnancies attributable to foetal resorptions, abortions or still-births (Larsen, 1947; Smith and Rous, 1948; Vesselinovitch *et al.*, 1967). In general, these mishaps have not been studied in detail but several observers have noted developmental anomalies and malformations in foetuses that survived long enough. Sinclair (1950), for example, described failure of dorsal closure of the central nervous system of mice when the mother received administrations of urethane beginning on the seventh day of pregnancy. This anomaly did not develop if the first dose of urethane was deferred until the eighth day but was replaced by another one, namely, degeneration of the basal plate of the brain. More recently, Di Paolo and Elis (1967) reported that a single dose of urethane given to mice on the eighth day of pregnancy led to a variety of malformations in their offspring and Druckrey *et al.* (1967) found malformations of the paws of the rats who had received injections of a carcinogenic nitroso compound on the fourteenth or fifteenth days of pregnancy.

Di Paolo and Kotin (1966) have recently maintained that there are many parallelisms between teratogenesis and carcinogenesis and that of the small number of compounds that have been assayed for both activities the majority have produced both malformations and cancer. One and the same compound may have different effects depending on the stage of development of the test organism at the time it is exposed to the compound. Di Paolo and Kotin suggest, therefore, that a compound found to be carcinogenic towards mature tissues might be teratogenic to immature, embryonic tissues, and they propose a close relationship between neoplasia and malformation for reasons to be discussed in the next section.

E. THE RELATIONSHIP BETWEEN CARCINOGENESIS AND TERATOGENESIS

The case for an important relationship between carcinogenesis and teratogenesis presented in Di Paolo and Kotin's review, containing 307 references, is based on two kinds of evidence: first the association of congenital tumours and congenital malformations in children and second that, to the limited extent to which they have been tested experimentally, most teratogenic agents are also carcinogenic.

Reference has been made already to the excessive incidence of congenital malformations in children with Wilm's tumour of the kidney. The excessive frequency of leukaemia in children suffering from Mongolism has also been quoted in this context but it is not clear to what extent the concurrence is found in the generality of congenital neoplasias of childhood (Di Paolo and Kotin, 1966; Kotin, 1967; Miller, 1963, 1965; Miller et al., 1964).

Several environmental factors have been recognized in the aetiology of congenital malformations in children; they include virus infections, notably German measles; exposure of the foetus to X-irradiation, and chemical substances administered to the mother during pregnancy. The importance of chemical teratogens is demonstrated, alarmingly, by the consequences of giving thalidomide during pregnancy. It should be noted, nevertheless that some congenital malformations are not attributable to environmental factors but to sporadic genetic mutations, which geneticists have examined in detail in the laboratory mouse. It is widely believed that the environmentally-induced malformations are phenocopies of the mutational ones but this interpretation is not unchallenged. A phenocopy has the same phenotypic characters as a mutant, from which it is distinguishable only by tests of heritability; mutant characters are heritable, phenocopies are not (Landauer, 1959).

The greater part of Di Paolo and Kotin's review deals with teratogenesis and carcinogenesis in experimental animals. They list twenty-six teratogenic agents that have been tested also for carcinogenicity, with positive result in twenty. Unfortunately, comprehensive parallel tests of teratogenesis and carcinogenesis have rarely been carried out by the same investigators; the assessment and comparison of results obtained by different investigators using different methods and different species and strains of animals are often difficult. Some of the twenty examples of substances that are claimed to be both carcinogenic and teratogenic are unconvincing. I am not prepared to accept oxygen as a carcinogen; the references cited in support do not justify this interpretation. I have misgivings, too, about calling galactose a carcinogen and although oestrogens are recognized aetiological factors in mammary neoplasia in mice it is in my opinion misleading to call them "carcinogens". The carcinogenicity of some of the accepted teratogens is, at best, of low degree. Thalidomide, whose teratogenic action on the human foetus is all too well established, is at most weakly carcinogenic. The isolated observation of choriocarcinoma in an armidillo which had received thalidomide (Marin-Padilla and Benirschke, 1963) is not acceptable proof of carcinogenicity, especially in view of the notorious difficulty of accurately identifying choriocarcinoma. Roe and Mitchley (1963) found a few sarcomas, after a long delay, at the site of subcutaneous injections of large doses of thalidomide into rats but emphasized that the carcinogenic action, if any, was very weak. Many experienced and cautious investigators are disinclined to accept the production of low incidence of subcutaneous sarcomas in rats as sufficient proof of carcinogenicity. It is instructive to note that the teratogenic action of thalidomide in laboratory animals is erratic; it is most pronounced in rabbits, less so in mice, only occasionally demonstrable in rats and not demonstrable at all in any laboratory primate so far tested (Di Paolo and Kotin, 1966; Grüneberg, 1963). The demonstration of teratogenicity and also of carcinogenicity depends to a regrettable extent on luck in choosing a sensitive test animals. Incidentally, the use of primates in cancer research so far has been disappointingly unrewarding.

The demonstrated correlation between carcinogenesis and teratogenesis is incomplete and it seems to me premature to suggest that the potential carcinogenicity of a chemical agent might be assessed by testing its teratogenicity in embryos (Di Paolo and Kotin, 1966). A few years ago, a similarly imperfect correlation between carcinogenicity and mutagenicity received a good deal of attention and its significance has not yet been clarified (Burdette, 1955). It might be more profitable to accept the limitations of the proposed correlations, at least for the time being, and concentrate attention on the overlap between the three activities. The overlap is sufficiently impressive to justify Kotin's (1967) opinion that the more we can learn about one, the better we shall understand the other. Some observations on teratogenesis that may be useful and illuminating in the study of carcinogenesis will, therefore, be discussed briefly in the following paragraphs.

One group of chemical agents, namely the alkylating agents, are of especial interest because they are mutagenic as well as teratogenic and carcinogenic and because their biological actions have been intensively studied. Their precise mode of action on tissue cells is not entirely clear but they are believed to act directly on nuclear DNA so as to change its chemical structure and produce what is, in effect, a mutation. The mechanism of this action will be referred to again in Chapter 8. Alkylating agents produce effects similar to those produced by X-irradiation and, on this account, have been described as radio-mimetic. X-rays, also, are mutagenic, teratogenic and carcinogenic and are known to damage chromosomes and, thereby, effect mutations. It is not to be inferred that all teratogens act in this sort of way; most of them probably do not.

The antibiotic actinomycin D is an effective teratogen in low doses; one quarter to one fifth or less of the doses used therapeutically are sufficient to evoke a big variety of malformations in rats (Tuchman-Duplessis and Mercier-Parot, 1960) The evidence for carcinogenicity is less decisive but Di Paolo (1960) reported induction of epidermoid carcinomas adjacent to the sites of subcutaneous injection of actinomycin into mice and Kawamata et al. (1958) had previously found a high incidence of sarcomas, four of which were successfully transplanted, at the sites of subcutaneous injections into mice of another actinomycin which resembled actinomycin A. Actinomycin D does not, like the alkylating agents, alter the structure of genetic DNA; its characteristic action is to inhibit transcription of genetic DNA to yield informational or messenger RNA which is the essential first step in "expression" of genetic information (Chapter 8).

The site of action of most other teratogens is unknown; there is no reason to suspect that they modify the structure of DNA or interfere with its transcription. It is more probable that the majority of teratogens act later at one or more of the links on the long chain of processes that are implicated in the final "expression" of the genetic information as an overt phenotypic character. According to this interpretation, most environmentally-induced malformations are phenocopies of those that result from genetic mutation (Landauer, 1954). It should be noted that the same phenotypic abnormality can be produced by altering the structure of DNA or by interfering with its transcription or with any link in the subsequent chain of processes leading to overt phenotypic expression; various teratogens may act on different links. It is difficult otherwise to account for the ability of teratogens of diverse chemical structures to produce similar malformations. For this and other reasons mentioned by Di Paolo and Kotin, it is highly probable that multiple intracellular mechanisms can be implicated in teratogenesis. According to the hypothesis advanced here they are divisible into three broad groups:

(1) alteration of the structure of genetic DNA by some chemical teratogens including the alkylating agents and by X-irradiation,

(2) interference with the transcription of DNA by actinomycin D and, perhaps, by other chemical agents as yet uninvestigated and,

(3) interference with some later step in the expression of genetic information. It seems at present that most teratogens operate in the third way.

6. THE HISTOLOGICAL ANALYSIS OF NEOPLASMS

To be acceptable, Di Paolo and Kotin's assertion that carcinogenesis and teratogenesis have in common "the biological expression of phenotypic change in the absence of convincing evidence for direct genotypic involvement" needs much qualification. There is evidence for the direct modification of DNA by alkylating agents and for inhibition of its transcription by actinomycin D; moreover, the genome is "involved", in a real sense, in the expression of genetic information and in the phenocopy phenomenon, as will be shown in Part III of this volume.

The effects of a teratogenic agent depend in part on its physical and chemical properties but conspicuously as a rule on the stages of development reached by the various tissues when they are exposed to the teratogen. The timing of a teratogenic stimulus is usually more important than the chemical structure of the agent in deciding what kind of malformation shall result. Each developing tissue seems to have a characteristic "susceptible" or "critical" period during which it is especially sensitive to teratogenic stimuli (Di Paolo and Kotin, 1966; Wilson, 1965). Different patterns of sensitivity are demonstrable either by applying different stimuli at a given stage of development or the same stimulus at different stages (Landauer, 1954, 1959). The pattern of sensitivity depends on the genotype and differs in different species and in different strains within a species (Gruneberg, 1963). Moreover, it changes with time and the time pattern is genetically determined (Wilson, 1965). There may be only a narrow time-zone during which a teratogen may interfere with a particular developmental process without killing the foetus. No teratogenic response has been observed during cleavage of the ovum. Sensitivity to teratogenic stimuli seems to develop somewhat abruptly at about the time the primary germ layers are established, which, in mammals, is usually several days after fertilization. The sensitivity waxes to a maximum, attained at different times in different tissues, and then wanes. The tissue-pattern of sensitivity may change from one day to the next. In general, sensitivity decreases as differentiation and organogenesis advance (Grüneberg, 1963; Landauer, 1954; Wilson, 1965).

F. THE SIGNIFICANCE OF EXPERIMENTAL STUDIES OF CARCINOGENESIS AND TERATOGENESIS DURING PRE-NATAL AND EARLY POST-NATAL LIFE

The studies of teratogenesis and of carcinogenesis are to a considerable extent complementary and in no obvious way conflicting. Both have disclosed periods of high sensitivity to inciting agents of various kinds. In general the sensitivity to teratogenic stimuli is highest during early pre-natal development whereas the sensitivity to carcinogenic stimuli is usually highest during later pre-natal, or early post-natal, life. It is possible, nevertheless, that the sensitive periods for teratogenesis and carcinogenesis may overlap. This overlap might account for embryonic tumours in general having some characteristics proper to malformations as well as others proper to neoplasms. It might also account for some remarkable nephroblastomas of the human kidney described elsewhere and in particular for the "bilateral nephroblastomatosis" of Hou and Holman (p. 185). The age incidences of the less bizarre embryonic tumours of man agree fairly well on the whole with the varied times at which their parent tissues are known to begin to mature.

The responses to teratogenic or carcinogenic stimuli do not depend on chronological age *per se*, but upon the stage of development towards maturity; the various tissues reach the critical stage of sensitivity to inciting agents at widely differing chronological ages. Although the type and location of the response seems to depend primarily on the responding tissues, it apparently depends also, to some extent, on the quality of the inciting stimulus. The evidence that some inciting agents are both teratogenic and carcinogenic, although not as a rule at the same stage of development, is impressive. Nevertheless it seems to me premature to conclude that every teratogen is a carcinogen and every carcinogen a teratogen, the type of response being determined only by the developmental state of the responding tissue, or to infer that there is no basic difference between a malformation and a neoplasm. It is gratifying that the latter much disputed inference is now coming within the range of experimental investigation.

The most important clinical implication of these findings is that chemical carcinogens introduced accidentally or deliberately into a pregnant woman might initiate neoplasia in her unborn child. This was formerly considered improbable (e.g. Willis, 1967) although Peller (1960) argued from statistics and from the anatomy of the foetal circulation that the tumours of early life are attributable to the action of carcinogens, mainly of extrinsic origin, reaching the foetus through the placenta. In the light of more recent observations, the possibility of transplacental transfer of potent carcinogens has to be considered seriously.

As yet there is little evidence that chemical carcinogens contribute to an important extent to neoplasia in early life. Certainly nothing comparable with the thalidomide tragedy has occurred but it may be well to be aware of the implied warning. Embryonic tumours are uncommon and some types are extremely rare; the only reason for some concern at present is the epidemiological evidence for a moderate increase of leukaemia in children. Some alarming reports of enormous increases in the incidence of childhood leukaemia are fallacious. They are based on big increases in the percentage of childhood deaths attributable to leukaemia; the relative increase is explicable mainly, but perhaps not completely, not by an increase in the incidence of leukaemia but by a dramatic decrease in the mortality of nearly all other diseases of childhood. According to recent information the relatively small but progressive increase in deaths from leukaemia during childhood has been checked, for no known reason (Case, 1965).

It is apparent that the qualitative response to carcinogens and teratogens depends to an important extent on the timing of the extrinsic stimulus but the reasons for the time-specific responsiveness are not known. Many factors are conceivably implicated in ensuring an adequate effective exposure to the extrinsic stimulus at particular times. Variations in the permeability of the placenta for example could be critically important in modifying the effective exposure of foetal tissues *in utero*. It has been shown that certain carcinogens persist longer in the tissues of newborn animals than in those of adults; this prolongation of the effective exposure may account for some examples of high sensitivity of newborn mice to carcinogens but it does not account for all of them (Shubik, 1967b). It does not seem likely that effective exposure alone can be the decisive factor in determining the type and localization of the response. Much more probably, the crucial factor in carcinogenesis, as well as in teratogenesis, is the stage of development of the target tissue.

Several causes or conditions of the sensitive state of target tissues have been suggested. Di Paolo and Kotin maintained that cell division is a prerequisite for both teratogenesis and carcinogenesis. Vesselinovitch and Mihailovich (1968b) mention this as one of several factors which might cooperate in modifying the inception or subsequent development of neoplasia; other possible factors include the degree of cell differentiation and the metabolic or immunological state of the tissues. These authors also inferred from the strain and species differences in response that the "intrinsic predisposition" of the target tissue might be a most important factor. A genetic factor is clearly implicated in the sensitivity to carcinogens and teratogens. I have already proposed, with especial relevance to teratogenesis, that the sensitivity to extrinsic stimuli is comparable with embryonic competence. The same hypothesis applies also to carcinogenesis. This "competence", equivalent to "intrinsic predisposition" resembles embryonic competence in changing with time. According to the concept discussed fully in Part III, competence has a genetic basis being determined by the facultative genome of cells of the target tissue.

VI. Teratomas

A. Teratomas in Man

Willis defined a teratoma as a true tumour or neoplasm composed of multiple tissues foreign to the part in which it arises. He accepted the name teratoma, meaning literally "a malformation that is also a true tumour" as entirely appropriate and made a division into benign and malignant teratomas, which, in his opinion, provided a terminology for their description that was both simple and adequate. Willis conceded, nevertheless, that the distinction between malignant and benign was not always clear cut. Both benign and malignant teratomas have the capacity for progressive growth and, by this criterion, they are, distinguished from simple malformations; their lack of regional specificity differentiates them from the embryonal tumours (Willis, 1951, 1958, 1967).

Most if not all teratomas are thought to originate during early embryonic life. Many are recognizable at birth or during early childhood but some ingenuity is needed to reconcile an early embryonic origin with the first recognition of some tumours, notably in the ovary, testis or mediastinum during the third decade of life.

The growth of teratomas, whether benign or malignant, like the growth of neuroblastomas and ganglioneuromas, depends on the proliferation of undifferentiated, immature cells. Some of the proliferating tissue, and most of it in benign tumours, differentiates into relatively mature quiescent tissue. The proliferating tissue is pluripotential and constitutes the essential parenchyma of the tumours. Teratomas might be redefined therefore as tumours of pluripotential, regionally non-specific tissue. Malignant tumours nearly always contain substantial amounts of embryonic tissue of varied degrees of immaturity, whereas benign tumours are composed in the main of mature, fully differentiated tissues. It should be noted that every ultimately benign teratoma must have been an embryonic, undifferentiated growth at the beginning just as every ganglioneuroma was once a neuroblastoma in the descriptive, histological sense.

The differentiation of many components of teratomas, especially the benign ones, is far advanced. In general the varied components are combined in a disorderly fashion but often they are associated one with another in characteristic ways to produce an organoid pattern. Respiratory cavities are often associated with cartilage; alimentary cavities are often encircled by smooth muscle, masses of central nervous tissue are sometimes encircled by sheaths resembling meninges and are often accompanied by cartilage or bone and teeth are often set in bony sockets. Evidently contiguous tissues exert inductive effects on one another similar to those that operate in normal organogenesis. Some portions of teratomas are highly organized so as to form, for example, well formed digits with nails, phalanges and metacarpals. Willis insists, nevertheless, that teratomas have no true organs or body regions; they have scattered patches of central nervous tissue but no brain, renal tissue but no kidney, teeth without a mouth and so on. Teratomas have also anomalous multiplicities of various tissues as, for example, hundreds of teeth. Furthermore, Willis maintains that teratomas essentially lack vertebrate organization and that they are not foetiform as supposed by proponents of the hypotheses that teratomas are attributable to twin-inclusions or to parethenogenetic development of ova. In summary, teratomas owe their complex structure to divergent differentations of a multipotent undifferentiated parenchyma; differentiation or maturation is often far advanced and interactions between contiguous regions produce an organoid structure but the basic organization of vertebrate organism is lacking. Willis follows Askanazy, Nicholson and Needham in drawing the inference that teratoma originates in a region of early embryonic tissue that has escaped from the control of the primary organizer. The force of this argument has been weakened considerably by changes in the concept of "organizer", which is no longer credited with the autocratic powers formerly attributed to it.

The basic difference between benign and malignant teratomas must be sought in their undifferentiated stem cells which alone proliferate and have the capacity for multiple divergent differentiations. Willis' contention that all the constituents of a malignant teratoma are, as a rule, malignant would be better revised to say that all the undifferentiated tissue is malignant. Dissemination of the multipotential undifferentiated cells adequately accounts for the structure of metastatic tumours; they are often less differentiated than the primary tumour but often composed of many or all of the tissues present in the primary. Separate dissemination of cells from each of the component tissues is much less easily credible. The predominance in the secondary growths of one or more of the several types of tissue that compose the primary is reasonably attributable to local conditions at the site of deposition or, perhaps, even to a chance selection of one of the several available paths of differentiation. The presence of a single malignant component, which is sometimes observed, is most easily and plausibly explained by focal progression within a benign teratoma.

B. Experimental Investigations

1. *Experimental Induction of Teratomas of the Testes*

Michalowsky (1928) induced teratomas in the testes of domestic fowls by injecting a solution of zinc chloride into both testes and several investigators have confirmed his findings. The procedure is effective only during the months of January to March, the result being strongly conditioned by hormonal factors. Zinc sulphate or nitrate or copper sulphate can be substituted for zinc chloride. The yield of teratomas is low; Guthrie, for example, found teratomas in only eight out of 111 cockerels that had been injected (Bagg, 1936; Falin, 1940; Guthrie, 1962, 1964).

It has proved extremely difficult to reproduce these results in mammals. Bresler (1959, 1964) found teratomas and other testicular tumours in 9% of mice that had received intratesticular injections of a cauterizing emulsion containing copper sulphate and simultaneous administrations of testosterone propionate. He also found teratomas in rats similarly treated. Rivière et al. (1960), injected zinc chloride solutions into the testes of 125 rats and found a teratoma in one of them. Other investigators have used similar methods without success.

The mode of action of the metallic salts on the testes is unknown. Their primary effect is more probably a destructive than a specific carcinogenic one. Champy and Lavedan (1938, 1939) reported that almost complete but not total removal of the testes from fowls led to the growth of embryomas and malignant seminomas. The injections of metallic salts inflict brutal injury on the testes and make satisfactory histological studies of the earliest stages of neoplasia impracticable.

2. Spontaneous Teratomas of the Gonads in Mice

Jackson and Brues (1941) described an embryoma in a C₃H mouse, which was transplanted through eleven serial passages without loss of pleomorphism. The tumours contained squamous epithelium with keratinizing pearls, pigmented and ciliated epithelium, epithelium with goblet cells and various glandular structures, which secreted watery or mucinous material and sometimes formed cysts. Often the tumours contained, also, smooth muscle, cartilage, bone and nervous tissues. The varied types of tissue were ascribed to divergent differentiations of a single tissue that was carried on by transplantation from one animal to another.

Fekete and Ferrigno (1952) made comparable observations on a transplantable ovarian teratoma of spontaneous origin in a C_3H mouse. Transplanted tumours, like the primary, were structurally complex but the proportions of differentiated adult tissue was greater in large tumours of long duration than in small ones of shorter duration. The disposition of the varied tissues was confused and organization was almost completely absent. The differentiated tissues contained few mitoses whereas other, undifferentiated, tissue contained many. The chromosome number was diploid, indicating that a popular hypothesis attributing teratatoma to parthenogenetic development from an ovum or a polar body was incorrect. The pleomorphic structure persisted through nine serial transplantations and the authors believed that the varied constituents could be produced only by the continued growth and differentation of pluripotent cells. They regarded the undifferentiated, most rapidly proliferating cells, which were present in all tumours, as the essential elements.

Stevens and Hummel (1957) extended the opportunities for studying teratomas of the testes experimentally when they found that these tumours developed spontaneously in the testes in about 1% of the male mice of an inbred strain "129". The primary tumours, as described by Stevens and Hummel, closely resembled malignant teratomas of the human and equine testes. They were easily detectable in mice more than one week old but most of the descriptions were based on tumours from mice twenty to thirty days old. By that time most of the tumours were composed of differentiated tissues of adult type but about one-third of them had embryonic and immature components adjacent to the differentiated tissues. The latter were of very varied types, which differed greatly in relative frequencies. Nervous tissue was present in 100% whereas hair follicles and sebaceous glands were found in only 7%. Certain types of tissue were not found, notably pigment cells, trophoblast, hepatic, renal and lung tissues and teeth. Consistently with the earlier observations of Willis and others on human teratomas, true organs were not formed and axiation and segmentation were not recognized but certain tissue relationships were noticed; nearly all smooth muscle was in apposition to an epithelium with goblet cells, skeletal muscle was often attached to bone or cartilage, sometimes by a tendon, and hair follicles and sebaceous glands were always associated with keratinized squamous epithelium.

Stevens, noting a tendency towards increasing complexity of primary tumours with advancing age, traced the developmental course of the teratomas. The most primitive tumours found in newborn mice were made up of clumps of undifferentiated cells, many of which were in mitosis. Next, cells took on an epithelial arrangement around pools of blood and tissue debris and two different kinds of epithelium became distinguishable signifying, as Stevens believed, the formation of the two primary germ layers, ectoderm and endoderm, in approximately equal amounts. Embryonal and mesenchymal cells surrounded and separated the ectodermal and endodermal vesicles. Thereafter, the ectoderm differentiated into neuroepithelium, the endoderm into alimentary and respiratory epithelia and the mesoderm into cartilage and muscle. It is noteworthy that the neuroepithelium was the earliest recognizable tissue, that the adult, differentiated elements of the tumours were produced by the differentiation of embryonal cells and that the development of a teratoma was, basically, an epigenetic process. Progressive growth was attributable to the proliferation of embryonal cells; teratomas usually stopped growing when they lost their embryonal component.

3. *Transplantation of Teratomas of the Mouse Testis*

Stevens (1958) attempted transplantation of nineteen primary tumours in strain 129 mice and succeeded in establishing four transplantable strains. The failure to establish the other tumours was traced to the use, in the first or second passages, of grafts composed wholly of tissues of adult type. The four transplantable strains differed amongst themselves in important ways, testifying to the uniqueness of individual tumours of one general kind, to which references have been made previously. One strain (strain 233) was an embryonal tumour from the first and remained so through seventy-six serial transplantations. The cells of strain 536 resembled those of mouse trophoblast until transplantation was discontinued after the ninth passage. Tumours of strain 1062 consisted predominantly of immature neural cells together with small regions of differentiated tissues. The fourth strain, strain 402, was studied in the most detail and provided material for further investigations by Stevens, as well as by Pierce and his colleagues.

After the second transplanted generation, strain 402 was continued in fifteen distinct sublines. Embryonal cells and immature epithelium were the most conspicuous components of all of them but each had as well some kind of epithelium, muscle was present in twelve, cartilage in four and bone in two. In subsequent generations, the morphological differences between sublines became stabilized. Stevens considered that most or all of the differentiated tissues in these sublines derived from a pluripotential stem cell and that the differences between the lines was based on differences in the developmental capacities of their undifferentiated stem cells. He regarded the predominantly cartilaginous structure of subline X, for example, as being inherent in the stem cell; attempts to separate cartilaginous or non-cartilagenous material for transplantation were not successful.

The frequency of neural tissue in the sublines was notable. The morphological features that characterized the individual sublines were heritable in the sense of being maintained through many serial transplantations. Some changes occurred during the course of transplantation and seemed to be irreversible and, usually, to be in the direction of decreasing pluripotency so that the capacity of the stem cells for differentiating into many types of tissue became restricted. Stevens thought that changes of this kind might be comparable with the restriction of potency that occurs during embryonic development. Each type of tissue seen in primary tumours of strain 129 mice was found in some tumours of one or another of the sublines. Thyroid tissue, which had not been seen in a primary tumour, was found in one tumour of subline 402. Several grafts of one subline failed to grow after the line had maintained through twenty-two transplant generations over a period of more than three years. Tumours of another subline differentiated completely and failed to grow progressively but, apparently, it could be transplanted in series by using incompletely differentiated tissue from young tumours (Stevens, 1958, 1959).

Pierce and his colleagues extended the study of one of Stevens' transplantable sublines (402 VI). Subcutaneous implants of the tumour produced undifferentiated embryonal carcinoma with the addition of elements from the three primary germ layers. Intraperitoneal injections of minced tumour produced solid tumours and, also, a haemorrhagic ascites in which many small cysts floated. The cysts were not found after the fifth intraperitoneal serial passage. Morphologically, they contained at most three types of cell, namely, cells closely resembling cells of the yolk sac, spindle-shaped mesenchymal cells and embryonal epithelial cells, which, at times, showed slight differentiation into, for example, early neural or squamous cells. When transplanted subcutaneously the cysts developed into fully-differentiated teratocarcinomas indistinguishable from tumours that had been maintained throughout by subcutaneous transplantation. The authors suspected that the cysts were foetiform derivatives of the teratocarcinoma and considered that their observations supplied definite proof that the differentiated elements of a teratocarcinoma developed from embryonic cells by a process believed to be analogous to morphogenesis of a normal embryo (Pierce and Dixon, 1959a, b).

The "cysts" described by Pierce and Dixon correspond with the embryoid bodies first recognized, it seems, by Peyron (1939) and studied, also, by Stevens (1959, 1960). Stevens found that, under certain conditions, strain 129 teratomas produced thousands of these bodies, which resembled early mouse embryos. He saw them in a metastatic growth of a primary tumour, but not in early primary tumours themselves, in the ascites resulting from the intraperitoneal injection of transplanted sublines and, occasionally, in subcutaneous grafts. Transplantable tumour sublines were established by grafting single embryoid bodies into the eyes of strain 129 mice. After two to four weeks the grafts contained undifferentiated embryonal cells, primary ectoderm and endoderm, cuboidal, columnar and pseudostratified, ciliated epithelium, neuroepithelium, muscle, cartilage and trophoblastic giant cells. One graft contained several formations with an unmistakable resemblance in structure to portions of normal embryo about nine days old. Stevens believed that the embryoid bodies resembled six-day mouse embryos and were, indeed, hemologous with them. Nevertheless they were composed of truly neoplastic cells since they grew progressively during serial transplantation (Stevens, 1960).

Pierce et al. (1960) produced embryoid bodies by injecting one of the transplantable derivatives of subline 402 intraperitoneally into strain 129 mice and transplanted the cysts subcutaneously into similar mice. Out of ninety-nine grafts, thirty-one were teratomatous by the sixty-fourth day, thirty-nine were cysts lined by endoderm and twenty-nine were not recovered. Even the poorly differentiated tumours contained at least ten somatic structures. The embryonal component of the embryoid bodies changed little before the ninth day after implantation. After the tenth day, the embryonal tissue developed through the successive stages of neural plate and neural tube to yield considerable masses of brain tissue. The authors inferred that embryonal carcinoma is the neoplastic equivalent of an ovum. In one cyst deep in the substance of a subcutaneous teratocarincoma they found embryonal carcinoma arranged as a neural fold separated from flattened visceral yolk sac by mesenchyme. Somites were present on each side. Because the structure was bilaterally symmetrical and showed evidence of primary organization they considered that there could be little doubt as to the embryoid nature of the teratocarcinogenic cysts.

4. *The Histogenesis of Teratomas of the Testis*

To study the earliest stages on the development of teratomas it is highly desirable, if not essential, to use material in which neoplasia occurs with a much higher predictable frequency than it does in strain 129 mice. Using a combination of genetic and experimental procedures, Stevens succeeding in producing material with the required high incidence of teratomas.

The liability to develop teratomas of the testis, which seems to be peculiar to strain 129 mice, probably depends on multiple genes but Stevens found that the normal incidence of about 1·0% could be doubled by introducing a single gene Steel (Sl^J), which was known to have a conspicuous effect on the development of normal germ cells. Stevens found also that certain environmental factors influenced the incidence of congenital teratomas; teratomas developed more often in the left testis than in the right and twice as often in mice of second or later litters as in mice of first litters. With the most favourable combination of genetic and environmental factors, the incidence of teratomas rose towards 10%. Stevens then found that teratomas developed with a still higher frequency in successful grafts of germinal ridges of 12·5 day embryos implanted in the testes of adult strain 129 mice. Implants in the spleen yielded only about one tenth as many teratomas as implants in the testis. Evidently some unidentified factor present in the environment provided by the adult testis influenced the result to an important degree. The genotype of the host mouse was of minor importance; teratomas developed satisfactorily in implants in alien strains of mice. On the other hand, the genotype and age of the genital ridge were crucially important. The ability of the genital ridges to yield teratomas after transplantation was lost between 12·5 and 13·5 days after the beginning of gestation. The importance of genotype was demonstrated in experiments with two stocks of strain 129 mice carrying different alleles of Steel, Sl^J and Sl^d. The two stocks were interbred. The genital ridges of mice heterozygous at the Sl locus (Sl^J Sl^d) yielded teratomas in 80% of the successful implants whereas ridges from homozygous mice (Sl^J SL^J or Sl^D SL^D) yielded teratomas in only 2·7%.

Histological examinations of genital ridges at the time of transplantation showed that the ridges from heterozygous mice, which yielded abundant teratomas, contained the normal number of primordial germ cells but the ridges from homozygotes, which yielded very few teratomas, were almost devoid of primordial germ cells. Stevens interpreted these findings as providing strong evidence that teratomas of the testis in mice originate from primordial germ cells (Pierce et al., 1967; Stevens, 1964, 1966, 1967; Stevens and Mackensen, 1961).

Pierce, Stevens and Nakane refer to Stevens' transplantation procedure as the "experimental induction" of teratomas. It is of more than semantic importance to enquire closely into the propriety of this description. At first sight the phenomena do not seem to correspond with what is ordinarily understood by the "induction" of neoplasia although the term might be acceptable in one of the senses in which it is loosely used in experimental embryology. In application to neoplasia, the term *enhancement* would be preferable; the difference in the incidences of tumours in implants in spleen and testis respectively may be a measure of the enhancement and the phenomenon as a whole may provide some badly-needed clarification of the nature of enhancement. The loss, between 12·5 and 13·5 days after the beginning of gestation, of the ability of genital ridges to produce teratomas after implantation into normal testes strongly suggests a loss of "competence" to respond to an "inducing" stimulus in the embryological sense and it recalls the loss of "sensitivity" of foetal and neonatal tissues to carcinogenic and teratogenic agents to which reference was made in the preceding section. There is an important possibility that responsiveness to enhancement is of the same general nature as embryonic competence. The identity of the enhancing stimulus provided by normal mouse testes is completely unknown; presumably it is not a carcinogen in the ordinary sense as most other enhancing agents are.

Stevens (1967) summarized other evidence in support of the contention that teratomas of the testes in mice are derived from primordial germ cells. In conformity with that interpretation, many of the teratomas seen in grafts examined seven or nine days after implantation were intratubular, as were some spontaneous tumours removed from fifteen to seventeen-day-old embryos. The tumours that developed in the grafts were histologically similar to the spontaneous ones in strain 129 foetuses but in the foetuses there were only one to six tumours, usually situated close together in a single tubule, whereas in the grafted genital ridges there were ten to twenty widely separated foci. Teratomas evidently originated at multiple unconnected foci. Teratomas in grafts examined about three weeks after implantation were composed of a variety of tissues, amongst which immature and adult neural tissue predominated.

Pierce and his colleagues came to the same conclusion from their studies of the ultrastructure of primordial germ cells and embryonal teratoma cells. Pierce and Beals (1964) compared the ultrastructure of primordial germ cells of the fifteen-day-old foetal mouse testis with that of embryonal carcinoma cells and found a resemblance so close as justify the presumption that the teratoma cells originated from the primordial germ cells. More recently Pierce et al. (1967) reported similar observations on transplanted genital ridges and drew the same conclusion from them.

The histogenesis of teratomas in man has been highly controversial in the past and is still disputed but over a narrower range of plausible interpretations. Teratomas are no longer attributed to parthenogenetic development of ova or to development of isolated blastomeres. Two main hypotheses are now argued. One supported in particular by Nicholson, Needham and Willis and, hence dubbed the "British" hypothesis proposes that teratomas represent regions of embryonic tissue that have escaped from the primary embryonic organizer (Pugh and Smith, 1964), whereas the other, "American", hypothesis maintains that teratomas are derived from multipotential germ cells (Melicow, 1965). As mentioned earlier the alleged "escape" from the embryonic "organizer" seems to me implausible. The experiments of Stevens and Pierce support the origin from primordial germ cells but do not entirely dispose of the criticism of Collins and Pugh (1964) that the concept of "primordial germ cells" is a hazy one and that the origin of germinal and seminiferous cells is still controversial. Collins and Pugh as well as Willis prefer to leave the exact histogenesis of teratomas an open question and to my mind, nothing much is lost by doing so. With a little modification of each, the two hypotheses are not so far apart as their more enthusiastic proponents imply. It is relevant to note that not all teratomas can be presumed, with confidence, to originate in primitive sex cells. There is no real dispute about the origin of many or most teratomas from multipotential cells present, ordinarily, in the early embryonic tissues of the genital ridges but so to derive all teratomas involves subsidiary hypotheses of varied, sometimes low, credibilities.

5. *Cloning Experiments*

Kleinsmith and Pierce (1964) used an enzymatic procedure to dissociate embryonal carcinoma tissue from tumours of the transplantable substrain 402 AIII and thereby isolated single tumour cells which they injected, subcutaneously, into strain 129 mice. In this way, they established forty-one clones of teratocarcinoma cells, which proliferated and differentiated into tissues proper to each of the three embryonic germ layers. The clones differed from one another consistently in the type and degree of differentiation. Stratified squamous epithelium, which was present in 79% of a series of tumours of the parent substrain 402 AIII, was found in over 50% of the tumours in six of the clones and in none of the tumours in six other clones. Tumours of one clone produced notochord tissue, which had not been seen in the parent substrain and most of the clones produced small embryoid bodies, likewise absent from that substrain. Each clone was carried through at least five serial passages during which it retained its distinctive characteristics.

Finch and Ephrussi (1967) confirmed and extended these observations. They started with the transplantable substrain 402 IIIa, as used by Pierce and obtained from him, and maintained it by serial intraperitoneal transplantations in strain 129 mice. From the transplanted tumours or mass cell cultures derived therefrom they obtained embryoid bodies and dissociated these to yield isolated cells which they used to establish cell clones by long-term continuous culture *in vitro*. From time to time, the capacity of the cultured cells for differentiating into a multiplicity of tissues was tested by subcutaneous or intraperitoneal injection into strain 129 mice. The first tests were made on twelve clones which had been cultivated through at least twenty-five to fifty generations *in vitro*. From four of these clones, twenty-three subclones were established and some of the clones and subclones were maintained through many more generations *in vitro*.

As in the experience of Kleinsmith and Pierce, the clones differed widely from one another in their capacities for divergent differentiation. Some differentiated excellently *in vivo* whereas others differentiated poorly but even the latter yielded tumours containing at least five different tissues, of which the most frequent were embryonal carcinoma, yolk sac epithelium, neuroepithelium, keratinized squamous epithelium and well-formed neural tubes. One clone (No. 8), in addition to these tissues, originally produced also, but in smaller amounts, ciliated epithelium and keratinized squamous epithelium. It was cultured for nearly a year longer, during which time it passed through at least 140 more cell generations *in vitro*. When tested at the end of this time, it produced nearly the same types of differentiation as it had done originally although in reduced amounts. Subclones were isolated after thirty-eight passages of clone 8 and tested after twenty-three to twenty-nine further cell generations *in vitro*; the resulting tumours, as a rule, were not as well differentiated as those produced by clone 8 at earlier stages. It is important to note that during their long period of culture *in vitro* the teratoma cells multiplied rapidly and showed no signs of differentiation.

These elegant experiments demonstrate that the individual, undifferentiated "embryonic" or "stem" cells of a teratoma carry all the diverse histogenetic and morphogenetic potentialities of the whole tumour and second that these potentialities are transmitted through many cell generations in the course of rapid proliferation in an undifferentiated state. Finch and Ephrussi remarked on the relevance of the observations to the important but poorly understood biological phenomena referred to as determination, de-differentiation and re-differentiation whose nature will be discussed in Part III when the significance of the observations now under discussion will become more apparent. Briefly, they indicate that isolated teratoma cells carry the "programme of development" of a whole teratoma; and this includes a histogenetic and morphogenetic "memory" that is stable and transmissible from cell to daughter cell through an indefinite series of cell generations in the absence of phenotypic expression. It is inferred that the undifferentiated cells carry a specific facultative genome that is stable and replicable.

6. *Implications of the Experimental Studies of Teratomas*

Pierce (1961) and Kleinsmith and Pierce (1964) considered that their observations on teratomas of the mouse testis provided strong evidence of the validity of the "stem-cell theory of cancer" advocated by several investigators (Hauschka and Levan, 1958; Makino, 1956; Yoshida, 1956). According to my understanding, none of the authors said or implied that the stem-line concept, as Hauschka calls it, is a "theory of cancer". The experimental basis for the concept is the demonstration of heritable variations from the modal number of chromosomes of ascites tumour cells in the course of their serial transplantation and the separation of distinctive sublines by cloning. Hauschka and Levan showed that ascites tumours are heterogeneous and that diverse components can be isolated as propagable clones. The stem-line concept provides an explanation of the heterogeneity of transplanted ascites tumours, and as Hauschka suggested, may explain the mechanism of tumour progression but this reasonable proposition has not yet been substantiated by factual evidence. The heterogeneity of transplanted teratomas as demonstrated by Stevens and more decisively by the cloning experiments conforms well with this usage of the stem-line concept. The demonstration of stable clones, with diverse developmental capacities, each an unique stem-line, is especially noteworthy.

Haushka and Levan proposed that tumour progression is faciliated by continuous selective population shifts within a randomly mutating multiple stem-line. This statement, whether it be true or not, is quoted here to bring out the fact that the stem-line concept is inseperable from the concept of random mutation and selection. Unfortunately, in the writings of Pierce and, more recently of Leighton (1967) this *stem-line* concept seems to have become mixed up with a *stem-cell* concept, applicable to normal development. Leighton, for example, writes of the fertilized ovum as a totipotential stem-cell and of the stem-cells that give rise to metaplastic tissues. According to this usage, the normal basal cells of the epidermis are multipotential stem-cells capable of differentiating in several ways. The phenomena of normal development and differentiation do not depend on the selection of random mutants but on the differential utilization of the genome, as will be discussed fully in Part III. In this book the term stem-cell will be used to refer always and only to developmental phenomena and stem-line will refer to the phenomena observed by Hauschka and others in transplanted ascites tumours which are mutational and not developmental. Both concepts are valid and useful and the stem-cell concept, which is not new in principle, will be used repeatedly in other contexts but the two concepts should be clearly separated in language and in thought. The confusion which now threatens seems to be attributable to the circumstance that the observations on teratomas have provided evidence of both stem-cells and stem-lines.

The stem-cell concept is implicit in the repeated references made in this chapter and elsewhere to divergent differentions of one kind of cell. It has been remarked also in other places that inferences in harmony with the stem-cell concept lack finality when they are drawn from observations on transplanted tumours in the absence of cloning on account of the known or suspected heterogeneity of the transplanted material. This applies for example to the histological analysis of mammary adenoacanthoma and organoid tumours in mice by the transplantation technique. The notable achievement of Kleinsmith and Pierce was to show decisively that a single neoplastic stem-cell carries all the capacities for divergent differentiation and morphogenesis that are needed for the development of a complex teratoma. Finch and Ephrussi added the important demonstration that the programme of development or histogenetic and morphogenetic "memory" that is carried in individual stem cells is stable and replicable being transmissible from cell to daughter cell through an indefinite series of a cell generation in the absence of overt differentiation or morphogenesis. In terms of the hypothesis adopted here, the individual teratoma stem cells carry a facultative genome that is stable and replicable in the absence of phenotypic expression of any part of it.

CHAPTER 7

Biochemical Characteristics and Biological Problems of Neoplasia

I. Biochemical Characteristics of Neoplasia

It is not the purpose of this chapter to try to review the manifold applications of biochemical methods to the study of neoplastic diseases (Busch, 1962); the chief aim is to discuss some of the concepts that are being used by biochemists in their search for a biochemical basis of neoplasia and their relevance to the biology of neoplasia.

A. Early Biochemical Hypotheses of Cancer

The first notable biochemical "theory of cancer" was advanced by Warburg who proposed that the initial and basic change in the conversion of normal cells into neoplastic cells was an irreversible injury to their respiratory mechanisms resulting in the continued production of lactic acid from glucose in the presence of oxygen. On this view the most characteristic feature of cancer cells is *aerobic glycolysis*. Greenstein emphasized the tendency of tumours to converge towards a common type of biochemical uniformity by losing those enzyme systems that conferred biochemical specificity and diversity on the various types of non-neoplastic tissue cells. Miller and Miller proposed a "deletion hypothesis" of neoplasia, according to which carcinogenesis results from the loss or permanent alteration of those cell proteins that are essential for controlling growth but not for maintaining life. This hypothesis was derived from studies of liver tumours induced by carcinogenic azo-dyes and in particular on the observation that the carcinogenic dyes could bind to the proteins of normal liver cells but not to the proteins of hepatoma cells. V. R. Potter introduced a variant of the deletion hypothesis, called the catabolic deletion hypothesis, embodying the proposal that the deleted enzymes might be responsible for catabolic rather than anabolic processes.

Of these hypotheses, the Warburg hypothesis has proved the most durable and despite repeated searching criticisms, notably by Weinhouse (1955, 1966) it is still accepted by some people as the only valid biochemical generalization about cancer but it is now becoming recognized that all these hypotheses, including Warburg's, lack generality. Some exceptions to the generalizations have been known or suspected for a long time but fairly recently they have multiplied to the extent that they now invalidate all the hypotheses that have been mentioned so far. Many of the most decisive refutations have been obtained from a series of hepatomas to be considered in the next section.

B. Minimal Deviation Tumours

H. P. Morris and his collaborators have accumulated a large series of transplantable hepatomas derived from primary hepatomas induced in rats by carcinogenic azo-dyes. The tumours are extremely varied in many respects; experimental studies of a wide range of tumours has shown that no two are identical (Morris, 1966). The range of growth rates is wide and some of the more slowly growing ones resemble normal liver so closely that they have become known as *minimal deviation tumours* (V. R. Potter, 1961).

One of the most thoroughly studied of these minimal deviation tumours, the Morris 5123 tumour, resembles normal liver in nearly every respect. The morphology and ultrastructure of the tumour cells are similar to those of normal liver cells and after twenty-five generations of transplantation the chromosome complements were still diploid. The enzyme-patterns of the tumour cells are remarkably similar to those of normal liver cells. The enzyme-deletions previously found in other hepatomas were not demonstrable in 5123 and, therefore, were apparently not responsible for biological malignancy. Moreover, 5123 contains soluble proteins that can bind carcinogenic azo-dyes. Essentially no aerobic glycolysis is demonstrable in 5123 cells. As Potter remarked, the Morris 5123 tumour seems to have destroyed a series of earlier biochemical generalizations about tumours and especially Warburg's generalization to the effect that aerobic glycolysis is characteristic of tumours and is the first change to be manifested when a normal cell is converted into a cancer cell (Pitot, 1963, 1966; V. R. Potter, 1962, 1964).

Many biochemical investigations have been carried out on the basis of a "minimal deviation hypothesis", which presumes that if a sufficiently wide range of these tumours be examined it might be possible to identify some common enzymatic changes that correspond with the minimal abnormalities essential for neoplastic growth (Morris, 1963). No such changes have yet been found.

C. Recent Biochemical Hypotheses of Cancer

To replace the earlier theories, which had to be abandoned in the light of the observations on the minimal deviation tumours, Potter (1954) proposed a feed-back deletion hypothesis based on the proposition that the cancer cell may be regarded as a cell in which one or more connecting links have been deleted or altered to the point of functional ineffectiveness. The exact point of the break cannot now be specified but Potter believed that mutations of regulator genes or operator genes are probably implicated. He suggested that the earliest crucial mutations might affect the regulation of enzyme-sequences concerned in DNA synthesis and mitosis without affecting any other enzymes. Potter admitted that he had never seen a hepatoma that conformed to that ideal and proposed a succession of mutational events, each taking place in a previously mutated cell.

Pitot (1966), also, has emphasized the probable critical importance of faulty regulatory mechanisms in neoplasia. Although enzymes present in the minimal deviation hepatomas closely resemble those present in normal liver tissue, Pitot maintained that a defect in the response to the environmental factors that normally control enzyme-levels and, probably, enzyme synthesis is demonstrable in every tumour whether primary or transplanted, in the large series of Morris hepatomas available for testing. Nevertheless the pattern of the control defect is virtually unique in each of the highly-differentiated hepatomas; each tumour thus has an unique phenotype. Pitot remarked that this situation presents a complete antithesis to the Greenstein hypothesis. Also it seems to discourage undue optimism about the outcome of searches prompted by the "minimal deviation hypothesis" for a single critical enzymatic abnormality common to all the tumours. Pitot does not draw this conclusion although he considers that the facts complicate the problem to a fantastic degree. The complexity of the problem has been evident to pathologists for several decades but biochemists have been slow to admit it. The uniqueness of individual tumours has been clearly recognized for nearly thirty years (e.g. Foulds, 1940) and the independent variability of different characters of tumours has been repeatedly emphasized for nearly twenty (Foulds, 1949) and both have been discussed in early chapters of this book.

The phenomena now under review led Pitot to question whether cancer is a genotypic or a phenotypic disease. The evidence given previously strongly suggests that cancer is not a "a genotypic disease", if that is meant to imply a defect or abnormality of the *total genome*; no such defects have been shown to be consistent and essential characteristics of neoplastic cells although they may be occasional or perhaps even frequent accompaniments or consequences of neoplasia. Pitot suggested a form of "phenotypic disease" described in terms of a theory of altered template stability which presumes the existence of extremely stable messenger or informational RNA. According to this theory polysomes bearing mRNA interact with the endoplasmic reticulum and it is proposed that a change in the mosaic structure of the endoplasmic reticulum could effect an alteration of the polysome-membrane complex resulting in a change in template stability.

One virtue claimed for the "template stability" hypothesis is that it allows for the "reversibility" of neoplasia which, in my view, is of extremely dubious value. The outstanding difficulty in any non-genetic hypothesis of neoplasia is plausibly to account for the high degree of stability and accurate replicability of the non-genetic mechanism. There is no unequivocal evidence of the "reversion" of neoplastic cells to normal cells; alternative and much more plausible interpretations of the alleged examples have been advanced in Chapter 5. The most impressive evidence for reversibility comes from studies of the crown-gall tumours of plants (Braun, 1961, 1963) and the relevance of these tumours to neoplasia in vertebrate animals is not beyond question. Pitot cites the reversibility of *promotion* of chemically-induced neoplasia of mouse epidermis as described by Boutwell (1964). The empirical facts are that promotion by croton oil does not take place if the applications are too widely spaced; if the applications are sufficiently closely-spaced, the effects of croton oil are additive but not necessarily irreversible until shortly before the emergence of visible tumours. The nature of promotion is extremely obscure and will need further attention in Volume 2. It is not clear what is being "reversed". There is no reversion of cells to normal cells because the preceding initiation of incipient neoplasia, according to all available information, is *not* reversible.

The irreversible initiation of incipient neoplasia in mouse skin by chemical carcinogens is effected extremely quickly and perhaps, it seems, almost instantaneously. This may not apply to all other types of induced neoplasia. The observations of Kidson and Kirby (1965), also cited by Pitot imply that the effect of a carcinogenic azo-dye on rat liver is to establish a changed pattern of transcription of DNA. During the first three or four weeks of administration the change is reversible but further administrations soon lead to an irreversible change which might be due either to a change in the DNA itself or to a permanent change in the pattern of its transcription.

The observations of Kidson and Kirby suggest another possible hypothesis not considered either by Potter or by Pitot; the basic fault in neoplasia might be a change in the *pattern of transcription* of DNA not necessarily accompanied by any change in the total genome or by any defect in the mechanisms of protein synthesis on the ribosomes or endoplasmic reticulum.

D. Biological Problems of Neoplasia

Cancer in man is usually a chronic disease with a long history extending back for many years before clinical signs are evident. The study of occupational cancers in man and of experimental carcinogenesis in animals has demonstrated the long time lag between the first exposure to a carcinogenic stimulus and the emergence of a recognizably neoplastic lesion. What happens during the interim is obscure. Until fairly recently it was widely presumed that a tumour came into being with all its characters established in definitive form, its subsequent behaviour being attributable wholly to the multiplication and dissemination of unchanging neoplastic cells. It is now apparent that when the whole history of neoplasms is taken into account, neoplasia is often or, as some believe, always a process of discontinuous development through qualitatively different stages as a consequence of progression, by which is meant a permanent, heritable change in one or more of the characters of neoplastic cells. The study of progression has shown that the structure and behaviour of tumours are determined by numerous characters that, within wide limits, are independently variable, capable of highly diverse assortments and liable to independent progression. Terms like "malignancy" refer to certain commonly observed assortments which nevertheless are subject to some major, and innumerable minor, variations. Some neoplasms, for example, lack one or more of the characters usually deemed necessary for a firm diagnosis of either a benign or malignant tumour but even within these two broad groups laboratory tests reveal subtle variations in reactivity or in biochemical or immunological characters. Neoplasms indeed are almost infintely various, and to emphasize the variety James Ewing titled his classical book on the pathology of tumours "Neoplastic Diseases".

The general idea of tumour progression is now widely accepted but often with overemphasis of two special cases, namely, progression from "benign" to "malignant" or from "dependent" to "autonomous" growth and with relative neglect of the basic concept of neoplasia as a dynamic process of epigenetic development. Tumour progression is an important mechanism in neoplastic development, but as originally defined, it is applicable only to changes in the overt characters of the cells of recognizable "tumours". Neither normal nor neoplastic development can be fully comprehended in terms of extant characters; both require attention to developmental criteria which do not apply to present characters but to what present things become at a later time (Grobstein, 1959). Initiation, the first, decisive step in neoplastic development entails no consistent change in overt characters; it effects a change in reactivity or capacity revealed only by subsequent behaviour. It can be described as Grobstein described the first step in normal differentiation as the acquisition of "developmental bias without overt signs"; it corresponds with what many biologists call *determination* except that it is less disputably permanent and irreversible.

In studying the earliest, clinically "silent" stages, it is necessary to dissociate the concept of neoplasia from the concept of proliferation or "growth". Even in late, clinical stages the basic problem is not merely a problem of *growth*, as was widely held until recently and as some still believe. The *rate* of neoplastic growth is not extraordinary; sometimes it is almost imperceptibly slow and at its maximum, in some transplanted tumours, it approaches, but rarely, if ever, exceeds that of embryonic tissues or of regenerating adult mammalian liver. In extent, both in space and time, it is characteristically but far from invariably indeterminate. Neoplastic growth is not, as often supposed, wholly "uncontrolled" or "autonomous"; hormones can "control" some tumours that otherwise show all the cardinal signs of lethal, malignant neoplasia. Tumours are not consistently "undifferentiated" or devoid of function; some of them including some of the malignant ones show a high degree of histological differentiation and functional specialization sufficient for them to perform normal functions and produce, for example, physiologically active hormones. Neoplastic growth is not, in general, "chaotic" or "anarchistic". Ordinarily, tumours are *not* composed of formless conglomerations of cells but of tissues with organized histological patterns, which pathologists use to name and classify them. Sometimes the histological structure closely resembles that of the normal parent tissue. Nicholson (1950) maintained that the majority of tumours are "organoid" in the sense that they are built on the same plan as the tissues in which they originated and that their cells are grouped to form tissues and organs in a way that is sometimes remarkable. He inferred that the central problem of neoplasia is not merely one of "growth" but also one of morphological *pattern*. The problem of neoplasia comprehends, indeed, all the partial problems of *development* in its broadest sense, namely, growth, differentiation, supracellular organization, morphogenesis, maintenance and, even, senescence and death. The central biological problem is one of *biological organization* in space and time. Neoplasia cannot be precisely defined. Its essential character can be briefly, if imperfectly, expressed by saying that it constitutes a breach in the integrated "wholeness" of the organism.

The student of neoplasia has to face several of the most fundamental and controversial problems that confront academic biologists and, being concerned with a subject of peculiar gravity and urgency to human beings, he is less able to evade or ignore them. Tumour induction may, or may not, be basically similar to embryonic induction but they are alike in at least one way: in each the result depends as much, or more on the reactivity of the responding tissues as on the nature of the inducing stimulus. The concepts of incipient neoplasia, residual neoplasia and progression are probably closely akin to those of embryonic competence or capacity, determination and developmental imminence. The concept of a "programme of development" is as necessary in neoplastic as in normal development and so also in that of the differential utilization of the genome. It is no longer reasonable to be satisfied with vague references to "genetic factors" or "a genetic basis" in the many neoplastic phenomena in which the genome is clearly implicated without taking notice of the newer knowledge and concepts of genetic materials and genetic actions. Similarly, it is no longer desirable to be content with allusions to the "controlling forces" from which neoplastic cells have, allegedly, "escaped" without some attention to the known regulatory mechanisms, implicating both genetic and dynamic systems and operating according to cybernetic principles, that have come to light. Finally, the ultimate integrative quality of life cannot be ignored.

Part III, which follows, represents an outsider's efforts to assess the present state and probable trends of knowledge about biological organization and normal development. The discussion is necessarily incomplete, imperfect and selective being circumscribed both by the writers limited knowledge and by his primary commitment to the study of neoplasia.

Part III

Biological Organization and Developmental Biology

CHAPTER 8

Biological Organization

Lwoff (1962) discussed biological *order* as a sequence in space and time with two complementary aspects, namely, structural, static order and functional, dynamic order. These are not two different and separate systems of order but two kinds of manifestation of one system. They are not separable without abstraction although for purposes of investigation and exposition, some division is scarcely avoidable. The most dramatic innovations in biological method and thought have come from electron microscopy and molecular biology, both of which are concerned primarily with the static aspect of biological order. These disciplines are essentially analytical and reductionist; without concurrent attention to the dynamic aspects they can lead to a biology with the life left out, which is not biology at all. The most important corrective influence has been supplied by the other outstanding innovation of recent times, cybernetics, which has taught that the most distinctive characteristics of living organisms result from the complexity of a dynamic organization in which all biological processes are in some way coupled together. Complexity is an essential condition of life and not merely an inconvenient elaboration.

The present section deals in a general way, first with the relatively static genetic materials, next with dynamic organization and then with the interplay of the static and dynamic organizations as a necessary preliminary to the discussion, in the section on Developmental Biology, of the four-dimensional sequential changes of biological systems in time.

I. Genetic Materials and Genetic Actions

A. GENETIC MATERIALS

For a long time, chromosomes have been credited with the role of transmitting hereditary characteristics. Disturbances of particular regions of the chromosomes regularly lead to specific heritable changes in phenotypic characters of the organism. The precise analytic methods of Mendelian genetics have shown that the effective genetic regions have a constant and precise linear arrangement along the chromosomes and have led to the concept of specific *genes* as hereditary units or determinants arranged in a constant and specific order along the chromosomes. The cytologist's *chromatin*, which is the most characteristic substance of chromosomes, is made up of deoxyribonucleic acid and protein and now the prevalent view is that the essential genetic components of the chromosomes are specifically-patterned macromolecules of deoxyribonucleic acid (DNA) having a molecular weight of the order of hundreds of millions. The widely accepted representation of the structure of DNA is the Watson-Crick double helix, which Crick (1957) describes as follows: "It consists of two polynucleotide chains running in opposite directions and twined round one another. The two chains are held together by hydrogen bonds between the bases, each base being joined to a companion base on the other chain. The pairing of bases is specific, adenine going with thymine and guanine with cytosine." The complementary pattern relationship between the two chains stabilizes the double helix and provides for its exact copying, by replication, of the individual chains.

The term replication is applied, and should be restricted, to the synthesis of a new unit in opposition to a pre-existing one that serves as a model, "primer" or "template" and imposes its own pattern or a complementary pattern on the new unit. Each chain of the DNA double helix serves as a template for the assembly of deoxyribonucleotides in a specific order determined by base-pairing between these nucleotides and the nucleotides of the pre-existing chain. An essential enzyme, DNA polymerase, joins the assembled nucleotides into a continuous chain through phosphate linkages. The pattern of the new chain is complementary to that of the template and therefore identical with the companion chain of the model double helix. When the two chains of a double helix, arbitrarily designated A and B respectively, replicate simultaneously, chain A orders the synthesis of a new chain B^1 equivalent to chain B of the model and chain B orders the synthesis of a new chain A^1 equivalent to chain A of the model. The template and its complementary copy stay together and intertwine, A with B^1 and B with A^1, so that copying of the double helix AB yields two exact replicas, AB^1 and BA^1, each consisting of one "old" chain (A or B) and one "new" chain, (B^1 or A^1). It is to be noted that the "new" units are at no time material parts of the "old" ones. In this basic way, replication differs essentially from binary fission in which the "model" grows to a limiting size and divides into two parts, which are not necessarily identical with one another or with the model. Under appropriate circumstances, a single-strand of DNA can "prime" the synthesis of a complementary copy with which it intertwines to form a double helix. Moreover, polynucleotides can be synthesized *in vitro* from nucleotide triphosphates in the presence of a "primer" DNA and DNA polymerase; he "primer" orders the sequence of the nucleotides and the polymerase links them together. These *in vitro* systems have yielded much of the available knowledge about nucleic acid synthesis but, although the replication of DNA is understood in outline well enough for our present purposes, many details remain obscure (cf. Jacob *et al.*, 1963; Peacocke and Drysdale, 1965).

If, as now supposed, the essential genetic material is a continuous linearly-patterned double helix of DNA, the idea that genes are discrete particles strung along the chromosomes like beads on a string is untenable; the concept of the particulate gene is obsolete (cf. Pontecorvo, 1959). The gene has become instead a particular segment or region of a continuously-patterned thread of DNA or, as Waddington (1956) proposed, "some sort of small fragment of a chromosome acting as a unit". The phrase "acting as a unit" is crucial; the definition of "gene" depends on the criteria used for assessing unitary action. At least three criteria are in current use for recognizing genes, namely, genetic *recombination* by "crossing-over", *mutation* of the kind usually known as point-mutation, and *function*. The units distinguished by these three criteria are different from one another. The unit of function, which refers as a rule to the synthesis of a specifically-patterned protein, is called a *cistron*. Cistrons, as identified by biochemical criteria, have been estimated to comprise some hundreds of times as many nucleotide pairs as the "genes" distinguished by genetic analysis (Benzer, 1957).

The gene is now a functional and not a morphological unit (Picken, 1960) but the definition by function is not satisfactory unless the particular function is precisely specified. There is good reason to believe that the genetic material, directly or indirectly, controls the synthesis of specific proteins and, in particular, of enzymatic proteins, and this "function", although assuredly not the only one, is the current favourite for study and discussion. Biochemical studies have been concerned mainly with the synthesis of enzymes and other proteins in micro-organisms and were guided formerly by the "one gene—one enzyme" theory according to which one gene is responsible for the synthesis of one specific enzyme which, usually, was presumed to be a "primary gene product". This theory has been severely criticized on many grounds and is no longer tenable (Bonner, 1958; Goldschmidt, 1955; Kacser, 1957; Lederberg, 1956; Picken, 1960; Pontecorvo, 1963; Sager and Ryan, 1961; Waddington, 1956, 1962).

The current view is that the genetic DNA carries the coded "information" for the syntheses of specific proteins and that the de-coding involves much more than the fabrication of a "primary gene product".

B. THE GENETIC CODE

Crick (1958) advanced two propositions about the genetic information in DNA under the terms Sequence Hypothesis and Central Dogma. The sequence hypothesis maintains that the specificity of a piece of nucleic acid is expressed solely by the sequence of its bases and that this sequence is the code for the sequence of amino acids in a particular protein. The central dogma affirms that once "information" has got into a protein it cannot get out again; information may be transferable from nucleic acid to nucleic acid or from nucleic acid to protein but not from protein to nucleic acid or to another protein.

In principle, the synthesis of protein could be coded or "programmed" on DNA or on magnetic tape; the practical difficulty has been to make the four nucleotide bases in DNA code for the sequence of twenty or so amino acids commonly found in proteins. The principles of "coding" in general exclude the coding of one of twenty amino acids by one of four nucleotide bases; two bases together are barely enough but three bases together are ample and could, indeed, code for up to sixty-four amino acids. Crick and his colleagues proposed a *triplet code*, with three bases coding for each amino acid, which has proved successful over a considerable range of tests and base triplets, or *codons*, have now been assigned to the twenty amino acids. The codon for alanine, for example, comprises two cytosine bases and one guanine base and is written CCG. The sequences of bases within the codons have not yet been determined. This code seems to be *universal*, or nearly so in that it applies, so far as yet tested, to animal and plant cells as well as to bacteria. The code is believed to be *non-overlapping* and it must be presumed to be *degenerate*, *ambiguous* or *redundant* in that some of the amino acids can be specified by more than one codon. Many difficulties in applying the triplet code remain to be overcome but provisionally at least, it supplies a satisfactory working hypothesis of genetic coding for protein synthesis. It should be noted that the one gene—one enzyme theory has been transformed into a one codon (or more than one)—one amino acid hypothesis, the codon consisting of three members of a continuous sequence of nucleotides in a nucleic acid chain. It should be noted also, as will be discussed in more detail later, that there is no justification, at present, for thinking that the whole of the genetic DNA is coded for protein synthesis (Crick, 1963; Crick *et al.*, 1961; Gros, 1964; Hechter and Halkerston, 1965; Matthaei *et al.*, 1962; Nirenberg *et al.*, 1963; Peacocke and Drysdale, 1965; Speyer *et al.*, 1962, 1963; Weinstein, 1963; Wittman and Wittman-Liebold, 1963; Ycas, 1962).

In the study of developmental processes, the genetic code itself is less important than the manner of its de-coding which, according to current ideas, entails two consecutive steps, *transcription* and *translation*. In the first step, the DNA code is transcribed onto ribonucleic acid (RNA) which, in the second step, orders the appropriate amino acids into a specific sequence. Transcription and translation have been studied most intensively in micro-organisms but there is now sufficient reason to believe that similar processes operate in animals and in plants although with differences in detail that may be highly important.

C. Transcription

In principle, transcription resembles replication; it entails appositional synthesis of a nucleic acid chain with a sequence of nucleotides determined by base-pairing between the new chain and a DNA template. It differs from replication in that the assembled nucleotides are of the ribose instead of the deoxyribose type, in that uracil replaces thymine as the pair for adenine and in that a different enzyme, RNA polymerase, is needed for joining up the nucleotides into a polynucleotide chain. The synthesis of RNA polymerase itself seems to require transcription of a DNA template and, hence, is described as "DNA-dependent". The base sequence of the RNA transcript is complementary to that of a sequence in the DNA template. In some *in vitro* systems both strands of the DNA are copied during transcription as they are in replication but it seems that *in vivo* only one strand of the double-helix is transcribed. It cannot yet be presumed that the companion strand is wholly inert or inessential for transcription of the other one (Ochoa, 1962; Spiegelman and Hayashi, 1963; Tatum, 1964).

The distinction of outstanding biological significance between the two forms of copying is that whereas the whole genome is replicated in continuity only discontinuous segments of it are transcribed. Transcription can begin at multiple points along the helix and advances to only a limited extent from each point so that the transcribed portions are not continuous and do not overlap. Geneticists (e.g. Pontecorvo, 1959) have recognized the problem of accounting for the evident discontinuities of genetic action without structural discontinuities along the sequence of genetic sites. To meet this problem Jacob and Monod have advanced a hypothesis that provides for functional, instead of structural, discontinuities along the genome. They propose that the bacterial chromosome contains "units of transcriptive activity", which they call *operons*. The operon comprises one or more *structural genes* that are coded for protein synthesis. The constituent genes are arranged in tandem and a single *operator* situated at the beginning of the operon is presumed to initiate transcription of all the attached structural genes in sequence. The operator consists of a nucleotide sequence that does not code for protein. Certain other *regulator* genes, spatially separated from the operon, produce *repressors*, which as Jacob and Monod now suppose, are most likely proteins formed in small amounts as a consequence of transcription of the regulator. The repressor formed by a particular regulator is credited with a specific affinity for a specific operator with which it tends to associate reversibly. The combination of repressor with operator blocks the transcription of the whole operon and prevents the synthesis of the proteins specified by all the structural genes of that operon (Jacob and Monod, 1961, 1963; Monod and Jacob, 1961). Some elaborations of the operon hypothesis will be discussed later. The inference to be stressed here is that the transcription of discontinuous sections of the genome, called operons, can be regulated independently of one another by chemical processes which seem to be linked with portions of the genome spatially separated from the operons. The *operators* can be thought of as punctuation marks in the genetic code that separate operons and allow discontinuous transcription of the genetic DNA (Spiegelman and Hayashi, 1963).

Transcription is believed to produce three physically-distinguishable kinds of RNA, namely, (1) informational or *messenger* RNA (mRNA) which carries the code specifying the sequence of amino acids in the protein to be synthesized; (2) *adapter*, *soluble* or *transfer* RNA (sRNA or tRNA) which carries the appropriate amino acids to the messenger for ordering into the proper sequence and (3) ribosomal RNA which in combination with protein makes up the ribosome particles upon which the synthesis of protein takes place. All these types of RNA have nucleotide sequences complementary to those of the segments of DNA from which they were transcribed.

It may be useful to mention briefly here an important technical procedure known as *hybridization*. If an RNA transcript is brought into contact, under appropriate conditions, with single-stranded DNA of the kind from which it was transcribed it will unite specifically with the particular segment of DNA that was responsible for the transcription. In general, hybridization reveals complementary relationships between specimens of DNA and of RNA or between different specimens of DNA. It is important in the present context for its contribution to the demonstration that the three kinds of RNA are transcribed from different segments of the DNA and, further, that transcription of all three of them together utilizes only a small fraction of the DNA molecule.

1. *Ribosomal RNA*

Ribosomal RNA makes up more than 80% of the total RNA in bacteria but is transcribed from less than 4% of the DNA molecule. Apparently it carries no code for amino acid sequences and its function in protein synthesis apart from contributing to ribosome structure is still mysterious.

2. *Messenger RNA*

Messenger RNA, described by Jacob and Monod (1961) as the "primary gene product", carries the nucleotide code for protein synthesis. Messenger

RNA is a mixture of molecules of varied sizes corresponding probably with the sizes of the proteins for which they are coded. In general, the molecules are large, as they must be to accommodate the protein code. Haemoglobin, for example, needs a messenger with 450 nucleotides for coding by triplets. The various messengers are transcribed from different segments of the DNA and each has a nucleotide sequence complementary to that of the segment of DNA from which it was transcribed. Different kinds of animal cells as, for example, rat liver, spleen and kidney cells, have differently patterned messengers. It seems likely that a whole operon is transcribed to yield one messenger and not several messengers corresponding with the individual cistrons composing the operon. It is noteworthy that transcription of all the messengers recognizable in bacteria seems to utilize less than 1% of the total DNA.

Bacterial messenger RNA, as originally described by Jacob and Monod, was characterized by metabolic instability and rapid renewal. It was effective in protein synthesis for only a few minutes after its transcription and, in bacteria, continuous transcription was necessary for sustained protein synthesis. Animal and plant cells contain recognizable mRNA that is short-lived and rapidly renewed. Sustained protein synthesis in isolated nuclei of calf thymus has been reported to require continuous transcription but, more generally, protein synthesis in animal and plant cells is not always and, perhaps, not usually dependent on continually produced and rapidly inactivated mRNA. Short life and rapid renewal are no longer acceptable as essential criteria for the recognition of mRNA.

The distinctions between the different kinds of RNA have become considerably blurred and the only distinctive characteristics of the messenger is its "informational" property. (Scholtissek, 1962; Sirlin, 1963). Observations on RNA viruses and bacteriophages, to be mentioned later, show that informational RNA can be stable and replicable.

3. Transfer RNA

Soluble or transfer RNA is a mixture of RNA molecules, there being one tRNA (or more than one) for each of the amino acids used in protein synthesis. The nucleotide sequences of tRNA's are complementary to those of the regions of the DNA from which they are transcribed. The molecules are much smaller than those of mRNA and comprise some seventy or eighty nucleotides. The molecule is believed to be doubled back on itself and coiled into a double helix with complementary base-pairing that extends over a sequence of about twenty-five nucleotide pairs. The two ends of the molecule and the loop in the middle are unpaired. The tRNA molecules are stable and, apparently, can be used repeatedly for transferring appropriate amino acids to the mRNA where they are ordered into the correct sequence. The tRNA molecule must have recognition sites for specific amino acids and, also nucleotide triplets on the mRNA but the nature and the location of these sites are not yet clearly known. Apparently the tRNA molecules themselves are not coded for protein synthesis (Cantoni *et al.*, 1963; Gros, 1964; Hultin, 1964; Tatum, 1964).

D. Translation

Translation entails the linking together of a large number of amino acids in a sequence determined by mRNA. Protein synthesis takes place on ribosomes which are scattered in large numbers throughout the substance of bacteria and attached to cytoplasmic membranes in the cells of animals and plants. Only a small proportion of the ribosomes probably 10% or less are engaged in protein synthesis at one time and these are associated together in groups of various sizes by attachment to a thread of mRNA. These groups are called polyribosomes or, more briefly, *polysomes*.

Protein synthesis starts with the amino acids being "activated" by reacting with ATP to form adenylic derivatives which then combine with tRNA. Each tRNA has an end group comprising two cytosine nucleotides followed by a terminal adenine through which it combines with the -COOH group of an amino acid to form an amino acyl-tRNA. The specificity of attachment seems to depend on the enzymes (amino acyl-tRNA synthetases) which couple the appropriate amino acid to the specific tRNA adapter. The tRNA carrying the appropriate amino acid next attaches to a ribosome which becomes attached to the end of an mRNA strand, presumably as a consequence of base-pairing between a triplet of nucleotides on the tRNA and the first coding triplet on the mRNA. According to the current intepretation the ribosome then moves along the mRNA to the next coding site and another tRNA carrying the next amino acid needed to make the correct sequence moves on to the ribosome and the first one detaches from it. The amino group of the second amino acid couples with the carboxyl group of the first so that the ribosome now carries a tRNA bearing a dipeptide. The ribosome moves on again to the next coding site on the mRNA and another amino acid is added to make a tripeptide. The ribosome, bearing a growing polypeptide chain, continues to move along to the end of the mRNA molecule and then drops off and the completed polypeptide chain separates from the final tRNA. Meanwhile other ribosomes have moved on to the other end of the messenger, each carrying a growing polypeptide chain. A messenger with about 150 coding sites can accommodate five ribosomes at the same time and these, together with the mRNA thread, constitute a *polysome*. It seems that one ribosome can repeat this performance many times and on variously-coded messengers. The enzyme and energy requirements of these processes are obscure, as are many other details. The mechanism for moving ribosomes along messengers is unknown. In animal cells, most ribosomes are attached to cytoplasmic membranes and are, presumably, stationary; to say that in these cells the movement is "relative", implying that the messengers move across the ribosomes, is reasonable but not illuminating. It is not easy to understand how the growing polypeptide chain is held to the ribosome whilst it is being transferred from one tRNA to another as each new peptide bond is formed. What actually happens on the ribosome itself is, in fact, mysterious and little is known about the structure of ribosomes or the function of ribosomal RNA (Brenner, 1961; Gierer, 1963; Gilbert, 1963; Hultin, 1964; Rich et al., 1963; Peacocke and Drysdale, 1965; Scholtissek, 1962; Tatum, 1964).

1. Enzyme-Synthesis

It has been known for a considerable time that bacteria synthesize certain *constitutive* enzymes under almost all environmental conditions that permit healthy growth whereas they produce other *inducible* enzymes only if appropriate *inducers* are added to the environment. Substrates can act as inducers for their corresponding enzymes but inducers are not necessarily substrates although it seems that they must share some elements of steric configuration with the substrates. Small molecules such as amino acids can act as inducers although the products of induction are macromolecules.

It has been asserted that the ability to synthesize an inducible enzyme is genetically determined and that the inducer adds no new pattern to the organism but only accelerates the synthesis of proteins by existing pathways. Hinshelwood remarked that, from one point of view, induction involves the gradual development of latent characters already potential in the bacteria rather than the acquisition of new ones but he believed, nevertheless, that true "adaptation" can and does occur, and others are convinced that new characters are, in fact, induced. The induction is at first reversible but eventually becomes "virtually irreversible". Cohn recorded persistence of an induced enzyme through 150 generations in the absence of inducer. Hinshelwood doubted if the permanence is ever absolute but found that the new enzyme persists almost indefinitely if the period of "training" by exposure to inducer has been sufficiently prolonged; in the early stages of training it is easy to reverse the induction but at some later stages reversion is incomplete. He concluded that although the environment can exert an indirect selective effect on random mutants it can exert also a direct effect on enzyme pattern that is *not* attributable to selection of random mutants. Cohn maintained that induction effects a permanent phenotypic change without altering the genotype and explained how a fraction of a uniform population of organisms can acquire a new character and how clones differing by a specific inducible character can develop from the uniform population without change in genotype (Cohn, 1958; Hinshelwood, 1953; Monod, 1956; Pollock and Mandelstam, 1958).

More recent investigations have to a great extent reconciled the apparently dissonant opinions about induced enzyme synthesis by showing how different portions of an invariant genome can be selectively activated or repressed. Induction does not, indeed, introduce any new pattern into the genetic *material* but it elicits a new pattern of genetic *activity* by bringing into operation genetic units that, formerly, were quiescent. It has become apparent that induction is but one element in a complex system of regulation. The formation or action of certain enzymes can be retarded or repressed by their substrates or analogues thereof, by the end-products of their action on substrates or by intermediary metabolites. Two main inhibitory mechanisms have been studied in great detail, namely, *enzyme repression* and *end-point inhibition*. Repression affects the formation of enzymes whereas end-point inhbition, which is the more rapidly effective, affects the action of enzymes on their substrates. Both mechanisms can be interpreted as negative feed-back devices for preventing the synthesis or action of enzymes in excess of demand (Maas and McFall, 1964; Various authors, 1961).

Monod and Jacob distinguish a general class of interactions which they call "*allosteric* effects", in which there is no direct chemical relationship between the inhibitory or activating substance and the substrate of the enzyme. They attribute allosteric effects to a selective increase or decrease of the specific activity of enzymes, or of their affinity towards their substrates, effected by agents that do not act by reason of their being analogues of the substrates, or actual intermediaries in the reactions, but by binding with the protein at a site distinct from the active enzymatic site and, by so doing, altering the molecular structure of the enzyme protein (Jacob and Monod, 1963).

Jacob and Monod have described regulatory systems in terms of their *operon* concept. They propose that *repressors* may be liberated from regulator genes in either active or inactive form and that they can react specifically with certain cytoplasmic metabolites, or *effectors*, of unknown composition but low molecular weight, in such a way that the repressor is modified. In an *inducible* enzyme system, the repressor liberated from the regulator is presumed to be active so that it inhibits the corresponding operator and blocks transcription of the attached operon. When combined with an effector, which, in this system is an inducer, it is inactivated and the operator, freed from inhibition, can initiate transcription of the operon and subsequent enzyme synthesis. By contrast the repressor is presumed to be liberated in an inactive form in *repressible* systems, but it is activated by combination with an effector;

transcription and consequent protein synthesis, therefore, take place in the absence of an effector but not in its presence (Jacob and Monod, 1961, 1963; Monod and Jacob, 1961).

The specific feed-back mechanisms are essentially cyclical and Danielli (1959a) and others have devised hypothetical systems that constitute autonomously and continuously acting cycles. Jacob and Monod describe several hypothetical model systems in which one or another of alternative states can become established without alteration of the genetic material itself. The elements of these regulatory systems, like the elements of electronic systems, can be organized into a variety of circuits fulfilling various purposes and both kinds of systems can be represented diagrammatically in the same sort of way. Jacob and Monod illustrated a system based on the presumption that the enzyme acts on the substrate to produce an effector that inhibits the repressor of the operon responsible for specifying that enzyme. Once established, the system would be autocatalytic and self-sustaining; transcription would proceed indefinitely in presence of the substrate (Fig. 65). Another system shown

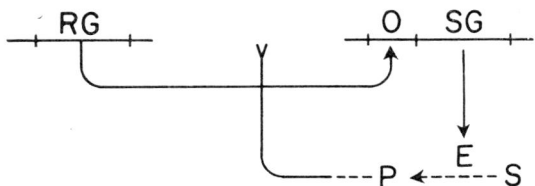

FIG. 65. Model I. Synthesis of enzyme E, genetically determined by the structural gene SG, is blocked by the repressor synthesized by the regulator gene RG. The product P of the reaction catalysed by enzyme E acts as an inducer of the system by inactivating the repressor. O: operator; S: substrate. (Jacob and Monod, 1963)

in Fig. 66 comprises two linked synthetic chains. Making the assumption that the end-product of each chain inhibits the first synthetic step in the other chain, alternative states could be established; either of the two pathways, once it had achieved a "head-start" or temporary metabolic advantage, would permanently suppress the other pathway. A switch from one pathway to the other could be accomplished in a variety of ways as, for example, by temporarily inhibiting any one of the enzymes on the active pathway. Jacob and Monod have described yet another system, which, making certain assumptions about the rate of decay of enzymes and other products, would oscillate from one state to another and provide a kind of biological clock. Many other variations are possible. One regulator might control several operons or it might control an operon that contains another regulator whose activity may control the first one ("cascade" regulation). Systems of various kinds can be interconnected so as to provide for the regulation and integration of extremely complex networks of synthetic activity (Goodwin, 1963, 1964; Jacob and Monod, 1963; Pontecorvo, 1963; Waddington, 1962).

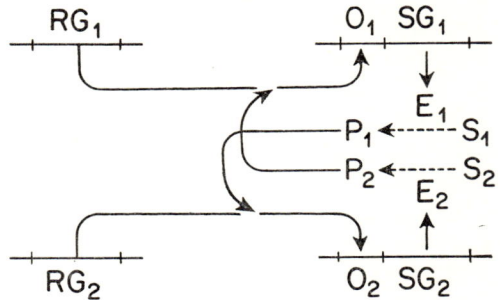

FIG. 66. Model II. Synthesis of enzyme E_1, genetically determined by the structural gene SG_1, is regulated by the regulator gene RG_1. Synthesis of enzyme E_2 genetically determined by the structural gene SG_2 is regulated by the regulator gene RG_2. The product P_1 of the reaction catalysed by enzyme E_1 acts as corepressor in the regulation system of enzyme E_2. The product P_2 of the reaction catalysed by enzyme E_2 acts as corepressor in the regulation system of enzyme E_1. O: operator; S: substrate. (Jacob and Modod, 1963)

The three major features of the foregoing concept of protein synthesis as proposed by Jacob and Monod (1963) are as follows: (1) The synthesis of proteins in bacteria is controlled by two kinds of genes namely *structural* genes, which determine the specific structure of the proteins and *regulator* genes, which regulate, negatively, the rate of information transfer from structural genes to proteins. (2) The bacterial chromosome contains "units of transcriptive activity", or *operons*, co-ordinated by a genetic element or *operator*. (3) Most importantly, regulation of protein synthesis operates at the genetic level by determining the rate of transcription of the code of the genetic DNA into mRNA. Jacob and Monod recognize, nevertheless, that other types of regulatory mechanisms interconnect metabolic pathways and regulate enzyme activity through allosteric effects on the configuration of certain proteins. More generally, it might be expected that many varied specific and unspecific factors will be able to modify protein synthesis by acting at any step along the complex sequence of processes, comprising transcription and translation, that leads from the genetic DNA to the finished enzyme or other protein. Synthesis can be inhibited not only by the mechanisms already mentioned but also by unspecific interference with the organized supply of the necessary raw materials and energy (Novelli *et al.*, 1961; Pollock, 1953). Evidence from several sources indicates that certain hormones regulate protein synthesis by interfering with the formation of peptide linkages during translation. The exact mechanism is unknown but it seems fairly clear that it operates *after* the amino acids have been activated and attached to tRNA molecules and conveyed to the ribosomes. Some authors refer the effect to regulation of the "function" or "activity" of ribosomes but these qualities of ribosomes are obscure. The primary action may possibly be on the production or utilization of mRNA (Hultin, 1964; Korner, 1960; Liao and Williams-Ashman, 1962; Wilson, 1962).

II. Changes in the Genetic Materials and in their Actions

A. MUTATION

Pontecorvo (1959) described mutation as any change in the quantity, quality or arrangement of genetic sites that is still compatible with replication. The recent trend has been towards a more explicit identification with changes in the genetic DNA. Herriott (1966) remarked that mutation can be defined loosely as an error or alteration in the genetic message or, in biochemical terms, as a change in one or more bases along the DNA strand. The term is ordinarily used to include many kinds of alterations ranging from change in a single nucleotide base to microscopically visible abnormalities of chromosomes. Point mutation, attributable to a copying error, is characteristically random in time and site; it may be restricted to a single nucleotide base on the DNA chain and each *muton* or mutable unit mutates independently of other units. Random "spontaneous" mutation by copying error takes place once in about 10^5 to 10^7 replications of the genetic material. The grosser changes in the genetic material include translocations, deletions, insertions, and inversions of genetic segments and recombination by "crossing-over" during meiosis. Recombination has been ascribed to breaking and re-joining of chromatid threads where they cross one another in the chiasmata but what actually happens is still obscure. An alternative interpretation is that recombination is due in whole or in part to a kind of copying error first described by Belling (1933) and now usually referred to as the *copy-choice* mechanism. In essence, copy choice means that during replication, copying can switch from one model to another as, for example, from one DNA strand to its partner in the double helix. There is thus an interchange of genetic *pattern* between the two strands but, according to the copy-choice-hypothesis, no interchange of genetic *material*.

A variety of physical and chemical *mutagens* can greatly increase the incidence of mutation. The mutagenic action of ultraviolet light is maximal at the wave length of maximal absorption by DNA and this has been widely interpreted to mean that ultraviolet light acts directly on the chromosomal DNA to induce mutation. The chemical mutagens are numerous and highly various in structure; they include both organic and inorganic compounds. It is difficult to understand how all the diverse mutagens can act in identical ways within cells; their ultimate effects may be the same but their primary actions may differ, being sometimes, perhaps, exerted on the synthesis of RNA or protein rather than on DNA (Auerbach, 1967; Sharma and Sharma, 1960; Straus, 1962).

In general, chemical mutations produce much the same effects as irradiation and are described as *radiomimetic* although the chemically induced changes are usually the more localized. At one time it was suspected that there was a close correlation between mutagenic and carcinogenic properties, the implication being that the primary action of carcinogens was mutagenic. The suggested correlation did not withstand critical examination (Burdette, 1955) but this does not finally dispose of the possibility that the intracellular actions of mutagens and carcinogens are similar in kind. Brookes (1966) has pointed out that the tests of mutagenicity and carcinogenicity are necessarily carried out on different organisms. Mutagenicity can be tested in microorganisms in which a frequency of mutation of 1 in 10^6 is easily detectable; a similar frequency is well outside the limits of recognition in mice, which are commonly used for the tests of carcinogenicity. A good deal of research is now being devoted to the direct primary action of mutagens and carcinogens on nuclear DNA. A few specific examples will be mentioned here; reference may be made to reviews by Herriott (1966) and Brookes (1966) for more detailed discussions.

One of the simplest chemical mutagens, nitrous acid, which is not demonstrably carcinogenic, oxidatively deaminates adenine, guanine and cytosine converting them to hypoxanthine, xanthine and uracil, respectively, and by so doing produces mutations. When nitrous acid acts on the transforming DNA of pneumococci which, as will be described later, can be considered as a fragment of a bacterial genome, it effects a mutation from streptomycin-resistance to streptomycin-sensitivity. The mutagenic action of certain alkylating agents which include the chemically simple sulphur and nitrogen mustards, have been studied in considerable detail *in vivo* and *in vitro*. These agents exert a variety of biological actions; they are cytotoxic, growth-inhibiting and mildly carcinogenic. They interact with DNA mainly at the N-7 position of the guanine moieties. There are two main hypotheses of how this alkylation leads to mutation. According to one of them the alkyl group on the N-7 position of guanine labilizes the glycosidic bond so that purine splits off, even at neutral pH. This loss of a base, constituting a genetic deletion, leaves a gap in the DNA template and during replication either the gap is ignored or a base selected at random is inserted into the new strand opposite to the deletion. There is a strong probability that a base different from that deleted will be inserted so that a mutation will result. According to the other hypothesis, the acidity of the -NH-CO-group is strengthened in the alkylated guanine so that as a consequence of its dissociation it tends to base-pair with thymine instead of cytosine during DNA replication. After such anomalous base-pairing, adenine-thymine would replace the alkylated guanine-cytosine pair in subsequent replications (Lawley, 1961, 1962; Lawley and Brookes, 1961, 1965).

The bifunctional or "two-armed" sulphur or nitrogen mustards have, in general, a stronger cytotoxic action than their monofunctional or "single-armed" counterparts. One possible reason for this is that the bifunctional compounds can effect "cross-linkage" between guanine bases on different strands of the DNA double helix and, by interfering with separation of the two strands, prevent replication. Evidence has been accumulating for some time that the genetic damage caused by cross-linkage can be repaired. The first step in repair seems to be deletion of the cross-linked bases followed by restitution of the continuity of the two DNA strands and consequent recovery of the ability to replicate. It is beyond the scope of this chapter to enter into the details of these repair processes but their importance may be considerable (Brookes, 1966; Lawley and Brookes, 1965). It is noteworthy that the majority of mutations in higher organisms have been attributed to deletion of, or damage to, comparatively large segments of DNA in contrast to the point mutations commonly observed in micro-organisms. The point mutations may not be recognizable in more complex organisms in which the tolerance of small mutations or the ability to repair them may be much higher (Brookes, 1966).

The carcinogenicity of the alkylating agents is relatively low. By contrast, the carcinogenicity of nitrosamines and nitrosamides is high and the induced tumours are notably widespread and varied. It is believed that the compounds are metabolized in the body after absorption so as to liberate an alkylating agent which is the effective carcinogen. The intracellular actions of these alkylating products have not yet been elucidated. It is more remarkable that despite long investigation the intracellular actions of the carcinogenic polycyclic hydrocarbons are still obscure. A direct action on nucleic acids is suspected but not yet demonstrated. These carcinogenic agents are not demonstrably mutagenic.

B. Phenocopy

The model regulatory circuits devised by Jacob and Monod are entirely imaginary but it is noteworthy that they yield the prediction that changes in circuits of this general kind could produce phenotypic effects closely similar to those produced by changes in the genetic material itself. The circuits supply indeed a plausible representation of the phenomenon of *phenocopy*, first described by Goldschmidt in Drosophilia. Phenocopies are phenotypic abnormalities similar to those that result from mutation in all respects save one—they are not heritable as mutations are. Almost any kind of "shock" effected, for example, by exposure of Drosophila larvae to cold, irradiation or many varied chemical substances, can induce phenocopies. Shocks of many kinds can induce the same sort of phenocopy and one kind of shock can induce a diversity of phenocopies. Goldschmidt maintained that every known mutant of Drosophila had its phenocopy and Stern predicted that a mutant could be found corresponding with every known phenotypic variation. These opinions are consistent with the presumption that phenotypic characters are end results of chains of developmental processes that are linked at one or many points with specific sites on the genome; the same phenotypic effect may be expected to result from alterations of the genetic sites as from interference with the chain of developmental processes linked thereto. There is good evidence that some phenocopies, although perhaps not all, result from disturbance of the same chain of processes as that deranged by the corresponding mutation (Dunn, 1941; Goldschmidt, 1955; Landauer, 1959; Stern, 1955).

The distinction between mutation and phenocopy is based on heritability. In Mendelian genetics heredity applies to the inheritance of characters through the germ cells of sexually reproducing organisms. In these organisms heritability is usually a reliable, if not infallible, criterion of change in the genetic material itself. Phenocopies do not entail a permanent change in the genetic material transmissible through the germ cells. Heritability has not the same significance in bacteria and in unicellular organisms that multiply by binary fission or in the individual somatic cells of metazoa. In all these there is cytoplasmic, as well as genetic, continuity between parent and offspring and although transmissible, seemingly "heritable", phenotypic changes *may* be attributable to mutation in the strict sense they may result also from transmissible self-maintaining changes in the pattern of utilization of an invariant genome. Monod and Jacob remarked that some of the changes of state described by them in their model systems, being dependent on control mechanisms operating directly at the genetic level, should closely mimic "true transmissible alterations of the genetic material itself". In other words the changes might be hard to distinguish from *mutation* mainly because the Mendelian test of heritability through the germ cells is not applicable. In these circumstances a rigorous proof of mutation seems likely to depend on the recognition of differences in the nucleotide sequences of the nuclear DNA or its RNA transcripts.

III. Genetic Fragments and Viruses

A. Bacterial Transformation and Transduction

Much of the important information about genetic materials and their action has come from studies of transmissible fragments of genetic material which in some sense "act as units" although they may comprise several or numerous cistrons and varied organized materials other than the genetic material itself. It is now commonplace to study viruses from this point of view but bacterial transforming factors provided the first, and still provide the simplest, example of transmissible genetic fragments.

1. Bacterial Transformation

The transformation of pneumococcal types, as first described by F. Griffith (1928), consisted of the imposition of a heritable capacity to synthesize a serologically-specific capsular polysaccharide upon unencapsulated pneumococci originally of a different serological type. Much later, Avery and his colleagues showed that the capacity could be transferred by a deoxyribonucleic acid extracted from type-specific encapsulated pneumococci. To quote from a valuable review by McCarty, the observations of Avery, McCarty and others showed that a purified deoxyribonucleic acid isolated from Type III penumococci had the property of inducing predictable type-specific modifications in unencapsulated pneumococci originally of a different serological type. The transformed pneumococci had the capacity to produce serologically-specific Type III capsular polysaccharide. The induced change was permanent and the transformed bacteria retained their capsular structure and type-specificity through repeated subcultures. Furthermore the active substance can be isolated from transformed bacteria in amounts far in excess of that originally used to induce the change, indicating that as well as imitating the process of capsule synthesis it is itself replicated in the cell. The accumulated evidence has established beyond reasonable doubt that the active substance responsible for the transformation is a specific deoxyribonucleic acid (McCarty, 1946).

It should be stressed that the specific DNA confers on the pneumococci the permanent heritable capacity to produce material of an entirely different kind, namely an immunologically specific polysaccharide. Many other characters of "markers" have been similarly transformed in influenza bacilli as well as in pneumococci but transformation of most kinds of bacteria has not been achieved. Apparently every defined character of a pneumococcus can be transferred by an appropriate transforming agent but not all by a single agent. Some linkages have been found but, in general, the various characters are transformed independently of one another, so that penicillin resistance, for example, is transformed without change in the capsular antigens. It is noteworthy that linked characters, transferred together, may bear no apparent biochemical relationship to one another. Resistance to penicillin, streptomycin and other agents noxious to the bacteria are especially valuable "markers" in transformation experiments because it is easy to separate the resistant organisms in pure culture by using the noxious agent to kill all the sensitive ones. To be transformed, the bacteria must be in a state of reactivity or *competence* that changes cyclically and reaches a maximum late in the exponential growth phase. The duration of the competent state is variously estimated but sometimes it seems to last no more than 15 minutes.

From these and other considerations it is inferred that the transforming agents are equivalent to nuclear genes and they either modify a corresponding portion of the genome of the recipient bacterium or substitute themselves for it at the time of replication. The more detailed interpretation is uncertain. The three main possibilities are: (1) the transforming agent becomes attached to a corresponding segment of the bacterial genome as an addition thereto; (2) the transforming agent becomes incorporated in the bacterial genome, a corresponding segment of the normal genome being ejected to make way for it; (3) the specific genetic pattern of the transforming agent is transferred to a corresponding segment of the bacterial genome by a "copy-choice" mechanism. Peacocke and Drysdale (1965) concluded that the second interpretation is the correct one, the transforming agent being incorporated in the genome of the recipient bacterium, presumably by recombination between the transforming DNA and the DNA of the transformed bacterium. The result, however achieved is, in effect a "mutation" consequent on the stable, heritable integration of the genetic *pattern* of the agent in the genome of the recipient bacterium.

The transforming agents provide the most convincing evidence that DNA alone can carry and transmit the genetic pattern, "information", or "code" of the bacterial genome. They show, too, that the bacterial genome can be fragmented, particular fragments bearing specific portions of the total "code". It is evident moreover that a substantial chain of unidentified processes must link the primary transcription of the genetic DNA with the phenotypic expression of a serologically-specific capsular polysaccharide.

2. *Bacterial Transduction*

Transduction is defined as a process in which a bacteriophage carries a fragment of a bacterial genome (other than genetic material specific to the phage itself) from one bacterium to another. The donor bacterium is destroyed during this process but the recipient survives and acquires heritable characteristics proper to the donor bacterium. Bacteriophage plays only a passive role as a vector in this process and apart from the special mode of transfer of genetic material, transduction is closely akin to transformation. It has been studied most intensively in the salmonella bacteria, and in particular in *E. coli*, where transformation comparable with that seen in pneumococci and influenza bacilli has not been demonstrated. In transduction as in transformation seemingly all characters can be transferred but as a rule only one at a time. The transduced characters are permanent and heritable and stable "integration" of the transduced genetic pattern into the genome of the recipient bacterium is inferred. The mechanism of the integration is probably the same as in transformation. (Hartman, 1957; Zinder, 1955).

B. Viruses

Viruses are organized infective particles having a characteristic size and morphology and containing a high molecular weight nucleic acid, either DNA or RNA and a protein as minimum components. Some of them, which have been called *minimal viruses* (Caspar and Klug, 1962), contain no other materials whereas some are much more elaborately composed. According to present beliefs, the nucleic acid component is wholly responsible for the genetic continuity of all viruses and for their actions within cells or micro-organisms, whereas all other components whether a single protein or complex organizations of proteins lipids and carbohydrates, are important only for supplying a protective covering for the nucleic acid, stabilizing it outside cells, and facilitating its transmission from cell to cell and from organism to organism. Within cells the nucleic acid replicates and in addition provides all the genetic information needed to direct the synthesis and assembly of all other components of a complete infective virus particle. The nucleic acid components are described by Lwoff (1960) as ". . . pieces of genetic material carrying the information for the reproduction of organized infective particles". Picken (1960) argues that because viruses are organized entities they can be called "organisms". Be that as it may, viruses are not *living* organisms; they can multiply only within living cells or micro-organisms upon which they are wholly dependent for the energy and materials necessary for their perpetuation and dissemination. In Piries' words, viruses exist only by courtesy of the synthetic mechanisms of a more elaborate host cell (Lindegren and Pirie, 1963). The essence of virus infection is the introduction of genetic material with a new genetic pattern into a living cell or micro-organism; replication of the new genetic pattern and the consequences of the infection depend on a linkage between the foreign genetic material and existing dynamic systems of the infected cell. The special feature of viruses is that in part at least they become incorporated in the material flux of living organisms and during their reproductive period they are part of a living organism. (Picken, 1960).

Even the "simple" or "minimal" viruses are composed of organized sub-units assembled according to physical principles which have been discussed by Caspar and Klug (1962). The minutiae of structure are not highly relevant to the purposes of this chapter, which is concerned primarily with the functional and dynamic aspects of virus infections. Caspar *et al.* (1962)

advanced certain "Proposals" for a nomenclature of viruses based solely on morphology. This may be useful to specialists in virology but unhelpful to biologists in general. The Proposals include the term *virion* for the complete infective particle and *virus* for "all phases of the viral life cycle". Unfortunately, the latter definition reflects the obsolete concept of viruses as independent living organisms; not being "alive" they cannot have "life-cycles". Even more unfortunately, the Proposals include no term for the biologically-decisive unit, the genetic nucleic acid or genome. As Wildy remarked in the Discussion of a related paper by Lwoff *et al*. (1962), the essential features and properties of a virus are, in the last analysis, the expression of the primary structure of its genetic material. The two outstanding biological properties of viruses are concerned with (1) the extracellular survival and dissemination of *infective* particles; without an infective phase a virus cannot be recognized (Lwoff, 1960), and (2) the intracellular operation of the genetic nucleic acid by entry into the organized biochemical flux of a living cell or microorganism. The specific pattern of the nucleic acid component accounts for the intracellular consequences of infection; the properties of the ancillary materials forming a "coat" for the nucleus account for extracellular survival and for the epidemiological and immunological features of the virus disease. The discussion of viruses and virus diseases in terms of static structures is artificial and unprofitable; it is preferable, if more difficult, to attempt to analyse them in terms of what Burnet (1957*b*) has called "a stream of biological patterns".

Until fairly recently the basic conceptions of viruses have been based almost entirely on studies of bacterial and plant viruses. The analysis of animal viruses is now contributing in an important degree to the understanding of the action of genetic fragments within cells and promises to illuminate also the action of the normal genetic materials within cells of the higher animals. The contributions of virology to the understanding of genetic materials and genetic action as well as the wide current interest in the possible virus aetiology of neoplastic diseases justify some further discussion of viruses although the discussion of the special case of the so-called tumour-viruses or oncogenic viruses cannot be pursued until the characteristics of neoplastic development have been surveyed. (Volume 2).

1. *Tobacco Mosaic Virus*

The virus of tobacco mosaic disease is, structurally, one of the simplest of the "minimal viruses", being made up of a core of ribonucleic acid and a "coat" or shell of protein. Within infected cells the virus increases enormously and the synthetic processes of the cells seem to be directed almost entirely to the replication of virus nucleic acid and the synthesis of virus-specific protein and the extreme disruption of normal metabolic processes kills the cells. Tobacco mosaic virus has been of outstanding experimental convenience in pioneer studies of the constitution of viruses partly because of its simple structure and partly because it can be extracted in large quantities and in a comparatively pure state from infected plants. Stripped of its protein coat, the nucleic acid core is unstable but it has been shown, with difficulty, that the nucleic acids alone is infective; it can induce typical mosaic disease in plants and infective virus particles complete with virus-specific protein coats are recoverable in abundance from the diseased plants. Furthermore, "synthetic" strains of tobacco mosaic virus have been made by combining the nucleic acid of one strain of virus with the protein of another strain. The synthetic viruses are infective and, again, complete infective virus particles are extractable abundantly from the infected plants. Both the nucleic acid and the protein of the extracted virus are identical with those of the virus that supplied the nucleic acid of the "synthetic" agent; the added protein from a different source stabilizes the nucleic acid and facilitates its transfer but the nucleic acid alone is responsible for the disease and for the type or virus that is produced. Experimental chemical alteration of nucleic acid bases of the virus has produced a mutant virus with a protein coat having a new and distinctive amino acid composition, showing that the virus RNA carries the information or code for the synthesis of the specific protein of the virus coat (Matthaei et al., 1962; Leslie, 1961).

The investigations of tobacco mosaic virus show decisively that the genetic continuity of the virus and its intracellular action depend on the nucleic acid component. Within infected cells the nucleic acid replicates precisely and through transcription of its genetic pattern, it directs the synthesis of the virus-specific protein coat and it diverts the metabolic activities of the infected cell almost completely to these ends. Tobacco mosaic virus provided, indeed, the first impressive evidence that RNA can behave like DNA as *genetic material capable of replication and transcription.* (Fraenkel-Conrat et al., 1957; Williams, 1956).

Ycas (1962) maintained that transcription of the genetic material for the synthesis of the virus-specific protein coat does not utilize the whole of genetic information available in the virus RNA molecule. The basic unit of the protein has about 157 residues. According to the triplet-coding hypothesis now in favour, this protein can be coded by 157×3 nucleotides of RNA. The virus RNA molecule contains in fact some 6000 nucleotides so that only a fraction of the genetic information can be used for the synthesis of the protein coat and, unless there is a flaw in the argument, the large excess of information is available for other purposes. Ycas noted that, consistently with this inference, some mutant tobacco mosaic viruses produce symptomatically differing diseases although their specific proteins are not demonstrably different. He suggested that the functional difference does not involve protein synthesis but may depend, perhaps, on a repressor action. The important indication is that the virus genetic material may participate in other actions as well as in its own replication and the specification of its own protective covering.

2. *Bacteriophage*

The bacterial viruses ("phages") are constructed on the same general plan as tobacco mosaic virus but they are more elaborately organized and their nucleic acid is usually of the deoxyribose instead of the ribose type. RNA phages exist but the vast amount of work on phages to which brief reference will be made here, refers to the DNA phages. Phages resemble tadpoles in shape having a "head" composed of DNA with a coating of protein and a "tail" composed wholly of proteins. The several proteins of head and tail are of virus-specific types. When a phage infects a bacterium the nucleic acid enters whereas nearly all of the protein of the head and tail stays outside. Such protein as enters the bacterium seems to play no important part in later events; the consequences of infection are attributable to the DNA. (Boyd, 1956; Cohen, 1954; Herriott, 1959; Luria, 1959*a*, *b*).

Phages are divisible into two main groups called *virulent* and *temperate*. Following infection with a virulent phage, there is an *eclipse* phase during which no infective phage particles are demonstrable. The metabolism of the bacterium alters quickly; the synthesis of many enzymes stops and it is no longer possible to induce adaptive enzymes; the synthesis of bacterial DNA stops completely and existing DNA is progressively degraded, but some protein synthesis continues. After a few minutes, phage DNA is detectable and replication of this *vegetative phage* continues until about fifty phage-equivalents of DNA are present, whereupon phage DNA units and phage specific protein, which meanwhile has been synthesized, combine to form infective particles complete with protein coat and tail. The infective particles increase to a critical number and then the bacterium lyses and sets free the mature phage particles. It seems that the DNA of virulent phage completely dominates the economy of the infected bacterium and redirects its dynamic biochemical systems almost completely towards the replication of phage DNA and the synthesis of phage-specific protein. Phage DNA thus carries out the same activities in infected cells as tobacco mosaic RNA does and, similarly, is wholly responsible for the genetic continuity of the virus and for the specific configuration of its proteins.

Temperate phages are less consistently destructive than virulent phages. They effect varied, but probably relatively small, shifts in the metabolic equilibrium of infected bacteria; the chief result may be the synthesis of phage DNA in preference to bacterial DNA but pre-existing DNA is not destroyed and many bacteria recover from the infection.

Temperate phages can establish a symbiotic type of union with their bacterial hosts which is known as *lysogeny*; no infective phage is demonstrable but the capacity to produce infective phage particles is retained as a permanent heritable character of the lysogenized bacteria and is exercised under certain experimental conditions (Bertani, 1958; Lwoff, 1953, 1962). The lysogenized bacteria carry all the genetic pattern or code needed to fabricate a complete infective phage particle. It is inferred that the lysogenized bacteria carry the phage genetic material in a non-infective form known as *prophage* and that the prophage is so linked with the bacterial genome that the genomes of prophage and bacterium replicate together, the prophage behaving as if it were a part of the bacterial genome. The nature of the union between prophage and bacterial genome is not certainly known but prophage is more likely carried as an addition to the bacterial genome than inserted into it by substitution or by a copy-choice mechanism. The attachment to the bacterial genome is not haphazard but restricted to certain specific sites, presumably on account of corresponding or complementary pattern relationships or on the degree of genetic homology between the prophage and a segment of the bacterial genome. It is a plausible hypothesis that prophage originates as fragment of a bacterial genome and becomes attached to the homologous segment of an intact genome. *Prophage is simply phage genetic material.* In lysogenized bacteria, as in the eclipse phase of infection with virulent phage, complete infective phage particles do not exist. During the eclipse phase the genetic material replicates as such; it also directs the synthesis of phage-specific protein. In lysogenized bacteria the phage genetic material seems to be carried as a harmless passenger but it probably exerts a more or less continual effect. Some lysogenic phages confer notable specific characters on their hosts. For example, certain phages confer toxigenicity on diptheria bacilli and the correlation between toxigenicity and lysogenization is complete; if the bacilli lose the prophage they lose also their toxigenicity. Under these circumstances the phage genetic material clearly directs dynamic systems of the bacteria towards ends unconnected with the replication of phage genetic material or the synthesis of phage-specific proteins. Phage particles are believed to carry a number of gene-like sites of mutation and one phage of *E. coli* contains, apparently, a thread of DNA with 50,000 pairs of nucleotides (Picken, 1960). The genetic information in the phage DNA is probably not completely utilized in the fabrication of new phage particles.

Jacob and Wollman (1959) cite prophage as an example of a class of "éléments genétiques ajoutés", which are to be distinguished from the classical genetic determinants, and which they call *episomes*. Episomes are dispensable components of cells; they may be absent and when present they may exist either in a free, independent, state or in an integrated state, attached to their hosts' genome. Prophage, as described by Jacob and Wollman, is an episome that becomes attached to a specific locus on the bacterial genome from which, under appropriate conditions, it can detach and replicate in a vegetative state. Other possible examples of episomes include the sex or "fertility" factor (F) of *E. coli*, the colicinogenic factors of some Enterobacteria and a sporogenic factor in *B. subtilis*. No clear examples of episomes have been recognized except in bacteria (Jacob *et al.*, 1960).

3. Animal Viruses

Historically, animal viruses have been studied primarily as disease-producing agents in animals and in man and a large and important body of information about the pathology and epidemiology of the virus-induced diseases has accumulated. Nevertheless, until recently, basic knowledge about viruses and their effects has come mainly from the more experimentally-amenable viruses of plants and bacteria. Many formidable technical difficulties have now been overcome more or less successfully and new information about animal viruses is being amassed rapidly. It has become evident that the basic general principles derived from the study of bacteriophage and tobacco mosaic diseases are applicable to animal viruses. In particular, the genetic continuity and specificity of animal viruses are vested in the nucleic acid components which carry the whole of the genetic code needed for the fabrication of complete infective virus particles; the nucleic acid component is probably solely responsible for the intracellular effects of animal viruses but the virus coats are important for their extra-cellular survival and dissemination and for their immunological properties. In some respects, however, neither bacteriophage nor tobacco masic virus provides a reliable "model" for the analysis of animal viruses and indiscriminate extrapolation from bacteria and plants to animals and man can be dangerous; animal viruses must be assessed on their own merits (cf. Luria, 1960).

4. Structure

Animal viruses are extremely diverse in structure and behaviour. All of them contain a specific nucleic acid of high molecular weight, which may be either DNA or RNA but never both, and a virus-specific protein; the smaller viruses seem to contain little or nothing else but the larger ones approach the bacteria in complexity and contain varied enzymes, antigenic proteins, carbohydrates and lipids. No single description can apply equally to a large, complex DNA-containing virus like vaccinia and a small, simple RNA virus like poliovirus. Even amongst the RNA viruses there are great divergences in structure and intracellular behaviour between, for example, the small poliovirus and the large influenza virus and other myxoviruses. Furthermore, intracellular behaviour depends on properties of the infected cell, as well as on specific characters of the infecting virus, so that a wide range of virus-cell interactions is found and viruses disturb the activities of the cells they infect to greatly varied extents and in varied ways (Green, 1962; Scholtissek et al., 1962). In this chapter it is not practicable to attempt more than an assessment of the general principles and trends that are emerging from a rapidly expanding and changing field of inquiry; details may be found elsewhere (e.g. Burnet and Stanley, 1959; Various Authors, 1962 summarized by Dulbecco, 1962).

5. Infection

Lwoff (1960) defines *virus infection* as the introduction into a cell of the genetic material of a virus. It is important, as this definition implies, to distinguish clearly between the process of entry of virus material and the consequences of the entry. *Entry* is probably governed mainly or entirely by properties of the virus coat and surface properties of the exposed cells. The first step of *adhesion* or *adsorption* may depend on electrostatic or on chemical properties of the surfaces of virus and cell; adsorption of the complex myxoviruses, including influenza virus is attributed to an interaction between virus agglutinin and receptors on the cell surface. Holland and Hoyer suggested that the presence or absence of surface receptors capable of binding poliovirus is a function of the differentiated state of various cells. Poliovirus multiplies in almost any human tissue cell that has been cultivated *in vitro* but cultured respiratory epithelial cells, which maintained their differentiated state as evidenced by the retention of cilia, were resistant to infection by poliovirus or by an adenovirus. The highly selective action of the poliovirus *in vivo* is more probably due to surface properties governing entry of virus than to differences in the intracellular behaviour of the virus. It may be noted in passing that the surface properties are much altered during culture *in vitro* and that observations on cultured cells may give a highly misleading representation of the properties of similar cells in their natural positions in the living animal. The second step in entry is *penetration* and it does not inevitably follow upon adsorption. The mechanism of penetration is not known. Animal viruses, unlike bacteriophages, have no mechanism for injecting their nucleic acid component into cells, leaving most of their other components outside. The coat of an animal virus is removed at or near the cell surface, during or soon after penetration, by a mechanism which is not known but which may require the synthesis of a new "uncoating enzyme". (Green, 1962; Holland and Hoyer, 1962; Hoyle, 1962; Joklik, 1962; Kerr et al., 1962). Infective nucleic acids have been extracted from RNA viruses and less often from DNA viruses. This artificially "uncoated" virus consisting of naked molecules of nucleic acid is weakly infective but, when derived from some RNA viruses at least, it infects a wider range of host cells than the complete virus particle can do, presumably because immune mechanisms directed against proteins of the virus coat do not come into play (Colter et al., 1957; Colter and Ellem, 1960; Dulbecco, 1962).

Infection, as defined by Lwoff, ends when virus nucleic acid is liberated within the cell. The immediate consequences of infection depend on the genetic pattern of the virus nucleic acid and on the susceptibility of the infected cell. It is evident from experiments with purified nucleic acids that the virus genetic material carries all the genetic information or code needed for the fabrication of complete infective virus particles, but the expression of its genetic capacities depends on the active metabolic processes of the infected living cells and the dependence is absolute (cf. Hoyle, 1962). Moreover, expression of the capacities is not an inevitable consequence of infection unless the infected cells are "susceptible". Burnet (1956) maintained that there must be a specific complementary-pattern relationship between the virus genetic material and some portion of the genome of the infected cell. The interaction between a virus and the cell receptors, whatever they may be, blocks or neutralizes the receptors so that they are no longer available to incoming viruses of the same kind or even of some different kinds and in consequence the cell is no longer "susceptible" to those viruses. This important phenomenon is called *interference* (Schlesinger, 1959).

It remains to note that the relationship between the virus genetic material and the infected cell is a dynamic one, not to be fully comprehended in terms of static materials.

6. *Eclipse*

Immediately after the virus has penetrated and has been uncoated, there is an *eclipse* phase during which no infective virus particles and no virus-specific proteins are demonstrable by any method now available. Eclipse seems to be a general feature of infection by animal viruses (Isaacs, 1959). During eclipse the virus nucleic acid participates in the two activities already described as characteristic of normal genetic materials, namely replication and transcription. Replication yields new units of virus nucleic acids whereas transcription yields the specific components of the virus coat or capsid. Replication and transcription may overlap in time, but they are separate processes taking place in different positions in the cell. The products of replication and transcription are assembled in the cytoplasm or at the cell surface to form complete infective virus particles. Eclipse, which begins with the liberation of virus nucleic acid from infective particles, ends with the assembly of new complete particles. It is clear that infective virus particles do not multiply *as such* within cells during eclipse; new particles are built up from the products of replication and transcription of the virus genetic material.

The details of replication and transcription in cells infected by animal viruses large and small and bearing either DNA or RNA as their sole genetic material are now being studied intensively but many remain obscure and, not surprisingly, they differ from one virus to another and from one type of cell to another (e.g. Green, 1962; Scholtissek, 1962). It is a reasonable working-hypothesis to suppose that the nucleic acid of DNA viruses replicates in the nuclei of infected cells in much the same way as the normal genetic DNA and that transcription is mediated by a mRNA, patterned by the virus genetic material, which becomes attached to normal ribosomes and there specifies the synthesis of virus proteins. In a general way, this hypothesis seems to be satisfactory. The behaviour of RNA viruses is not so easily interpreted. Normally new cell RNA is formed by transcription of DNA and not by replication of pre-existing RNA and, in particular, not by replication of the short-lived mRNA, with which virus-RNA has been compared. Evidence obtained in part from RNA phages (e.g. Spiegelman and Doi, 1963) and in part from small, animal RNA viruses (reviewed by Martin, 1967), indicates that virus RNA is replicated in the strict sense and not synthesized on a DNA template. Martin observes that this is the only known example of the direct transfer of genetic information from one molecule of RNA to another, without the intervention of the otherwise universal genetic material, DNA. On this account and in view of its possible significance in many other biological processes, the phenomenon merits some further attention here.

The first important clue to the nature of the replication of virus RNA was the observation that persistent synthesis of virus RNA was possible in the presence of concentrations of actinomycin D that inhibited normal RNA synthesis, implying that normal RNA and virus RNA were synthesized independently by different mechanisms (Franklin and Baltimore, 1962). More specifically, the observation indicates that replication of virus RNA does not require transcription of DNA, since the action of actinomycin is to block such transcription. Otherwise expressed, the synthesis of virus RNA is not "DNA-dependent". Other evidence in the same sense has accumulated. Virus RNA can replicate in the absence of DNA synthesis and even in enucleated animal cells but it cannot replicate without prior synthesis of protein. The sequence of events in virus replication as now understood and recounted by Martin is summarized in the following paragraphs.

After the virus has infected a cell and shed its coat, the released RNA begins to direct the synthesis of a number of virus-specific proteins. Some of the proteins apparently act as metabolic inhibitors that actively and profoundly repress the synthesis of host cell RNA and host cell protein. Other proteins are needed for the virus coat and others are RNA polymerases without which RNA cannot replicate. The normal genetic mechanisms of protein synthesis being suppressed, the virus RNA must carry the genetic code for RNA polymerases as well as for the proteins of the virus coat. Virus-specific protein synthesis takes place on cytoplasmic polysomes which are larger than those characteristic of uninfected host cells and which replace the normal ones, and it proceeds by the usual mechanism of peptide-binding of amino acids in a sequence specified by mRNA which in these particular circumstances is derived from the virus RNA (Attardi and Smith, 1962; Franklin and Baltimore, 1962; Kerr *et al.*, 1962).

The synthesis of new virus RNA probably begins as soon as the necessary RNA-polymerases have been formed. The precise mechanism of RNA synthesis has not yet been determined. According to one hypothesis it is a two-stage process; in the first stage appositional synthesis of a complementary copy of the single-stranded virus RNA takes place and a double-stranded RNA template is formed. In the second stage, only the negative strand is transcribed. In one variant of this hypothesis, displacement of the positive strand of the duplex is presumed; in another variant displacement is not presumed. Both variants assume that the two stages require different enzymes (Martin, 1967). The main alternative hypothesis presumes that only a single enzyme called a *replicase* is implicated and that the replicase catalyses the synthesis of progeny RNA in one step using the parental virus RNA directly without prior synthesis of a complementary strand (Spiegelman and Hayashi, 1963).

Probably none of the RNA first produced is incorporated into infective virus particles; instead, it attaches to ribosomes to direct the synthesis of more enzymes and more coat protein. Martin suggested that RNA synthesis and protein synthesis are coupled together in an integrated system and that replication entails a cyclical phenomenon in which virus RNA directs the synthesis of polymerase which makes more virus RNA and so on. Interruption of either virus synthesis or protein synthesis will stop both processes. The "maturation" of infective virus particles does not begin until a sufficiently large pool of components has been accumulated and then it is probably a spontaneous process requiring no energy.

Martin concluded that virus RNA has three distinct functions: (1) to act as *messenger* for the synthesis of virus-specific proteins, (2) to act as a template for the synthesis of RNA polymerase and (3) to become the RNA of the progeny virus. The wider implications, which may be of considerable importance in a variety of biological phenomena are that *informational* RNA can be both stable and replicable and that it can provide a stable *source* of information as well as an unstable *transmitter* of information vested in DNA. The term *messenger* is appropriate for the transmitters of information but is not synonymous with *informational* which is more widely applicable.

7. Defective Viruses

The mechanism of assembly of the independently synthesized components to make a complete infective particle is poorly understood. It is not clear how the components come together but once in apposition, the arrangements of nucleic acid and protein of the simpler viruses to form a precisely organized infective particle seems to be governed mainly by crystallographic principles (Caspar and Klug, 1962; Dulbecco, 1962). Larger, more complex viruses, notably influenza virus and other myxo-viruses are much less precisely organized. The assembly of these viruses seems to be an imperfect, not to say haphazard, process allowing wide variations in the relative amounts of nucleic acid and protein included in the complete particle. It is probable moreover, that these viruses acquire a final coating derived from the cell membrane at the time of their extrusion (Burnet, 1956; Hirst, 1962). Many of the particles are demonstrably, in some sense and in some degree, imperfect or *defective*. Recent studies of RNA viruses have shown how qualitative deficiences of the genetic RNA itself can lead to a *defective* virus.

Reichman (1964) described a satellite tobacco necrosis virus which was not by itself infective although it could multiply in the presence of most strains of infective tobacco mosaic virus. He showed by detailed analysis of the RNA and protein components that, within the limits of experimental error, the RNA contained sufficient genetic information to code only for its protein coat leaving none available for the synthesis of materials necessary for its own replication. Presumably tobacco mosaic virus replicating in the same cell provided the RNA-polymerase essential for the replication of the RNA of the "satellite" virus. Other unrelated viruses could not support proliferation of the satellite.

Hanafusa *et al.* (1963, 1964) showed that a strain of Rous sarcoma virus was *defective* in the opposite sense; the virus RNA was capable of replication but incapable of coding for the production of a virus coat and on this account infective particles were not produced. In this example of defectiveness again the defect could be corrected by a "helper virus" replicating in the same cell and, again, the "helper" had to be a closely related virus, namely, one of the viruses of the chicken leukosis group. The authors showed that the defective Rous sarcoma virus became infective by acquiring a protein coat identical with that of the leukosis virus although, after infecting cells, it produced the specific effects characteristic of the Rous virus.

8. Intra-cellular Actions

Burnet (1956) proposed that the intra-cellular action of animal viruses consists primarily in the taking-over of the host cell's synthetic mechanisms which instead of producing host components are directed to producing virus components. Replication and transcription of the virus genetic material is completely dependent on the host cell and necessarily competes with normal cell processes but the consequent derangement of cell activities ranges from slight to lethal. Two factors seem to contribute to the dominance of the virus genetic material. In the first place replication and transcription of the virus genetic material appear not to be subject to the normal regulatory mechanisms and feed-back controls which restrain the replication and transcription of normal genetic materials. In the second place the synthesis of the normal nucleic acids, DNA or RNA or both, seems to be repressed or inhibited in an active way albeit to greatly varied extents by different viruses. Both factors encourage the predominance of virus-directed processes at the expense of normal processes. The repression of the production of normal mRNA may be an important cause of disturbance of normal cell activities and of cell damage (Attardi and Smith, 1962; Dulbecco, 1960, 1962; Kerr et al., 1962; Luria, 1959b; Lwoff, 1962).

In some viruses the amount of genetic information in excess of what is needed to code the formation of genetic constituents and accessories is probably small but in other viruses it may be substantial. There is evidence to show that viruses confer new properties seemingly unconnected with the production of virus components. The "spare" information may provide the information for the production of a new kind of mRNA and consequently of proteins and enzymes not normally present in the virus or in the host cell. This clearly happens sometimes but it remains to be determined how often and to what extent it disrupts the normal activities of cells or damages or kills them. It is possible, too, that viruses can disclose new cell properties, in a less direct way, by de-repression of normally-inactive regions of the genome of the cell. Dulbecco thinks that each of these mechanisms may effect important changes in cell surfaces (Attardi and Smith, 1962; Franklin and Baltimore, 1962; Kerr et al., 1962; Dulbecco, 1962; Luria, 1959a, b, 1960).

It is pertinent to note that the maturation and release of complete infective particles does not entail lysis or notable damage of animal cells as it does of bacteria infected by virulent phages.

9. *Pathological Manifestations*

The manifest consequences of intracellular disturbances caused by viruses in one or other of the ways already discussed are limited in variety although greatly varied in degree. Viruses can damage or kill cells as a result of their "cytopathic" or necrotizing action and they can incite cell proliferation; not infrequently one virus can evoke both cytopathic and proliferative responses. Also, viruses can persist in cells for a long time in a *latent*, apparently symbiotic state, without producing recognizable changes.

It is surprisingly difficult to detect any pecularities in the cytopathic effects, by which they can be distinguished from cell-damage inflicted in many other ways or to distinguish, except in degree, between the overt effects of different viruses (Morgan et al., 1957; Rose and Morgan, 1960). In certain virus diseases cytochemical disturbances of RNA metabolism in the nucleus are conspicuous; similar disturbances are seen, although much less frequently in cells degenerating from other causes. Also, there is often evidence of hyper-function of the Golgi region which it has been suggested may be concerned in the transport of virus-specific proteins from the site of synthesis to the site of assembly of complete virus particles or in mechanisms of intracellular defence against viruses (Adams and Prince, 1959; Love, 1959; Marcus, 1962).

It is still disputed to what extent the proliferation of cells in some virus diseases is a direct consequence of the intra-cellular action of the virus and to what extent it is a secondary response of healthy cells to materials liberated from cells damaged by the intracellular action of viruses (Morgan, 1959). The "inflammatory reactions" evident in many virus diseases are almost certainly non-specific inflammatory responses to dead or damaged tissue and not direct responses to the viruses themselves. Even in the "virus-tumours" there are indications that cell proliferation is not a direct or primary consequence of the intra-cellular action of the inducing virus.

The specificity of the *diseases* caused by viruses is not easily explained by peculiarities in the intracellular actions of particular viruses; as shown in the preceding paragraphs, such peculiarities are hard to find. More probably, the specificity is referable to the selective action of viruses on cells of particular types in particular locations, the selective action being governed mainly by the factors of virus entry and cell susceptibility already discussed.

The *latency* of virus infections that produce no overt disease or cell changes is well established and of great epidemiological importance but it is not well understood. A possible analogy with lysogeny in bacteria has been much in the minds of investigators but they have found little evidence to support it. The only impressive example of an animal virus existing as "provirus" in a state comparable with lysogeny is the virus *sigma*, which produces sensitivity to CO_2 in Drosophila (Luria, 1960). Some tumour-inducing viruses seemingly become "integrated" into the cells they infect but it is not thought at present that the integrated state is closely comparable with lysogeny. An alternative explanation based on immunological tolerance has been proposed for latent infections by the viruses of visceral lymphomatosis of fowls and of lymphocytic choriomeningitis (Hotchin, 1962). Hotchin suggested that the virus of lymphocytic choriomeningitis can enter into a form of cell-virus symbiosis that does not depend on virus "masking" or on lysogeny but on the host developing immunological tolerance against the virus. He suggests further that overt disease caused by the virus is not attributable to the kind of direct intracellular actions to virus discussed in earlier paragraphs but to secondary immunological reactions. On this interpretation the acute disease which the virus may produce in mice is, in reality, an acute hypersensitivity reaction to an intra-cellular antigen. Incomplete tolerance or the breakdown of tolerance by cumulative stresses lead to a hypersensitivity reaction comparable to the homograft response and according to the severity of the response the host may suffer from acute shock or from sustained tissue damage constituting auto-immune disease. It is not clear to what extent these ideas about the intervention of immunological tolerance and its breakdown can be generalized but it is relevant to note that immune reactions like the well-known inflammatory reactions are secondary manifestations of virus infection as it is seen in diseases of living animals; they are not seen in the infections of isolated cells and they are not referable to direct and specific intracellular actions of viruses.

IV. Genetic Materials and Actions—Summary

The foregoing sections are intended to provide no more than a general idea of the impact of molecular biology on the study of genetic materials and actions. It should be noted here that "genetics" no longer deals only, or even mainly, with Mendelian "heredity" and that "genetic materials" are not to be identified with "hereditary determinants". The current view is that the genetic materials operate continuously throughout the lives of cells; they do not become ineffective or redundant once hereditary characters have been established.

The basic innovations are that the essential genetic components of chromosomes are specifically patterned macromolecules of deoxyribenucleic acid (DNA) having a molecular weight of the order of hundreds of millions, that they have the capacity for precise *replication* of their own structure and that they carry a "code" for the synthesis of specific proteins. The outstanding value of DNA as a "hereditary material" depends on its chemical stability and its precise replication. Its functional importance depends on its further capacity for *transcription*, whereby the genetic code is transferred to RNA which carries a complementary copy of the sequence of the nucleotides of the DNA. In another stage of *translation* the nucleotide sequence is decoded during protein synthesis; the nucleotide sequence of the RNA assembles the appropriate amino acids into the correct order for the synthesis of a specific protein. The whole of the DNA molecule engages in replication but only discontinuous segments of it are transcribed. Strictly speaking, translation of the genetic code yields only a polypeptide chain. It is widely, if somewhat too readily, presumed that secondary and tertiary protein structure develops automatically as a consequence of the amino acid sequence in the polypeptide chain. The relevance of the genetic code to developmental processes will be discussed later.

Much of the preceding account is based on studies of bacteria and various genetic fragments and viruses which have disclosed indispensable information about genetic materials and actions that could hardly have been obtained from complex metazoa. Although similar general principles are probably applicable throughout the animal and vegetable kingdoms the *detailed* extrapolation of inferences from observations on bacteria to higher animals and plants requires caution and close attention to certain peculiarities and limitations of bacteria. Bacteria have no nuclei, properly so called. The endoplasmic reticulum of metazoan cells and its specialization to form a nuclear membrane are entirely missing from bacteria. There being no clear separation into nucleus and cytoplasm it is hardly justifiable to call bacteria *bacterial cells*, which, incidently, occupies seven more spaces of type; as proposed by Ris and Chandler (1962), it is preferable to call them *prokaryocytes* to distinguish them from true cells or *eukaryocytes*. The designation of bacterial DNA as a *chromosome* is equally questionable and may be seriously misleading. The genetic material consists apparently of a single molecule of DNA. In *E. coli* the DNA has a molecular weight of about $2 \cdot 8 \times 10^9$ and is thought to form a closed ring. The DNA is believed to be a naked thread lacking the RNA and the histone protein which accompany the DNA in all animal cells except spermatazoa where protamine replaces histone protein. Bacteria contain no morphological chromosomes and no process comparable with mitotic division has been unequivocally demonstrated. Bacterial genes form a single linkage group and are sometimes linked in a sequence corresponding with sequentially operating enzymes. This is notably untrue of animal cells where the genes controlling one synthetic system may be scattered widely through the genome and on different chromosomes (Brenner, 1959; Ezekiel, 1962; Murray, 1960, 1963; Pontecorvo, 1963; Ris and Chandler, 1963; Waddington, 1962).

The operon concept of Jacob and Monod, based on studies of bacteria, has been criticized on various grounds, especially in its application to the higher animals and plants. Jacob and Monod themselves remarked that there was no direct proof of the basic presumption that mRNA carries "structural information" to the sites of protein synthesis. Waddington (1962) and Pontecorvo (1963) have noted the difficulty of applying the operon concept to the sequential operation of genes that are widely scattered through the genome in animal cells. Pontecorvo, indeed, goes so far as to say that "The functional operon is conspicuous for its almost complete absence outside the bacteria and no wishful thinking will create it." The term *regulon* has been proposed to the collection of spatially separated genes acting together on a single synthetic process but, whilst descriptively useful, it is not conceptually an alternative substitute for the operon. (Maas and Clark, 1964; Maas and McFall, 1964.)

Stent (1964) has questioned one of the basic premises of the regulatory mechanisms described by Jacob and Monod by suggesting that regulation may be effected not, as they proposed, at the genetic level by control of messenger *production* but at a later stage by control of messenger *function*. Messenger function on the ribosomes is poorly understood. The "polysome" hypothesis has been disputed (Roberts *et al.*, 1963), and the structure of ribosomes and the function of ribosomal RNA remain obscure. Reference has been made already to the increasing evidence that the informational, messenger RNA in animal cells is not necessarily unstable or short-lived as it seems to be in bacteria; in the RNA bacteriophages and viruses it is plainly stable and replicable.

Although the Jacob-Monod concepts may need substantial modification in detail if they are to be usefully applicable to the cells of higher animals and plants and despite all the criticisms to which they have been subjected, they provide a working hypothesis which remains indispensable, if only because there is no alternative hypothesis of comparable scope, internal consistency and plausibility. Detail apart, the hypothesis embodies several general principles that seem likely to be durable and widely applicable and therefore deserving of enumeration here.

The operon concept provided the first plausible explanation of discontinuities of genetic function along chromosomes in the absence of evident structural discontinuities in the genetic material. The survival of this concept in its original form seems doubtful at present and for the purposes of developmental biology is less important than the demonstration of the more general principle of selective utilization of an invariant genome. The whole of the genome is not in effective use at any one time or place; it can be selectively "activated". Selective utilization is now widely accepted as the genetic basis of *differentiation* in the higher animals and will need repeated attention in later chapters. Meanwhile, a useful distinction may be made here between (1) the *total genome* comprising all the potentially utilizable genetic patterns available in the whole of the genetic material and (2) the *effective genome* comprising the genetic patterns that are in effective use at a particular time and place. In general "effective use" will mean being engaged in transcription (Foulds, 1963).

Another highly important feature of the Jacob-Monod concept is the distinction between *structural* genes and *regulatory* genes. Mendelian genetics has been concerned almost exclusively with structural genes. So far as can can be judged at present, the regulator genes outnumber the structural genes even in bacteria, and their preponderance is the greater the more complex the organism. It may be mentioned here, in passing, that there has been some difficulty in finding enough work to keep the whole of the genetic material usefully occupied. It has been mentioned already that only a fraction of the DNA in bacteria engages in transcription of RNA. The regulatory genes and the principle of selective utilization account for some of the apparent excess of genetic information, but it is not yet clear that they account for all of it. *Redundancy*, in the cybernetic sense, may account for some more. At present, it seems that there is a real problem to be faced.

Perhaps the most important and distinctive feature of the Jacob-Monod concept is the linking together, by biochemical processes, of the structural and regulatory genes into *systems* operating according to cybernetic principles. It combines the static and dynamic components in a single system of biological order and by so doing puts some life back into molecular biology. The study of viruses has shown unequivocally that their genetic material is inactive and inert unless it is incorporated into the biochemical flux of a living cell. This applies to all genetic materials; they generate no activity but direct or impose a pattern upon dynamic biological systems into which they are incorporated. The dynamic order of living cells, which has been less studied than static order, probably because it is more difficult to study, will be considered in the next chapter.

CHAPTER 9

Dynamic Organization

Genetic action must depend on some kind of linkage between the genetic material, whose essential characteristic is stable, replicable, chemical pattern and dynamic biochemical systems whose essential characteristic is chemical flux. Chemically and metabolically DNA is relatively inert; its plausible and indispensable function is to impose *pattern* on dynamic systems that supply all the materials and energy without which DNA would persist only as inert fibres or, as indeed it is said to exist in sperm heads, in essentially crystalline form. The existence of relatively independent dynamic systems has been repeatedly presumed and discussed under a variety of terms; some authors refer simply to "metabolism" and others to self-maintaining reaction cycles, self-perpetuating biochemical patterns, self-regulating metabolic patterns and the like. None of these concepts has proved of much operational value. The problem, as Kacser (1957) and others have pointed out, is the investigation of complex systems.

Biological organization, Weiss (1958c, 1962, 1963) has said, depends essentially on inter-dependency of component reactions rather than any fixed framework and life depends on the *order* of the inter-actions. Metabolism and dynamic activity are essential at all levels of biological organization and chemical processes are coupled with an immense variety of molecular and atomic conformations in which, it is thought, information is stored in coded form (Elsasser, 1958). One outstanding feature of biological organization is its great complexity. It is becoming recognized that complexity is an essential condition of life. The recognition of cyclically linked reactions contributed to the realization that the whole metabolism, to quote Bernal (1962), is a construction of a dynamical kind and of an altogether different order of complexity from anything man has ever conceived in his chemical operations. The complexity is responsible for several of the remarkable properties of organisms, notably the relative invariance of the whole cell or the whole organism in spite of the relative indeterminacy of the component processes and the continual flux of materials. The ability of organisms to *regulate* must also depend on complexity.

Kacser (1957) and von Bertalanffy (1960) amongst others have discussed the chemical kinetics of dynamic biological systems. Their discussions indicate that the laws of chemical equilibrium are inapplicable to the state of the living organism as a whole. The living organism remains approximately constant in a state far from equilibrium; an approach to equilibrium means death. Thermodynamically, says Bertalanffy, the living organism is "a system of fantastic improbability". In spite of the irreversibility of some processes, the organism is maintained in a state of high order and even advances during ontogeny and evolution to states of still higher order. The constant ratio between the components of a biological system is maintained in a state far removed from equilibrium by a steady flow of matter through the system. Unlike most physical systems, biological systems are *open systems* sustained by continual chemical flux and passage of materials into and out of the system. Some of these systems reach a time-invariant *steady state* in which the outflow balances the inflow. The steady-state is re-established after disturbance or biological stimulation and, remarkably, it is independent of the initial conditions and of the previous course of the process, being determined only by the system parameters of reaction and transport. Consequently the same final state can be reached from different initial conditions or in different ways in accordance with what has been called the "principle of equifinality". The theory of open systems accounts also for the phenomena of "falsestart" and "overshoot" familiar in many physiological processes and can be accommodated to development towards states of higher order.

The extreme complexity of biological open systems raises great conceptual difficulties which have not yet been overcome. Northrop (1947) proposed a concept of biological systems in terms of a complicated relationship between chemical entities in continual motion and flux. In his opinion, although chemistry can account for the materials and thermodynamics for the energy, chemistry and thermodynamics, separately or together, cannot account for the particular relatedness into which the energy organizes the moving chemical materials. Northrop believed that the additional requirement is a postulational theory of physics, namely field physics, that will prescribe an irreducible relatedness between physico-chemical entities. More recently, Elsasser (1958, 1961) applied cybernetic principles to the study of biological organization and believed that a specialization of quantum mechanics is needed for its study. Several mathematical approaches were valuably summarized by Waddington (1962) but as yet none of these theoretical discussions seems closely relevant to current practical biological problems. A simpler empirical approach is probably more immediately useful at the present time. The "molecular ecology" of Paul Weiss (1950, 1953) has some valuable features and, despite adverse criticism, has not been displaced by a more satisfactory concept.

It is not likely that complex open systems can ever be adequately specified or described in terms of chemical *substances* or orthodox chemical equations. The following paragraphs attempt to convey some of the most important features of complex open systems with the help of conventionalized diagrams. They cover much of the same ground as "molecular ecology" but they substitute chemical *processes* for chemical *substances* and emphasize the *pattern* and inter-relatedness of systems established by materials in flux.

Orthodox chemical equations reflect little or nothing of the pattern or flux of chemical activity in living organisms but they can be converted into the more adaptable "biological" equation shown in Fig. 67. The diagram

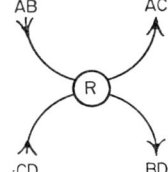

FIG. 67. Basic dynamic chemical equation.

shows two reactants or inputs AB and CD entering a *reaction region*, R, whence emerge two products or outputs, AC and BD. The basic presumption is that the materials in the reaction region are in a state of indeterminate flux. This presumption is consistent with the argument of Elsasser who maintained that, when two systems interact on the physical level, it is no longer possible to discriminate between them as logically distinct parts throughout the reaction, which is governed by uncertainty relations. If two molecules collide and part again resulting in the same or two different species of molecule, one cannot consider either the molecules or their constituent atoms as logical entities during the collision, which results in an increase in the total statistical indeterminacy (Elsasser, 1958).

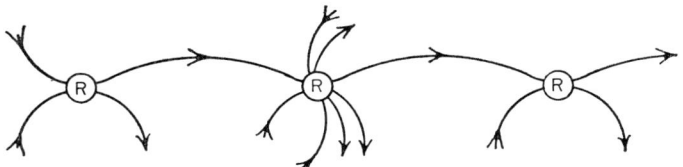

FIG. 68. Chain reaction.

The dynamic formulation is convenient for representing multiple inputs and outputs and for chain reactions (Fig. 68) and, more importantly, several reactions can be combined together to form *cycles*. The cycle shown in Fig.

69 represents an *open system* whose organized pattern is maintained wholly by the *flux* of materials, the flux being sustained by the continual inward and outward flow of materials at the periphery. The inputs and outputs are identifiable as chemical substances but the materials in flux in the cycle itself are not. In notable confirmation of this formulation, no measurable amount of any "intermediate" of the Krebs cycle is detectable during the reactions and, more generally, the intracellular concentration of most intermediary metabolites is vanishingly small (Monod and Jacob, 1961; Picken, 1960). It is predictable from Fig. 69 that the intermediates will exist only transitorially, if at all, *between* reaction regions and will be detectable only by breaking the cycle. Biochemical methods can identify the materials passing in and out of the cycle but cannot determine the flux and relatedness of materials within it, because the methods of analysis necessarily disturb or break the cycle.

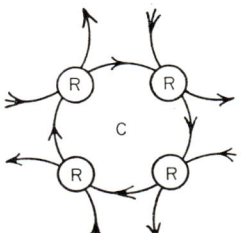

FIG. 69. Cyclic reaction.

Cycles can be combined with one another and with chain reactions to make complicated organized patterns of chemical activity. The example shown in Fig. 70 is an arbitrary arrangement and, for reasons that will be mentioned later, essentially artificial but it is convenient for illustrating some important biological concepts that depend upon, or are at least closely related to, the "open-system" organization. The whole system of Fig. 70 is an open system. The diagram illustrates how dynamic systems within the whole can be interlinked so that, for example, they can be coupled together by shared substrates and by the output of one system acting as the input of another (Kacser, 1957; Monod and Jacob, 1961). It illustrates also some basic conceptions of biological organization that need fuller discussion.

The central cycle (C) of Fig. 70, in a sense the "keystone" of the whole system, exists only as a pattern of materials in motion; if the motion stops, the pattern falls to pieces just as cells, when they die, lose their organized structure, autolyse, and disintegrate. This formulation suggests a new approach to an old problem by implying that biological organization is maintained *by and because of* flux of materials and not, as it was formerly expressed, *in spite of* that flux. It illustrates also the elements of a biological "field" theory. It represents a "field" whose organized pattern is established by, and wholly dependent upon, the relatedness of materials in motion. Biologists have often invoked "field" theories to account for the organized unity of the living organism but, as Waddington (1956) remarked, it is all too easy to use "field theory" as a sort of "joker" in the pack by means of which almost any situation can be dealt with. Some biologists seem to think of a "field" as an extrinsic system of forces that constrains materials to arrange themselves

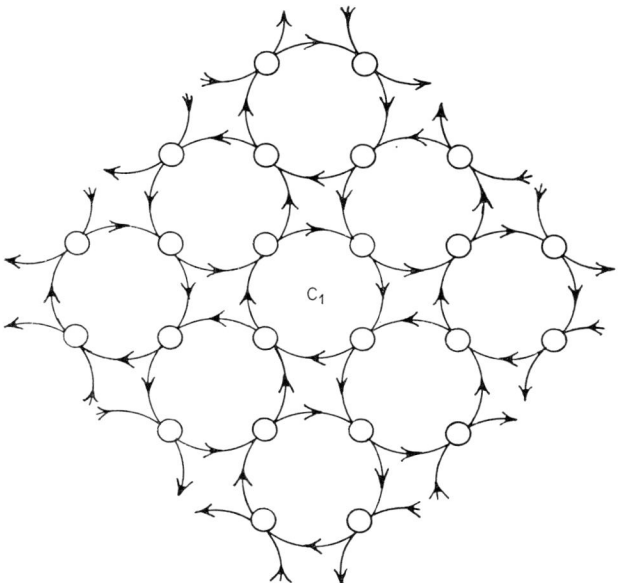

FIG. 70. Hypothetical dynamic system with "network" characteristics.

in an organized pattern, like iron filings between the poles of a magnet, the source and nature of the forces, which seemingly are presumed to be external and superior to the material constituents, being unspecified. A biological "field" becomes more significant when used for example by Waddington and by Weiss to imply that a number of processes interact with one another in such a way that they take up definite relations in space or by Northrop (1947) to refer to organized patterns established by materials in motion. The "field" properties of Fig. 70 are the consequences of the motion of the material parts and not of an extrinsic system of forces imposing order on those parts. The organized pattern and the "field" are co-extensive and inseparable; also, they are self-perpetuating so long as materials and energy pass in and out of the system in appropriate quantity and form.

The discussion suggests that the "control" of organized growth and structure should be viewed as an *intrinsic*, necessary characteristic or consequence of the dynamic pattern of open systems and not as an imposed mechanism for checking a natural trend towards disintegration. "Life is integration" said Sinnot; protoplasm seems to be inherently integrative and there is a persistent urge to bring all parts together into a unity (Sinnott, 1939). It may be added that there is a "persistent urge" to restore the unity when it has been disturbed by disease (e.g. Cameron, 1952). It will be necessary to discuss at a later stage the proposition that neoplasms have "escaped" from normal "controlling forces" of unspecified nature but it is worth mentioning here the possibility that the proposition is based on a superficial or mistaken interpretation of biological organization.

It is worth mentioning also that "information" can be carried in dynamic systems as well as in static genetic material. Hotchkiss (1958) has referred to the kind of information "derived from the impetus of a moving system in an active state" and King (1961) has compared the dynamic information with the storage information that disappears from a computer when the current is shut off. As yet, the significance of dynamic information has not been far explored but it may be substantial.

The idealized diagram of Fig. 70 is far removed from reality in at least two respects: first, it ignores non-cyclical chain reactions and irreversible reactions and second it omits the stable macromolecular constituents of all living organisms. Both Kacser (1957) and Elsasser (1958), who deal primarily with the kinetic aspects of organisms, emphasize the essential importance for "biotonic" behaviour of the linkage of dynamic systems with stable macromolecules. Kacser states that many of the macromolecules of proteins, nucleic acids and polysaccharides in organisms comprise a network of interlinked units and are stable and persist for a long time because of their extremely slow rate of change and not because they are being constantly renewed. Such systems are not in dynamic relationship with the components of dynamic systems but react through their surfaces with those components. When a macromolecule exceeds a certain size, it may gain the property of adsorbing smaller units. In view of the nature of the forces involved, mainly short-range coulombic and van der Waal forces, adsorption to the surface will be more or less selective. Surface configuration will determine which kinds of molecule will be able to approach close enough to be adsorbed. The effect of adsorption of small molecules is to alter their reactivity; their rates of reaction are increased so that the adsorbent molecule acts as a catalyst (Kacser, 1957). To bring Fig. 70 into closer correspondence with reality, it is necessary to introduce the relatively stable macromolecular component. Probably all the hypothetical *reaction regions*, symbolized by the small circles in Fig. 70, have a macromolecular basis which serves to catalyse the interaction of the small molecules of the dynamic systems and to stabilize the spatial relationships of interacting dynamic systems. The macromolecular component, in truth, forms a complex arrangement of fibres and membranes. It is proposed to describe such systems, comprising dynamic biochemical systems linked with a macromolecular skeleton as *biotonic* systems. The term "biotonic" with the special connotations defined by Elsasser (1958) to my knowledge, has not been used again in biological literature during the ensuing ten years. It seems permissible therefore to revive it now for use in the much broader sense here proposed. The term still implies that the dynamic systems have peculiarities restricted to those of living organisms but without specifying in detail the nature of the pecularities.

The hypothetical system represented by Fig. 70 conforms well with tho conclusion, drawn from pharmacological studies by McLean *et al.* (1965), that the structure and biochemistry of the cell can be likened to a net. If a net is pulled at any point, all the links will be distorted. Some links, being weaker than others, will tend to break irrespective of the point at which the original stress is applied. Each tissue has its characteristic network but although the initial reactions to toxins certainly vary from one tissue to another, possible owing to variations in metabolizing enzymes, there is a basic similarity in the pattern of response to a great many toxic agents.

E. A. Needham (1964) reached a similar conclusion, again from a completely different starting point. Needham pointed out that the complete pathway from a substance A to B and back to A by another route constitutes a cycle which combines the virtues of one way traffic with potentially indefinite persistence. The network may intermesh with other nets to form complex systems, the whole of which can be controlled at one specific point. Cyclic systems of this kind seem to be the essence of life; in them lies the dynamic permanence which alone can be equated with life.

Needham drew attention also to the probable importance in embryogenesis of the degrees of dependence and independence between the components of such systems. The important principle of *multiple assurance*, first proposed by Huxley and de Beer (1934), requires flexible connections between the components of a process and a degree of independence between them. The main virtue of this organization is that it prevents any one component of the process from being too vulnerable to fortuitous errors in a single controlling factor. The resulting flexibility is a conspicuous feature of living cells. The flexibility, and the network organization, it may be added, are probably at the root of the capacity for *regulation* and *adaptability* which is a characteristic feature of living organisms.

The diagram clearly provides for feed-back regulations of complex kinds. It may be recalled here that *complexity* is an essential condition of life. This is relevant to a recent discussion by B. E. Wright (1966) of the multiple causes and controls in differentiation. Variety and interdependence of causes, may be essential to the stability of morphogenesis. In dealing with complex phenomena brought about by varied and interdependent causes, a search for a single cause or trigger mechanism, as Wright says, can only delay our understanding of the problems involved. Embryonic induction, for example, is certainly *not* a trigger event and it is almost impossible to tell where it begins.

No doubt, at the cost of some labour, Fig. 70 could be elaborated to represent the relatively static macromolecular components of the systems and the compartmentalization of cells as illustrated by the various organelles which will be considered in the next section, but is doubtful if this exercise would be worth while. The present intent is to emphasize the basic organization of dynamic processes into complex open systems which has a far higher claim than DNA to be called the "secret of life", and which, unfortunately is much more difficult to study. A fundamental difficulty in biological investigation is to analyse living processes without destroying life. At the risk of tedious repetition, the complexity of biotonic systems is an essential condition of life; the reductionest analytical methods designed to evade the complexity can succeed only in producing an illusory simplicity which means death.

CHAPTER 10

Cytoplasmic Organization

On several grounds, it has been necessary for a long time to credit "cytoplasm" with a high degree of organization although most of the detail could not be detected without the help of electronmicroscopy. The concept of a homogeneous cell cytoplasm "corresponds with no aspect of reality but only with a particular level of analysis" (Picken, 1960) and it is becoming less and less possible to regard the cell as "a bag of enzymes milling around in search of each other" (McQuillen, 1965). The basic ultrastructure of cells is the same in all organisms from the protista up to man and its main features will be next considered briefly. For more detailed information and discussion of the many variations in special cases, reference may be made to reviews and atlases by Fawcett (1966), Porter and Bonneville (1963) and Rhodin (1963).

I. The Ultrastructure of Cells

A. THE ENDOPLASMIC RETICULUM

In addition to the long-familiar mitochondria and Golgi apparatus, the cytoplasm has a complicated system of membranes, vesicles, cisternae, tubules and fibrils making up the *endoplasmic reticulum* or *ergastoplasm*. (The terms are not strictly synonymous but in the present context the rather subtle distinctions that they imply are not important.) The membranes enclose spaces so as to separate an internal phase of the cytoplasm from an outer or continuous phase which constitutes the cytoplasmic matrix of the cell. In Porter's opinion, the membrane-limited units probably represent various and varying packets of metabolites and enzymes whose movements do not conform with the usual laws of diffusion. Two kinds of membrane, called rough and smooth, are distinguishable. They are continuous with one another. The *rough* membranes are studded with particles of ribonucleoprotein known as *Palade granules* or *ribosomes*. The granules are always situated on the outer side of the membranes in contact with the cytoplasmic matrix. The fibres and membranes, as well as the granules, contain RNA. The rough membranes and, in particular, their ribosomes seem to be the most important sites of protein synthesis; they are most conspicuous and abundant in growing cells, in secreting cells and, in general, in cells evidently synthesizing proteins. The *smooth* membranes are conspicuous in cells that maintain a constant form; they predominate in striated muscle cells and are common in cells that synthesize steroids, in liver cells storing glycogen and in pigmented epithelial cells of the retina.

Electron microscopy has settled many age-long disputes about the structure and even the very existence of the Golgi apparatus. It consists of vesicles limited by membranes free from granules and is reasonably interpreted as a localized specialization of the endoplasmic reticulum. It probably serves for the transportation, segregation and eventual secretion of materials formed elsewhere in the cell as some cytologists had inferred many years ago.

The endoplasmic reticulum is continuous with the nuclear membrane and the nucleoplasm is continuous with the cytoplasmic matrix through pores in that membrane. The endoplasmic reticulum, the Golgi apparatus and the nuclear membrane are now thought to constitute a single system of membranes of which the nuclear membrane is the most stable portion. Porter (1961) summarized the probable functions of the endoplasmic reticulum under four headings: (1) the synthesis of proteins and their segregation for export from the cell, the rough membranes and the Golgi apparatus being especially concerned in these activities; (2) the intracellular transport of metabolites by a mechanism that awaits clarification and the widespread, but patterned, distribution of enzyme-rich surfaces or membranes; (3) other synthetic mechanisms in which the smooth membranes carry enzymes and metabolites that are important in physiological events taking place within localized regions; the synthesis of steroids and the storage of glycogen probably belong to this category; (4) the conduction of impulses within cells which is a possible function in muscle cells. Porter suggested that the form of the endoplasmic reticulum is a rather precise indicator of cell "differentiation" for certain recognizable functions and, no doubt, for others not yet recognized. He suggested, further, that the endoplasmic reticulum constitutes a mechanism of differentiation rather than being a product of it.

Palade (1956) suggested that intracellular integration depends on a suitable and precise disposition of enzymatic units and that the structures revealed by electron microscopy provide the support needed for establishing an arrangement of functionally interdependent units in such a way as to facilitate, co-ordinate and control their activities and to guide metabolites through appropriate channels and prevent interference amongst different synthetic activities. Some of the best evidence for the general validity of this interpretation has come from studies of mitochondrial membranes and will be discussed in the next section.

The endoplasmic reticulum is not a rigid or particularly stable system as electron photomicrographs might suggest it to be. It varies with physiological activity of the cell and, therefore, must be considered to respond to a form of feed-back regulation. In some circumstances, as, for example, after prolonged fasting, it is absent or vestigial. Moreover, the endoplasmic reticulum including the nuclear membrane becomes fragmented and much of it disappears during mitosis. It is not clear how it is reconstituted at the end of mitosis. Fragments of the old nuclear membrane probably contribute to the new structures but to an uncertain extent and the process seems to be a rather random one. Porter has suggested that the system possesses a degree of self-determination in that the pattern it assumes preferentially in any particular type of cell may depend on properties such as chemical composition and organization within the system itself. (Anfinsen, 1959; Porter, 1961; Porter and Machado, 1960; Weiss, 1958c, 1962).

B. MITOCHONDRIA

Mitochondria are not, like the Golgi apparatus, specialized portions of the endoplasmic reticulum but distinct organelles visible in the light microscope. In electron photomicrographs they are sharply outlined by a double membrane. The outer membrane is smooth whereas the inner is involuted and the infoldings or *cristae* protrude into a central matrix. Mitochondria are present in all eukaryocytes but not in bacteria. The present account is limited to two or three of their features that are of especial interest and importance in a general study of biological organization; other details are readily available in general text-books or in more specialized reviews (e.g. Novikoff, 1961; Parsons, 1965; Wilkie, 1964).

The principal function of mitochondria is to generate a supply of adenosine triphosphate (ATP) that can be used in many enzyme reactions and active transport mechanisms throughout the cell. The adenosine triphosphate is produced by oxidative phosphorylation of the diphosphate (ADP). The energy for the synthesis of ATP is obtained by the oxidation of certain substrates including some intermediates produced by the Krebs citric acid cycle, the oxidation proceeding through a series of cytochromes, flavoproteins and dehydrogenases. The respiratory, phosphorylating and active transport systems seem to be localized on or in the inner mitochondrial membrane (Parsons, 1965). Palade (1956) suggested that repeated oxidative and phosphorylating enzymatic chains or patterns are woven into the texture of mitochondrial membranes as repeated decorative patterns are woven into a sheet of damask. From the biochemical view-point, Keilin (1959) maintained that enzymes are not free within cells but are linked by more or less stable bonds to other components of catalytic systems of which they form integral parts. An enzyme within a complex catalytic system is much more stable than when it is isolated and kept *in vitro*. Physical disintegration of mitochondria yields a typical cytochrome system whose activity depends not only on the integrity of each of its components but also on the structural integrity of the particle that supports them and assures their proper spatial orientation and mutual accessibility.

Several other authors have similarly emphasized the importance of the spatial organization of enzymes by their incorporation into intracellular structures for ensuring their integrated functioning and they have rejected the old idea that enzymes operate as freely diffusible molecules in a "bag of enzymes" (Lehninger, 1956; McQuillen, 1965; Novikoff, 1961; Siekevitz, 1962). The mitochondrial system demonstrates that it is not necessarily possible to infer the properties of an organized unit from a knowledge of the properties of fragments of it (Picken, 1960).

The finding of RNA and, more particularly, of DNA in mitochondria has led to a reconsideration of several long-discussed problems. Several years ago DNA was demonstrated in preparations of mitochondria obtained by differential centrifugation of cell extracts but its significance was not at first apparent; it was small in amount and the method of extraction did not decisively exclude contamination with nuclear DNA. It now seems to have been established beyond reasonable doubt that the DNA under consideration is *not* a contaminant but is an intrinsic component of mitochondria themselves.

The mitochondrial DNA differs from nuclear DNA in several of its physical properties and in its composition. It is now believed to be single-stranded and to have a circular configuration and it seems to be unaccompanied by histone protein. In these respects it differs from the DNA of all eukaryotic cells but closely resembles the genetic DNA of bacteria and viruses, and Nass, Nass and Afzelius have proposed that the mitochondrion is a homologue, or even a derivative of a prokaryotic organism, which is partly under control in a eukaryotic cell. Other observations are more consistent with a derivation from nuclear DNA. In recent experiments, du Buy and Riley (1967) studied the possibility of hybridizing mitochondrial DNA with nuclear DNA. Their observations indicated that mitochondrial DNA carried sequences of nucleic acids similar enough to those of fragments of nuclear DNA to allow those fragments to hybridize with mitochondrial DNA 40–70% as effectively as with homologous nuclear DNA. The findings indicate a close relationship between the two kinds of DNA but not complete identity. The physical properties of the mitochondrial DNA imply that it had a higher degree of homogeneity than nuclear DNA and a lower molecular weight. The available information seems to be consistent with an homology or at least closely similarity between mitochondrial DNA and a portion of the nuclear DNA (du Buy and Riley, 1967; Horowitz and Metzenberg, 1965; Kroon *et al.*, 1966; Nass, 1966; Nass *et al.*, 1965; Wilkie, 1964).

The function of mitochondrial DNA had not yet been demonstrated unequivocally but observations made in at least three different laboratories show that actinomycin D inhibits the synthesis of RNA and of protein in isolated mitochondria. Presumably, therefore, the genetic code of the mitochondrial DNA undergoes transcription and translation. The phenomena dealt with in the next paragraph imply that it also replicates.

1. Proliferation of Mitochondria

At fertilization the mid-piece of the spermatazoon carries male mitochondria into the ovum and the "diploid" number of mitochondria per cell remains approximately constant through all successive cell divisions. Clearly the mitochondria must double in number at each cell division and the DNA must double in amount, but it is not clear how these results are achieved.

The multiplication of mitchondria as a result of division by transverse splitting has been described repeatedly but it has not been decisively shown that in the generality of cells all mitochondria originate by fission of pre-existing mitochondria. In the course of cell division the mitochondria break down, just prior to nuclear division, into fragments with a reduced optical density which are randomly shared out between the daughter cells; new mitochondria begin to form at the end of nuclear division. Despite the lack of decisive evidence of a material continuity of microscopically-visible mitochondria from mother cell to daughter cell, structural *pattern* must somehow be transferred (Picken, 1960). The DNA component, at least, can hardly be generated *de novo*. In view of the distinctive characteristics of mitochondrial DNA, it is difficult to belive that it can be increased in quantity by any process other than replication of pre-existing mitochondrial DNA. This being so, it is reasonable to suppose that a specific DNA is transmitted from cell to cell and that the transferred DNA carries genetic information for the fabrication of a new mitochondrion. It may be recalled that the structural pattern to be reconstructed includes an elaborate organization of enzymes on mitochondrial membranes. It is doubtful if the genetic information carried in mitochondrial DNA can or should be credited with the ability by itself to order a complex process of *morphogenesis*. Some observations by Luck (1965) indicate that, in fact, there is a continuity of material other than DNA from cell to daughter cell.

Luck's observations were made on a choline-requiring mutant of *Neurospora crassa*. Radioactively-labelled choline was fed to the cultures for a while and then replaced by unlabelled choline. The label was retained in the mitochondria as a stable marker of "old" mitochondria. It persisted through three subsequent doubling-cycles at the end of which it was found to be distributed randomly through *all* the mitochondria; there was no division into "old" mitochondria with labels and "new" mitochondria without labels. This result is inconsistent with the *de novo* generation of mitochondria during cell proliferation. Luck's hypothesis is that mitochondrial mass grows by the addition of new material to an existing framework and the mitochondrial population increases by division.

Luck notes that his observations do not indicate the site of synthesis of the macromolecules needed for the growth of mitochondria or the form in which they are added to the pre-existing framework. Luck does not discuss the nature of the "pre-existing framework" or the ordering of the materials to make organelles with organized structure. Wilkie (1964) was of the opinion that the available evidence indicates that the formation of mitochondria requires the interaction of nuclear genes with a cytoplasmic template. In his view, therefore, the mitochondria are not "self-reproducing" in the same sense as genes are considered to be "self-reproducing". Nevertheless, one constituent of mitochondria, namely the DNA, is "self-reproducing" in that sense; more accurately, it is a replicating component of mitochondria and in theory, at least, it provides for a genetic continuity between successive generations of mitochondria and it may carry a genetic code for the synthesis of other components. In most animal and plant cells mitochondria clearly differentiate along with the cell and acquire different morphological characteristics in different tissues (Swift, 1965). These differences are apparently transmissible to daughter mitochondria and may, but need not necessarily, have a genetic basis.

The assembly of the components to make a structurally ordered and functional mitochondrion is an example of intracellular *morphogenesis*, the development of structure and form. Bacteria do not show even the rudiments of morphogenesis (Pontecorvo, 1963) and molecular biology has not yet contributed significantly to its study. Weiss (1958*c*) has maintained that structure is inconceivable in the absence of structuring processes. At least three possible types of structuring processes may operate in the morphogenesis of mitochondria. The most likely process is appositional growth on a pre-existing model which determines the *pattern* of deposition of new material (Sonneborn, 1966). To over-simplify, a structural element may "grow" rather like a crystal can "grow". This kind of process seems to be implied by Wilkie and by Luck in their references to a "cytoplasmic template" and a "pre-existing framework". Other possible structuring processes are the patterned biotonic systems of the cell cytoplasm regulated by the nuclear genome and similar biotonic systems directed by intrinsic genetic DNA of the mitochondrion itself. For a detailed examination of the complex problems of morphogenesis in cells and tissues reference may be made to Waddington (1962).

The reconstitution of the endoplasmic reticulum after its disintegration during mitosis and in certain other circumstances poses similar morphogenetic problems. It may be remarked that the stability of structure, which is more pronounced in the mitochondria than in the endoplasmic reticulum, is a dynamic rather than a static one. It is well known that the contents of cells can be stirred up and their interrelationships severely disturbed by microsurgical procedures without apparent effect on the future life of the cells. Inferentially, the cell contents maintain their relative positions under the action of continually-acting forces and not because they are attached to a rigid framework. It is wished to emphasize here only that formidable problems of *morphogenesis* are encountered at every level of analysis of biological organization; they are often insufficiently heeded in embryology and almost completely ignored in the experimental study of neoplasia.

C. CHLOROPLASTS

Chloroplasts, the organelles that carry the photosynthesizing apparatus of green plants, closely resemble the mitochondria in several important respects; they have been investigated along similar lines and with similar results. It seems to be undisputed that chloroplasts develop always and only by division of pre-existing chloroplasts. They contain DNA and RNA and can synthesize RNA and proteins. The chloroplast DNA seems to be a partial replica of nuclear DNA; it replicates in the same way as nuclear DNA but not synchronously with it. This DNA has been called the "genome" of the plastid and presumed to be responsible for the transmission of specific properties of chloroplasts from generation to generation. Differences can be detected between the chloroplasts from closely related species and these, being stable and transmissible through successive generations, are presumably determined by the chloroplast "genome". The morphogenetic aspects of chloroplast development do not seem to be receiving much attention (Chiang and Suedka, 1967; Gibor, 1965; Granick, 1963; Swift, 1965; Wilkie, 1964).

Most authors assert that chloroplasts are self-reproducing and autonomous or semi-autonomous organelles. These assertions may need some qualification. Without doubting that the chloroplasts of algae carry genetic information, Wilkie maintains that they cannot be considered autonomous organelles because they do not possess the complete "blue-print" for their own reproduction; action of the nuclear genome is also necessary for some details. In higher plants, at least, the hypothetical genetic units of the chloroplasts are not independent of the nuclear genome. Some other evidence relevant to the discussion of the "autonomy" of mitochondria and chloroplasts is given in the next chapter.

CHAPTER 11

Biological Organization in Unicellular Organisms

Biological organization, as already argued, depends on an interplay of static genetic elements, nuclear and extranuclear, and dynamic biotonic systems. It is convenient, as well as logical, to begin a discussion of this interplay with a consideration of the organization of unicellular organisms whose special features have been exploited to illuminate some principles of wide generality. After a short discussion of enucleation and nuclear transplantation experiments, observations on the remarkable ciliated protozoa are reviewed in some detail.

I. Enucleation and Nuclear Transplantation

A. ENUCLEATION

Experiments on unicellular animals, as well as on cells of vertebrate animals, have shown consistently that a nucleus can be transplanted successfully from one cell to another on condition that the nucleus is moved directly from cell to cell without exposure to the ambient medium; even momentary exposure kills the nucleus. An isolated nucleus is not viable (Brachet, 1961; Briggs and King, 1959; Lorch and Danielli, 1953). Cytoplasm, by contrast, can live and function for a while and sometimes for a long time.

Enucleated protozoa behave variously but biological activities never stop immediately and completely. Metabolism continues for a longer or shorter time. Some enzymes disappear but others persist or even increase in amount and some synthesis of RNA and protein continues. Cell movements may continue but usually to a limited extent. Enucleated amoebae or anucleate halves of amoebae may survive nearly as long as nucleated controls provided that the animals are kept short of food. Respiration continues indefinitely, owing, no doubt, to the persistence of the mitochondria which seem to be little affected by the removal of the cell nucleus. To this extent, mitochondria, once formed, are "autonomous" and independent of the nucleus. The endoplasmic reticulum is much less stable and is severely damaged by enucleation. There is no evidence that cytoplasm ever completely lacks powers of synthesis. The intrinsic activity and stability is substantial but no unlimited; in general, constructive synthesis soon stops in enucleated cells, growth and adaptive modifications become impossible and eventually the cells die (Brachet, 1961; Picken, 1960; Wilkie, 1964).

Many years ago, Hammerling (1953, 1963) described the long survival, sometimes for as long as several weeks, of anucleate fragments of the green alga, *Acetebularia*. Moreover, the fragments had a surprising capacity for morphogenesis; differentiation of the apical region could produce not only a stalk but also one or more whorls and one or even two caps. Hammerling ascribed the morphogenic capacity to "morphogenetic substances" which, he believed, were produced by nuclear genes and stored in the apex of the alga. The observations have been fully confirmed and extended and recently they have been re-evaluated in the light of the more recent knowledge about genetic materials and their actions and about the presence in chloroplasts of DNA that can be transcribed and translated for the synthesis of RNA and proteins.

Recent observations show that photosynthesis and respiration continue for a long time in anucleate fragments and provide carbohydrates and energy-rich phosphates that are available for use in anabolic processes (Keck, 1961). Protein synthesis, including the synthesis of several enzymes, can proceed for two or three weeks in the absence of a nucleus. Morphogenetic activity covers a similar period. The long survival and sustained photosynthesis are due, no doubt, to persistent functioning of the chloroplasts. Brachet and his colleagues attributed the morphogenetic capacity to an accumulation of informational RNA which they suspected was equivalent to Hammerling's "morphogenetic substances". They thought that the measurable but small amount of DNA in chloroplasts might serve as a template for the synthesis of chloroplast RNA and proteins but that the informational RNA responsible for morphogenesis was transcribed from nuclear DNA. They recognized, nevertheless, that this informational or mRNA had to be credited with a long life of two or three weeks, which is now easily credible in view of recent evidence already mentioned for stable, informational RNA and that its origin by transcription of chloroplast DNA might have to be considered.

In the course of these experiments Brachet and his colleagues found that in *nucleated* portions of *Acetabularia*, actinomycin-D inhibited growth and morphogenesis, and more surprisingly, that *momentary* suppression of protein synthesis by puromycin led to frequent heteromorphoses. Interference with the *orderly* synthesis of proteins apparently caused serious developmental abnormalities, a finding which may be significant in the study of morphogenesis (Baltus and Brachet, 1963; Brachet, 1963, 1964; Brachet et al., 1964; Denis, 1964a, b, c; Keck, 1961; Wilkie, 1964).

The evidence that has been reviewed indicates that "cytoplasm" has a substantial measure of intrinsic stability and activity; some cells at least can grow, regenerate and differentiate to a limited degree in the absence of a nucleus. There is no evidence that "cytoplasm" ever completely lacks intrinsic powers of synthesis. The inference from observations on bacteria that protein synthesis in the cytoplasm depends always on a continual supply of short-lived mRNA from the nucleus is no longer tenable. Nevertheless, none of the evidence controverts the opinion that, *in the long run*, the nucleus is indispensable and irreplaceable (Picken, 1960). None of it permits the inference that phenotypic characters can be established in the first place or substantially altered afterwards, except in the sense of degradation, in the absence of a nucleus. On the other hand, as Picken maintains, it is not justifiable to refer the generation of all attributes of a differentiated cell in a multicellular organism to the nucleus; the operative system is a nucleus in a cytoplasm.

It is worth noting that the *maintenance* systems essential for life are more stable than the *differentiative systems* and are differently located. Important maintenance processes take place in the mitochondria, whereas differentiative processes are located in the more labile endoplasmic reticulum.

B. Nuclear Transplantation

By transplanting nuclei of *Amoeba proteus* into enucleated cytoplasm of *Amoeba discoides* and *vice versa*, Lorch and Danielli produced "hybrid" amoebae having characters intermediate between those of the two parent species. Evidently both nucleus and cytoplasm powerfully influenced certain morphological and physiological characters and, if either tended to predominate, it was the cytoplasm with the exception that the capacity to adapt to antiserum seemed to depend entirely on the nucleus. The results were not considered inconsistent with other evidence to the effect that the nucleus determines the range of macromolecules that can be synthesized, whereas the cytoplasm and the environment influence the decision as to which macromolecules within that range are actually present at a given time (Danielli, 1959*b*, 1960, 1963; Lorch and Danielli, 1953).

Clones of some of the "hybrids" were maintained for many years. The influence of the cytoplasmic component seemed to stabilize within a few divisions following nuclear transplantation; thereafter, phenotypic characters showed no further tendency to approximate to those of the species that supplied the nucleus or the cytoplasm even after many generations extending over as long as eight years. At the end of this time further transplantation experiments indicated that both the nucleus and the cytoplasm had become relatively incompatible with the cytoplasm and the nucleus of the parent species. There had been a drift away from the properties of each parent, and no drift towards one or the other. The observations were not explicable by a change in either nucleus or cytoplasm alone (Danielli, 1959*b*; 1960).

These observations point to a mutual conformance of nuclei and cytoplasms that were originally disparate. Nuclear transplantation and hybridization are possible only if the nuclei and cytoplasms are mutually conformable; they need not be identical. More generally, and at a deeper level of analysis, the problem of "nucleo-cytoplasmic" relationships resolves into one of the mutual conformability and conformance of genetic and biotonic systems. Although the chemical pattern of the total genome is presumed to be invariant, the pattern of genetic activity as represented in the effective genome is subject to mutual conformance with biotonic systems. It is proposed here that the general principle of the mutual conformability and conformance of genetic and biotonic systems is an important key to the understanding of developmental processes and, in particular, of the mechanism of differentiation.

II. The Ciliated Protozoa

The especial relevance of ciliated protozoa to a general discussion of biological organization depends on two cirumstances: first, they have provided evidence of varying credibility about extra-nuclear genetic mechanisms and "cytoplasmic inheritance" and second, all but the most primitive of them are unique in the animal kingdom in having two kinds of nuclei, micronuclei and macronuclei, which provide exceptional opportunities for studying the interaction of genetic and biotonic systems in eukaryotic cells.

A. Extra-nuclear Genetic Materials

1. Cortical Organization

A ciliated protozoon is equivalent to an entire metazoan organism rather than to one of its constituent cells. The extraordinary specializations of structure within one unicellular organism are conditioned by the demands of unitary existence and affect especially the ectoplasm or cortex, which elaborates a pellicle, simple or compound cilia, trichocysts, skeletal plates and contractile fibres.

The cilia have been most thoroughly studied. At the base of each cilium there is a granule called a basal body or *kinetosome*. Kinetosomes are linked together in rows by *kinetodesmal fibres* to form a *kinety* and the whole system constitutes the *infraciliature*. The system has a high degree of intrinsic stability but in physiological reorganization, in division or regeneration, in encystment and excystment, the cortical structures de-differentiate and later re-differentiate, whereas the kinetosomes persist and duplicate before the onset of visible mitosis in dividing cells. The kinetosomes are directly implicated in the morphogenesis of new cilia and the associated fibrillar system but apparently not in the formation of other cortical structures notably the trichocysts. All but the most primitive protista contain mitochondria, a Golgi apparatus and miscellaneous cytoplasmic membranes (Lwoff, 1950; Pitelka, 1963; Trager, 1964; Weisz, 1954).

One of the remarkable outcomes of electron microscopy has been the demonstration that, in all organisms above the bacteria, mobile cilia and flagella are organized on the same pattern. Each cilium or flagellum is covered by a unit membrane continuous with the plasma membrane. The membrane encloses a matrix of low density in which longitudinal fibres are embedded in a distinctive pattern. Nine double fibres are evenly spaced around the periphery and a pair of single ones are placed in the centre. In addition, nine very thin secondary fibres, more difficult to demonstrate, are situated midway between the nine peripheral fibrils and the central pair. The fibrils originate from the kinetosome, a cylindrical body in the cortex which has nine longitudinal fibrils situated peripherally and continuous with the nine peripheral fibrils of the cilium. The two central fibrils of the cilia spring from the distal end of the kinetosome. Kinetodermal fibrils, also, originate from the kinetosome. In protozoa the replication of kinetosomes commonly precedes any other sign of morphogenetic activity; it is not known how the characteristic pattern of the longitudinal fibrils is thereby reproduced (Pitelka, 1963).

No clue to the significance of the 9/9 +2 pattern of fibrils in cilia has yet been found. The universal occurrence of this unvarying geometric configuration at such a relatively coarse multi-macromolecular level is unparalleled. The pattern seems to be an indispensable factor in the mechanics of flagellar locomotion and to be one of the very few successful designs for a cytoplasmic centre of genetic activity (Pitelka, 1963). Moreover, the cortical mechanisms in general and the kinetosomal mechanism in particular were surely amongst the earliest morphogenetic systems to evolve (Weisz, 1954). The morphogenetic properties of the cortex have been especially emphasized in Tartar's studies of the ciliate *Stentor caeruleus* (Tartar, 1956, 1960, 1961).

Provisionally, it is reasonable to interpret kinetosomes as an evolutionary experiment in the decentralization of genetic action and the delegation of certain controlling functions to structures in outlying parts of the cell. The experiment has been successful in providing a mechanism for the action of cilia and flagella which has not been superseded during the evolution of all the metazoa up to man, but there is little reason to think that this particular kind of delegation of nuclear control to cytoplasmic constituents operates in the higher animals except in the special case that has been discussed.

2. The "killer" Phenomenon in Paramecium

Certain paramecia, designated "killers", liberate into the medium a material that can kill other paramecia called "sensitives". The "killer" trait is due to the presence of microscopically-visible κ particles, which are absent from sensitives. If a paramecium receives no κ particles from its progenitors or if it loses them, it cannot make new ones. Moreover, if it has received them, it cannot maintain them or support their multiplication unless its nucleus carries a dominant gene K.

The κ particles are described as measuring about $0\cdot2$–$0\cdot8\mu$ in diameter in stained sections; in fresh material studied by phase-contrast microscopy, they commonly measure $1\cdot5$–$2\cdot5\mu$ in diameter or even as much as $10\cdot0\mu$. The particles contain DNA and protein. They are too big to be viruses. In electron photomicrographs they look like sectioned bacteria and κ is now widely believed to be a parasite allied to the bacteria or to the rickettsiae. The κ particles multiply in the cytoplasm. About half are transmitted to each cell at fission and the population level is restored by multiplication before the next fission. A single paramecium may contain hundreds or thousands of κ particles. Occasionally one or more escape from the cytoplasm into the culture medium and infect and kill any sensitive paramecia that are present. Similar particles called μ particles are responsible for the mate-killer trait in paramecia. The mate-killers, unlike the killers, do not liberate μ particles into the medium. Transmission of the particles to sensitive paramecia requires direct and prolonged physical contact such as occurs during conjugation and the sensitive mate is not killed until conjugation has been completed. The μ particles persist and multiply within the cytoplasm only if the nuclear genome carries either of two dominant genes M_1 and M_2 (Beale, 1954; Nanney and Rudzinska, 1960; Sonneborn, 1947, 1950, 1951; Wilkie, 1964).

It has been proposed that κ is important as a formal model of a certain type of nucleo-cytoplasmic system of wider distribution but apart from the μ particles and other close relatives of κ found in paramecia, few recognizable copies of this model have been described. The sensitivity of *Drosophila* to killing by carbon dioxide resembles the killer phenomenon in being attributable to a transmissible cytoplasmic particle σ but σ is probably a virus and it is not, like κ, notably dependent on a nuclear gene. Beale (1957) described the killer phenomenon as "something of a freak"; its general significance as a model of "cytoplasmic inheritance" has been exaggerated.

3. *Metagons*

During studies of the mate-killer phenomenon, Gibson and Beale (1962) encountered a special case of *phenotypic lag* similar to one previously noticed by Sonneborn in his observations on the killer phenomenon due to κ particles. Sonneborn had previously coined the term *phenotypic lag* to refer to delay in the phenotypic expression of a genetic change. In their first paper,

Gibson and Beale reported that if, by appropriate genetic manipulations, the dominant genes M_1 and M_2 were replaced by the recessive alleles m_1 and m_2 to give the genotype m_1 m_1 m_2 m_2, the μ particles responsible for the mate killer phenomenon disappeared from the paramecia but only after a surprisingly long phenotypic lag extending over about eight to fifteen fissions. The μ particles then seemed to disappear abruptly; animals carried either large numbers of μ particles or none at all. To account for these findings, Gibson and Beale proposed that, as well as containing visible μ particles, mate killer paramecia contained other particles designated *metagons* which were responsible for maintaining the population of μ particles; in the absence of metagons, μ particles were eliminated. The metagons were not demonstrable microscopically. They appeared to be stable but not, as a rule, replicable; it was presumed that they were formed only in the presence of M_1 or M_2 genes. The experimental findings were consistent with the supposition that about 1000 metagons were present in a mate killer paramecium, that no more were produced after exclusion of M genes and that those already present were shared out passively, but not completely at random, between the daughter cells. On these suppositions it was calculated that animals completely free of metagons would appear with appreciable frequency after about nine fissions which seemed to agree well with the observed phenotypic lag.

Further experiments (Gibson and Beale, 1963, 1964) indicated that metagons contained RNA as an essential constituent and probably as the only constituent. The metagons in mate-killer paramecia could be destroyed, or irreversibly inactivated, by exposing the animals to RNase; the μ particles persisted until after the animals had divided once and disappeared before the second division. Metagons could be regenerated at a later time in paramecia carrying M genes. Metagon activity was demonstrable in extracts of paramecia carrying the dominant genes M_1 or M_2 or both and in fractions of those extracts freed from microsomal membranes and ribosomal proteins. Both kinds of preparation were inactivated by RNase. The evidence indicated that the metagons were composed solely of RNA and Gibson and Beale thought that the metagon might be regarded as a very stable and repeatedly-functioning form of messenger RNA The investigations up to this point were summarized by Beale (1964) and by Sonneborn (1965) who discussed also the extension of the investigations by Gibson and Sonneborn to be discussed next.

Gibson and Sonneborn (1964) studied the behaviour of metagons in other genera of ciliates, namely *Didinium* and some species of *Dileptus*, that prey on *Paramecium*. Didinia were able to maintain μ particles which they acquired from ingested paramecia and the maintenance depended on the presence of metagons similarly acquired. *Didinium* did not otherwise contain metagons and could not produce them but the metagons acquired by ingestion persisted indefinitely, being followed for more than six months, covering about 1000 cell generations. Moreover, the metagons increased in number indefinitely to keep pace with the multiplication of the didinia. By contrast, metagons multiply extremely slowly or not at all in their native environment, the *Paramecium*.

Hybridization experiments were carried out to study the provenance of metagon RNA. RNA, extracted from paramecia containing metagons, was tested against DNA from *Didinium*, from another ciliate *Tetrahymena*, from a bacterium *Aerobacter aerogenes* and from two stocks of *Paramecium*, 540 and d-200. The two stocks were largely isogenic except at the M-m locus, where 540 had the M_1 M_1 M_2 M_2 genotype and d-200 had m_1 m_1 m_2 m_2. The outcome was that the metagon RNA hybridized only with the *Paramecium* DNA and eighteen times as effectively with the DNA from the stock carrying M_1 M_2 genes as with that from the stock with the recessive m_1 m_2 genes.

The first important inference from this experiment is that metagon RNA is complementary to the DNA of the M-m locus of *Paramecium* and more especially with the M_1 M_2 alleles. This leads directly to the hypothesis that the metagons are RNA transcripts of the M_1 and M_2 genes. The failure to hybridize with DNA from *Didinia* indicates not only the absence of M-like genes from these ciliates but also that the multiplication of metagons in them is effected by a different mechanism from that responsible for the multiplication in paramecia. It is inferred, briefly, that metagons replicate in didinia but are multiplied by transcription of M genes in paramecia. In support of this conclusion, actinomycin D, which inhibits transcription of DNA, suppresses the formation of new metagons in paramecia but not in didinia whereas guanidine hydrochloride, which represses RNA-dependent synthesis of RNA, does not affect the production of metagons in paramecia but inhibits it in didinia.

The general conclusion that the metagon is an RNA transcript of a localized region of the nuclear genome seems to be well founded but Sonneborn (1965) pointed out that serious difficulties and obscurities have still to be resolved. In particular the basic observations on the phenotypic lag are not consistently reproducible; the disturbing factor remains to be identified and controlled. It is still not clear whether or not the genetic code for the synthesis of the polymerase is needed for its own replication. It is not evident why the synthesis of a metagon extends over about two cell generations and the manner in which metagons protect μ particles is obscure.

4. *The Significance of Metagons*

Apart from a few possible examples in *P. aurelia*, no other metagons seem to have been recognized (Sonneborn, 1965). It would be as premature to infer that they do not exist as to presume that they are reliable "models" of ubiquitous "units of cytoplasmic inheritance". Their immediate interest and importance is in demonstrating the existence of stable, replicable RNA transcripts of small localized sections of the nuclear genome. In their native habitat in Paramecium, they do not replicate or do so minimally; it is reasonable to suppose that instead they engage in translation to produce the unidentified factor that directly or indirectly preserves the μ particles. In the alien host *Didinium*, they replicate and, as Gibson and Sonneborn say, they provide suggestive evidence about the origin of RNA viruses including the RNA tumour-inducing viruses, which is one of the reasons for discussing them at some length here. So far, at least, there is no evidence that they can direct the synthesis of a protein coat and Sonneborn (1965) classes the metagons in *Didinia* as *defective viruses*. In certain Didinia, according to Sonneborn, the metagons behave as plasmagenes or cytogenes, and Granick (1965) claims that the metagons show that genes may consist of either DNA or RNA, which seems to be an undesirable extension of the term *gene*.

B. Genetic and Biotonic Systems in the Ciliates

The division of the genetic material of a single cell between two morphologically and functionally different kinds of nuclei is an evolutionary experiment peculiar to the ciliates but highly relevant to the study of biological organization and development of the higher animals because they have used essentially the same device for the same purpose but in a different way. The ciliate micronucleus corresponds with the nucleus of metazoon germ cells and has the same function, namely, to preserve, replicate and transmit the *total genome*. The ciliate macronucleus corresponds with the nucleus of the somatic cells of metazoa; its genetic material is in active use as an *effective genome* but in diverse ways under various circumstances and in different cells. The main theme of this chapter is the selective utilization of an invariant total genome as illuminated by studies of the ciliated protozoa carried out in particular by Sonneborn, Nanney and Beale and their collaborators. Most of the factual basis of this chapter has been drawn from reviews by these authors, although their interpretations are not always followed (Beale, 1954, 1957, 1964; Nanney, 1957, 1958, 1963, 1964; Nanney and Rudzinska, 1960; Sonneborn, 1947, 1950, 1951, 1965).

1. Organization of the Nuclear Material in Ciliates

The most thoroughly studied organism, *Paramecium aurelia*, is cigar-shaped and commonly about 150μ long and 50μ broad. Ordinarily it contains two micronuclei measuring about 3μ in diameter and one elongated macronucleus about 50–60μ long and 20–30μ wide. The micronuclei are believed to be diploid. The macronuclei are in some sense multiple or polyploid nuclei. Sonneborn and Nanney believe that they are made up of numerous diploid subnuclei. Estimations of the amount of DNA present in macronuclei imply that about 860 haploid sets of chromosomes are present (Grell, 1964). Unfortunately, electron-microscopic examination of protozoan nuclei has yielded disappointing results and many details of nuclear structure and nuclear division are obscure (Pitelka, 1963).

Paramecia multiply vegetatively by binary fission and also following a mating between two individuals known as *conjugation*. During binary fission the micronuclei divide by mitosis, the details of which are not well known. The macronucleus elongates, constricts across the middle and then divides into halves. The genetic evidence, according to Sonneborn, indicates that the division is precise and effectively equal. In general the division of the macronucleus effects no Mendelian segregation but Nanney (1963, 1964) believes that the subnuclei of some macronuclei, newly formed as described in the next paragraph, are heterogeneous and that the diverse subnuclei are sorted our during subsequent fissions.

During the complex phenomenon of conjugation two individuals (conjugants) attach to one another. The macronucleus in each conjugant disintegrates into fragments numbering forty or more. The micronuclei undergo two successive divisions including a reduction division. Each conjugant now has eight, presumably haploid, micronuclei and a disintegrated macronucleus. Seven of the eight micronuclei disappear and the one haploid nucleus surviving in each conjugant divides by mitosis to yield two haploid nuclei, often called "gametic" or "germinal" nuclei. One of these two, the "stationary" or "female" one, remains in the same conjugant but the other, the "migratory" or "male" nucleus passes into the cytoplasm of the other conjugant. After the reciprocal interchange the two haploid nuclei in each conjugant fuse to form one diploid nucleus containing one haploid set of chromosomes from each of the conjugants. The diploid nucleus now divides twice in succession by mitosis to yield four diploid nuclei in each ex-conjugant, the two mating animals by now having separated. Two of the four transform by condensation into micronuclei; the other two increase in size and eventually "mature", presumably after successive replications of their genomes, into macronuclei. At the first binary fission after conjugation, the two new micronuclei divide by mitosis but the macronuclei do not divide and one of them passes into each of the daughter cells. The original content of two micronuclei and one macronucleus is now restored and is maintained indefinitely during vegetative multiplication, as already described. It is noteworthy, first, that fragments of the old macronucleus do not degenerate completely until the new macronucleus is fully developed and, second, that as a rule there is no significant interchange of cytoplasm between conjugants. Under certain conditions that prolong the duration of pairing, the transfer of cytoplasm may be substantial. Under certain conditions, also, a new macronucleus fails to develop in the usual way and the deficiency is made good by *macronuclear regeneration*, a new macronucleus developing from the persisting fragments of the old one. Apparently each of the forty or more fragments can regenerate a new, complete, functional macronucleus.

In the remarkable phenomenon of *autogamy*, complex nuclear changes take place exactly as in conjugation but in unpaired animals. There is no interchange of material between mating individuals and the two "germinal" micronuclei fuse together in the same animal to form a diploid nucleus from which two new micronuclei and one new macronucleus are derived as described for conjugation. One consequence of this most intense form of "inbreeding" is that during succeeding fissions the animals become genetically homozygous. Another and probably more important consequence of autogamy is that the animal acquires a new macronucleus. Autogamy occurs in most strains of *P. aurelia* at intervals of about ten days; it may be delayed for a long time by supplying food abundantly but eventually, failing either autogamy or conjugation, the clone will die out. The interchange of genetic material between individuals during conjugation is credibly beneficial but it is hard to understand the advantage of the homozygous state resulting from autogamy. The common feature in autogamy and conjugation is the formation of a new macronucleus and the reasonable inference is that the macronucleus must be renewed periodically for a clone to live indefinitely.

2. Functions of Micronuclei and Macronuclei

Paramecia can survive for a long time without micronuclei but the absence of a macronucleus is lethal within one or two fissions. No evidence of growth or morphogenesis has been found in the absence of a macronucleus. The phenotype of a paramecium depends on its macronucleus and not at all on its micronuclei. Macromolecular regeneration during conjugation results in purely "maternal" inheritance; the phenotype of the animal corresponds entirely with that of the parent supplying the macronuclear fragments and is uninfluenced by the genotype of the micronuclei. A regenerated macronucleus can support the normal functioning of a paramecium even in the complete absence of micronuclei. The micronuclei serve mainly or entirely as "germinal" nuclei and have no direct effect on the organism; the macronucleus, derived from a micronucleus is the "somatic", physiologically active, nucleus (Beale, 1954; Sonneborn, 1947; Nanney, 1963; Nanney and Rudzinska, 1960).

The unique feature of the ciliated protozoa is the material separation within a single cell of the *total genome* vested in the micronuclei and the *effective genome*, vested in the macronucleus. In metazoa the separation is achieved in different cells but not within single cells. The micronuclei of ciliates are comparable with the nuclei of the germ cells of higher animals and, like them, their function is accurately to preserve, replicate and transmit the *total genome*. The macronuclei are comparable with the nuclei of somatic cells of the higher animals and carry an *effective genome*, which is implicated in the differentiation and behaviour of cells. The macronuclei provide for the selective utilization, as effective genomes, of genetic patterns preserved in the micronuclei from which the macronuclei are derived and from which it seems they must be renewed periodically.

3. Selective Utilization of Genetic Patterns

A culture of paramecia comprising the descendants of a single individual collected from its natural habitat is called a *stock*. The various stocks are divided into several *varieties* by their ability to conjugate with one another to yield viable offspring. There is a notable diversity of phenotypes amongst the individuals of a particular stock but there are good grounds for believing that the genetic information for all the varied phenotypes is present in the micronuclei of every individual of the stock. In other words all individuals of the stock carry identical total genomes in their micronuclei although different effective genomes in their macronuclei. All the possible patterns of genetic activity of a stock are represented in the total genome of each individual paramecium but only one of them is used at a particular time, the others being repressed. There is a general resemblance, but with differences in detail, among the total genomes of different stocks of one variety and still greater differences among stocks of different varieties. The especial interest and importance of the ciliated protozoa derive from the opportunities they provide for studying the mechanisms of selection and maintenance of diverse effective genomes from an invariant total genome.

The two phenotypic characters that have been most thoroughly studied depend on surface properties; they are first, the *mating types* which depend possibly but not certainly on complementary configurations on the surfaces of cells that are able to conjugate with one another (Nanney, 1963) and second, the *serotypes* which depend on the presence of specific antigens believed to be localized on the cilia. The discussion of these properties which follows is based mainly on reviews by Beale (1954, 1957), Nanney (1957, 1958, 1963), Nanney and Rudzinska (1960) and Sonneborn (1947, 1950, 1951) to which reference may be made for additional details.

4. Mating Types

Paramecia are separable into *mating types* by their abilities to conjugate with one another; two animals cannot conjugate unless they belong to different mating types. In the most-studied varieties of *P. aurelia* only two mating types are distinguishable but up to eight types are present in some ciliates, *Tetrahymena pyriformis* having seven types. Exceptionally, all animals of one stock of Paramecium belong to the same mating type so that conjugation is impossible but the stock can be maintained indefinitely without loss of vigour; autogamy, without introducing new genetic material provides the periodic "rejuvenation" that seems to be needed. Apparently the differentiation of mating types is not crucial for the survival of a species although it may be more advantageous in nature than it seems to be in the laboratory.

Varieties of *P. aurelia* showing the more usual differentiation into two mating types have been divided into two groups, A and B, by supposed differences in the regulation of mating type. These two groups will be described separately but the differences between them are not so sharp as formerly supposed and are still more blurred when observations on other ciliates are taken into account.

Group A varieties are distinguished by what is described as "caryonidal inheritance". A *caryonide* is a clone of paramecia comprising individuals whose macronuclei all are derived by fission from a single macronucleus newly-established during conjugation or autogamy. The description of conjugation already given indicates that four new macronuclei are formed, two in each ex-conjugant, so that after the first vegetative division, four caryonides are established and persist until new macronuclei are established by conjugation or autogamy. As a rule all members of the same caryonide belong to the same mating type but the four caryonides may be of the same mating type or of different types in all possible proportions. Sister caryonides from the same ex-conjugant do not resemble one another in mating type more often than do caryondies from different ex-conjugants. The distribution of mating types amongst the caryonides seems to be random except that the predominance of one or another of the types is influenced by the environmental temperature prevailing whilst the new macronucleus is being formed. Once formed, the macronuclei are stable and retain their characteristic properties unchanged through repeated vegetative fissions until eventually they disintegrate during autogamy or conjugation. Various ingenious experiments, which need not be detailed here, show that the mating type of an individual paramecium is determined by its macronucleus and that the diversities of mating types result from differences in macronuclei established during conjugation or autogamy. Autogamy yields diverse mating types even when the micronuclei are genetically homozygous.

Group B varieties are distinguished by what is called, misleadingly, *cytoplasmic inheritance*. The two caryonides from the same ex-conjugant are alike in mating type and the mating type corresponds with that of the conjugant in which the new macronuclei developed. The distribution of mating types amongst the ex-conjugants and their progeny is *not* random. As a rule the two clones from the same conjugant are of like mating type and autogamy does not establish different mating types. The choice of mating type is attributable to cytoplasmic factors operating whilst new macronuclei are being formed and to environmental factors operating through the cytoplasm at that same time. Sonneborn described an experiment in which conjugating paramecia were exposed to a high environmental temperature as a consequence of which there was a massive interchange of cytoplasm between the conjugants and varied degrees of interference with the development of new macronuclei. When macronuclei developed in the normal way, the mating type of the ex-conjugants and their progeny corresponded with that of the alien cytoplasm, received by interchange. When normal development of new macronuclei was retarded or suppressed, macronuclear regeneration from fragments of the old macronucleus took place; the mating type of the ex-conjugants and their progeny then corresponded with that of the conjugant providing the macronuclear fragments and not with that of the alien cytoplasm in which macronuclear regeneration took place; no change in mating type occurred immediately, during subsequent fissions, or even after autogamy. These observations justify three important inferences: (1) Changes in mating type during conjugation were due to the action of foreign cytoplasm on the macronuclei newly-developing in it; (2) Regenerating fragments of macronuclei were *not* modified by the same foreign cytoplasm; (3) Regenerated macronuclei modified the cytoplasm in which they were present in such a way that new macronuclei developing in a subsequent autogamy were not altered, as they should have been if the alien cytoplasm had retained its original qualities.

The behaviour of group B varieties demonstrates in a vivid way the operation of the general principle of the *mutual conformance of genetic and biotonic patterns*; it also illustrates the limitation of genetic conformance to the stage when new macronuclei are being formed from micronuclei or, in other words, when a new effective genome is being derived from the total genome. A developing macronucleus "conforms" with the cytoplasm but the cytoplasm "conforms" with a mature macronucleus. The stable system comprises a mutually-conforming macronucleus and cytoplasm. Inheritance of mating type is "cytoplasmic" only to the extent that biotonic patterns of the cytoplasm establish a conforming pattern of genetic activity within the limits of choice permitted by the total genome and within limits of time corresponding with the derivation of a new effective genome from the invariant total genome. In the Group A varieties, by contrast the mating type depends mainly on the random selection of one of the two possible mating-type effective genomes available in the total genome. Mutual conformance in Group A seems to be a one-sided adjustment, the cytoplasm doing nearly all the conforming. It should be emphasized that in both groups the effective genome of the *mature* macronucleus is intrinsically stable during vegetative fission. The difference between the two groups, on this interpretation, is restricted to the degree to which biotonic systems of the cytoplasm influence the *selection* of an effective genome from the total genome. Even in this respect the distinction is not sharp; environmental temperature influences the choice in both groups. Moreover, there are many exceptions to the general rules of behaviour as stated above. Group A varieties, for example, include some "unstable caryonides" liable to frequent changes of mating type during vegetative reproduction.

The intrinsic stability of the effective genome is not absolute. In some ciliates the mating type changes when the clones grow old. Still more remarkably, *Paramecium bursaria* has a diurnal rhythm of mating behaviour; the capacity to mate appears in the morning, reaches a peak about noon and disappears towards evening. The initiation and the maintenance of the ability to mate are dependent on sustained protein synthesis continuing throughout the mating period (Cohen, 1965). In view of the proposition that circadian rhythms in general depend on feed-back control systems, it may be inferred tentatively that the ability to mate is sustained by specific protein synthesis subject to feed-back control systems.

It is clear that the determination of mating type in *Tetrahymena* is not due to the mere presence or absence of a single factor, as it might be in ciliates with only two mating types; there are seven choices and the "expression" of any one of them excludes the "expression" of the other six. Furthermore, although the choice is made during conjugation, it is not expressed in mating behaviour until from 50 to 100 negative fissions have been accomplished; the maintenance of mating-type "differentiation" is not dependent on its "expression" (Nanney, 1963).

5. *Serotypes*

Antisera produced by injecting paramecia into rabbits immobilize living paramecia by reacting specifically with antigens believed to be situated on the cilia. Immobilization tests and genetic analysis indicate that each *stock* of *Paramecium* carries a large number of antigens controlled by different genetic loci scattered through the chromosomes and designated A, B, C, etc. Different stocks have different alleles at one or more of these loci and the corresponding antigens differ to a greater or lesser extent from one stock to another. Genetic analysis indicates that all the paramecia carry in their micronuclei the genes corresponding with *all* the antigens found in that stock although as a consequence of the *mutual exclusion* phenomenon, only one of the antigens is demonstrable in any single paramecium. The diversity of antigenic types in stocks and varieties of Paramecium is extremely complex but experiments have been interpreted to indicate first, that the genome rigidly controls the specificity of antigens and the variety of antigens that *can* be manifested within a stock and, second, that one of many possible mutually-exclusive "cytoplasmic states" selects and establishes *one* of the possible antigens in any individual paramecium. Differences between stocks are attributed to differences in the total genomes; differences between individuals of the same stock are attributed to differences in their "cytoplasmic states". The serotypes are much more labile than the mating types and more conspicuously modified by environmental factors acting, it is presumed, through the "cytoplasmic states" even at times when no new macronuclei are being formed. None the less, there is little doubt that the serotype depends on the interplay of utilizable patterns in the macronucleus and patterned biotonic systems in the cytoplasm. The serotype and mating type systems are basically alike but some important differences in detail require attention (Beale, 1954, 1957).

The diverse environmental conditions that can alter the serotype include temperature and salinity of the medium, the presence in it of homologous antiserum, the rate of growth of the organisms, irradiation, and exposure to various enzymes and other substances. Changes of temperature are notably effective in determining the antigenic type. By manipulating the temperature of the medium, the antigenic type of a clone could be changed at will and once a new type was established it was maintained for a long time during vegetative reproduction under standard conditions but could be altered again by a new change in the environment. None of the alterations in antigenic type in response to environmental factors was irreversible or completely permanent. The long persistence of some of them has been attributed to "cytoplasmic inheritance" which, like the "cytoplasmic inheritance" of mating types, probably depends on the mutual conformance of effective genomes in the macronuclei and biotonic systems of the cytoplasm.

The transformation of antigenic type effected by temperature changes is sometimes demonstrable after a few fissions or almost immediately but in some stocks evident transformation may be long-delayed. It is noteworthy that the same environmental conditions control the expression of the various alleles at the same locus.

In the serotype system as in the Group B mating type system, biotonic systems of the cytoplasm are important in selecting an effective genome from the total genome whilst a new macronucleus is being formed and the phenotype is stabilized by mutual conformance between cytoplasmic biotonic systems and genetic patterns in the macronucleus. The outstanding difference seems to be referable to the much lower *intrinsic* stability of the serotype effective genome and to adaptations of the *mature* macronucleus to changing biotonic systems in the cytoplasm, although even in these properties wide variations are encountered. Certain antigenic types are unusually stable or more than usually unstable and the unusual degree of stability persists through considerable periods of binary fission but disappears after conjugation. Furthermore, certain antigenic types tend to change abruptly at autogamy and the new type is stable. Beale infers that the stability of the "cytoplasmic states" is, in part, under genetic control; according to the present interpretation it depends on the degree of *intrinsic stability* of the effective genome in the macronucleus. The serotype effective genome, it seems, can be replaced in response to environmental factors operating on a mature macronucleus during vegetative reproduction. Transformation of antigenic type by temperature changes or by homologous antiserum is completed as a rule only during periods of growth, active synthesis of new material being essential. Exceptionally, transformation is effected quickly and without fission (Austin et al., 1956; Beale, 1957). Beale emphasizes the extreme variability of the duration of the *phenotypic lag* preceding the manifestation of a new serotype in response to environmental stimuli; it may be completed within two days or delayed for more than fifty fissions. By contrast, new serotypes, depending on alterations in the genotype effected by conjugation are detectable fairly consistently after five fissions. Apparently biotonic systems operate more regularly and consistently to establish a new effective genome in a developing macronucleus than they do to change the effective genome in a mature macronucleus.

The regulation of a particular serotype depends not only on the existing environmental conditions but also on the physiological state of the organism, this in turn being related to previous conditions. The reaction to homologous antiserum may vary with the state of nutrition. Homologous antiserum does not alter the serotype of well-fed animals but when it is applied to severely starved animals it can bring about successive changes in the serotype. Different serotypes of the same stock transform preferentially to diverse types when exposed to homologous antiserum. As an example, a proportion of type B paramecia of stock 172 transformed to types D and E, type C transformed frequently to type B, type D transformed sometimes to type B but more often to types C or E and to type I which resulted from the transformation of no other type. One type (E) was not altered by antiserum. Although all the paramecia had the same genotype and were exposed to the same environment and in the same degree to the homologous antiserum, each type behaved in its own characteristic way by transforming preferentially to certain other types or by remaining stable (Beale, 1957; Skaar, 1956). These observations show that *alternative patterns of genetic activity* are available in the mature macronucleus and can be brought into use by environmental conditions without new macronuclei being formed. They show, also, that the utilization of a particular genetic pattern as an effective genome influences in a specific way *which* of the other multiple patterns will be substituted if the utilization of the original pattern is suppressed. The utilization of one genetic pattern thus determines in a specific way the *preferential availability* of an alternative pattern. It is useful here and, as will be shown later, almost obligatory in studying development in the higher animals, to distinguish between the *effective genome*, signifying the genetic pattern in actual use and the *facultative genome*, comprising all the genetic patterns available for use as effective genomes under various circumstances.

6. *Selective Utilization of the Genome in Ciliated Protozoa*

The studies of ciliated protozoa are particularly relevant to the subject of this book insofar as they illuminate, at a surprisingly low level in the evolutionary scale, certain aspects of the complex interplay of nuclear and cytoplasmic systems that underlies developmental processes in the higher animals, including the conspicuous process of differentiation.

Beale (1954) has said that the mating-type system of Paramecium is irrelevant to the study of differentiation in the higher animals because it depends on the macronucleus, which those animals do not possess. Functionally, nevertheless, the macronucleus is equivalent to the single nucleus in somatic cells of higher animals. The multiplicity of genomes in the macronucleus introduces some complications which must be taken into account. With this reservation, a paramecium differs from a somatic cell most significantly in having, as well as a macronucleus, two micronuclei which correspond functionally with the nuclei of the germ cells of higher animals. This circumstance is more of an asset than a liability in some studies of the differential utilization of the genome to which reference will be made in this section.

In this discussion of the pertinent evidence obtained from the ciliates, the terms *total genome, effective genome* and *facultative genome* are used freely. They were designed to refer conveniently and as precisely as present knowledge allows, to certain concepts of basic importance in developmental biology that derive from the newer knowledge of genetic materials and genetic actions (Foulds, 1963, 1964). The first two are scarcely controversial and can be derived from observations on bacteria and defined with some precision. The facultative genome is a more complex concept not as yet so easily definable. It has no counterpart in bacteria, its particular application being to developmental processes in which bacteria are notably lacking. It should be said that the terms do not, in general, refer to three distinct material entities but to three states of the genetic material. These three states are next discussed in the light of information obtained from the ciliates.

(a) THE TOTAL GENOME. The concept of a total genome is a chemical one, not a biological one. The total genome carries the whole of the inherited genetic information of the organism coded, as is now believed, in DNA. It is a precisely replicating entity and in the absence of mutation it is invariant. As exemplified in the micronuclei of ciliates and in the germ cells of higher animals, it is physiologically inert and engages in scarcely any activity other than its own replication. For most practical purposes the total genome can be equated with the inactive genetic DNA of the zygote nucleus of metazoa.

(b) THE EFFECTIVE GENOME. The effective genome comprises those discontinuous sections of the total genome that are in effective use in a particular cell at a particular time. Briefly, it is genetic DNA *in action*. In the ciliates, it is located in the macronucleus and it is the physiologically active genome. The activity, as already argued, depends on its linkage with dynamic biotonic systems. The operative system in the cell comprises a specifically patterned effective genome linked with a biotonic system of conforming pattern. The two constituents are not separable without abstraction although the genetic pattern endures without linkage as a *potential effective genome*. Although differently expressed, these assertions are consonant with the statement of Nanney and Rudzinska (1960) that "Specific nuclear activity is required for the perpetuation of the cytoplasmic state as unequivocally as a particular cytoplasmic state is required for the manifestation of a particular gene. Nucleus and cytoplasm operate in a single system of perpetuation." This was written with special reference to the ciliate serotypes but it is more widely applicable.

The effective genome comprises many sub-divisions implicated in diverse cell activities. The sub-divisions do not all behave alike and in some contexts it is necessary to specify which of them is under consideration. The two requiring specification here are the mating-type and the serotype effective genomes. The mating-type effective genome of a mature macronucleus is extremely stable although the stability is not absolute. As a rule a new effective genome cannot be established until the mature macronucleus disintegrates and is replaced by a new one during conjugation or autogamy. Once established the mating-type genome persists through an indefinite series of regetative fissions. Moreover, it persists in a macronucleus in an alien cytoplasm whose biotonic systems apparently conform with the effective genome of the macronucleus. Superficially, it seems as if the effective genome is both stable and replicable as such. More probably the stable and replicable basis is the *potential effective genome*, a constituent of the facultative genome whose properties will be described in the next section. Under standard conditions, the stability of the serotype effective genome is substantial but it does not withstand variously-produced changes in the environment.

(c) THE FACULTATIVE GENOME. The facultative genome comprises all the genetic patterns that are *available* for use as effective genomes under various conditions and at various times. It comprises, in short, all the *potential effective genomes*. It corresponds with the genetic system described by Nanney (1958) as maintaining the complete "library of specificities, expressed and unexpressed". For embryologists, the facultative genome provides a genetic basis for *capacity* and *competence*; in pathology it provides for *metaplasia*; in general, it provides for alternative patterns of utilization of the genome. It differs from the total genome in several important respects as will emerge from the following discussion. In particular, it is not an invariant entity and its properties are not accounted for solely by the structure of the genetic DNA. The facultative genome must be conceived as a total genome modified in some way by the ancillary materials that invest the DNA in the chromosomes of all eukaryotic cells.

In some ciliates the mating-type changes with increasing age showing that an alternative potential effective genome is present in the macronucleus but in general the high stability of the established mating-type effective genome makes the demonstration of alternative ones extremely difficult. By contrast the sensitivity of the serotypes to environmental change allows the relatively easy recognition of multiple alternative potential effective genomes in the mature macronucleus. Sufficient evidence for this has been summarized in earlier paragraphs.

The number of potential effective genomes in the facultative genome may be substantial but it is less than might be inferred from Mendelian genetic analysis or from the chemistry of the genetic DNA. Nanney (1958) remarked that the genetic potentialities of cells are expressed in integrated patterns; the expression of one specificity may involve the expression of a series of specificities, a phenomenon that he describes as *simultaneity of expression* or, as a consequence of *mutual exclusion*, it may debar the expression of other potentialities.

In all types of "differentiation", whether in protozoa or in vertebrate animals, an essential early step is the *selection* of one of the several alternative potential effective genomes present in the facultative genome for effective use. The important question for consideration is how the selection is made. Two choices of mating type are available in *P. aurelia*. In the Group A varieties, the choice seems to be mainly a matter of chance except that environmental temperature influences the probability that one or the other will be preferred. In the Group B varieties temperature is again a factor but the selection is clearly effected mainly by the cytoplasm in accordance with the principle of mutual conformance of genetic and biotonic patterns. An effective genome is selected in the newly-developing macronucleus to conform with biotonic systems in the cytoplasm, whereas the cytoplasmic biotonic systems conform with the effective genome in a nature macronucleus. The selection, whether random or directed, is effected whilst a new macronucleus is being formed during conjugation or autogamy.

The relative instability of the serotype genomes is disclosed when environment conditions are changed in a variety of ways, the act of changing being apparently more decisive than the quality of the new environment (Beale, 1957). Skaar (1956) suggested that the significant agency might be the generalized physiological shock produced by altering the environment and Nanney and Rudzinska (1960) wrote, in a similar vein, of "environmental stress". The effect of the shock or stress is to switch one effective genome "off" and a different one "on". Presumably the shock or stress produces its effects by modifying biotonic systems of the cytoplasm but the selection of a new effective genome does not seem to depend on mutual conformance with the altered biotonic systems. The environmental change apparently exerts a trigger-like action rather than a selective or directive one. Nevertheless, the triggered changes are not random; certain of the available potential effective genomes are used in preference to the others as though they were in some way more immediately "available" or "accessible" for effective use. Observations on the serotypes, as already noted, indicate that the utilization of a particular effective genome influences in a specific way *which* of the other available patterns will be substituted for it if the environment changes. By manipulating the environmental temperature a predictable sequence of serotypes can be induced. It is remarkable that in this way animals so low in the evolutionary scale as the ciliates can show the rudiments of a process of epigenetic development.

There is no evidence to suggest that the sequential or other changes of serotype entail any alteration of the total genome or that repressed potential effective genomes are expunged from the facultative genome. None of the environmentally-induced serotypes is completely permanent and none is irreversible under environmental stress. The only discernible alteration of the facultative genome is in the probability that one of the potential effective genomes will be preferentially activated. Except in this respect, the facultative genome of its macronucleus is highly stable and replicable.

As already mentioned, Beale noted that the manifestation of a new serotype induced by environmental stimuli might be delayed for more than fifty fissions, which implies that some replicable change in the macronucleus persisted during that time without phenotypic expression. Nanney recorded a comparable delay in the manifestation of mating-type in Tetrahymena. Nanney's observations showed that the "differentiation" of mating-type was effected during a short period following conjugation but it was not expressed until fifty to 100 regetative fissions had taken place and the animal had become mature and able to mate. Nanney insisted that the decision as to which mating-type eventually will be expressed is independent of the actual expression of that type and believed that the *expression* of mating-type is relatively easily disengaged from mating-type *differentiation*. Comparable delays in the expression of serotypes have been mentioned already in the discussion of serotypes.

In Nanney's discussion, "differentiation" corresponds closely with the phenomenon in vertebrate embryogenesis to which the term *determination* is often, if controversially, applied. It signifies a *decision in advance of performance*. Apparently a genetic pattern is selected for use as an effective genome a long time before it is actually so used. This is conveniently described as the *preselection* of an effective genome by analogy with the preselection of a gear in advance of its engagement with working machinery. Preselection is an important phenomenon in vertebrate development and will need further discussion in later chapters. It is relevant to note here that the generality of the phenomenon precludes any interpretation or explanation invoking peculiarities of the ciliated protozoa.

The remarkable feature demanding explanation is the persistence of a stable, replicable but unused pattern through fifty to 100 vegetative fissions. Preselection implies that the chosen genetic pattern is not engaged with biotonic systems which, therefore, cannot be expected to maintain the chosen pattern according to the mutual-conformance principle or through feed-back circuits of the Jacob-Monod type. The more probable explanation is that a segment of the facultative genome has undergone a persistent replicable change not in the genetic DNA itself but in the availability of genetic sites for transcription as a consequence of changes in the ancillary constituents of the chromosomes. This is not inconsistent with Nanney's (1963) proposal that the critical event in mating-type differentiation is a modification of the sub-nuclei whilst a new macronucleus is developing during conjugation or with other attributions to "macronuclear differentiation" (Siegel, 1961; Skaar, 1956), but it is more generalized, being applicable as well to determination in vertebrate embryogenesis and it is more specific in identifying the facultative genome as the site of change. There is no evidence that the structure of the DNA of the total genome is altered.

It will be more convenient to defer further consideration of the material nature of the facultative genome until relevant evidence from higher animals has been dealt with and to emphasize here that the evidence obtained from ciliated protozoa indicates that the material basis of the facultative genome is some kind of "differentiation" of the nucleus, more specifically of the chromosomes. The facultative genome is the expression of an intra-chromosomal *organization* not wholly explicable in terms of DNA or of dynamic regulatory systems. The organization is notably stable in the absence of effective use and it is replicable.

7. *Epigenetic and Biotonic Systems*

Reasoning mainly from observations on ciliates, Nanney proposed that two types of cell-regulating mechanisms maintain persistent cell characters but by different means. One system, the *genetic system*, depends on a primary genetic material that replicates by a template mechanism; the other, the *epigenetic system*, operates, perhaps, through self-regulating patterns. The genetic system maintains the complete "library of specificities, expressed and unexpressed"; the epigenetic system, comprising auxiliary mechanisms with different principles of operation, is involved in determining which specificites are expressed in a particular cell. The epigenetic systems are conceived as devices responsible for rhythmic patterns of activity and for "hereditary" but "non-genetic" mechanisms. The observations on ciliates imply that cells with the same genotypes, existing and reproducing in the same environment may manifest different traits and suggest that a cell may store information about past experience in forms other than nucleotide sequences. Further, they indicate that the time-honoured distinction between genetic and non-genetic traits is of limited validity; persistent differences may be controlled by either genetic or epigenetic mechanisms (Nanney, 1958; Nanney and Rudzinska, 1960).

The concept of *biotonic* systems advanced here so evidently corresponds in a broad way with Nanney's concept of *epigenetic* systems that I have been reluctant to use a different name. The chief reason for doing so is that in a book devoted primarily to developmental biology it is hardly possible to avoid using "epigenetic" in the sense in which embryologists have used it for several decades; to use the same word in two different senses would be seriously confusing. Apart from that, there are some differences, at least in emphasis, between the two concepts as well as between individual interpretations of *epigenetic* (c.f. Abercrombie, 1967*a*). The term *biotonic* applies primarily to the dynamic component of biological organization as the essence of life and only secondarily to its controlling or regulating function. Some other differences may be attributable to changes in the knowledge of genetic materials and genetic actions that have taken place since Nanney's publication.

In a passage already quoted Nanney remarked that "nucleus and cytoplasm operate in a single system of perpetuation". At a deeper level of analysis, genetic and biotonic mechanisms operate in a single system of perpetuation and it is becoming increasingly difficult to separate them even in thought. The steadily, if slowly, increasing recognition of extra-nuclear DNA and of informational RNA that is both stable and replicable accentuates the difficulty by blurring the distinctions between "nucleus" and "cytoplasm" and between "genetic" and "non-genetic". If it be accepted that the observations on metagons show that "genes" may consist of either DNA or RNA, as Granick (1965) maintains, then it is but a short step to extend the term "gene" to informational RNA carried by the endoplasmic reticulum and by the mitochondrial cristae. This step being made, all cytoplasmic processes will have to be credited with a "genetic" basis. The concept of biotonic systems as three-dimensional networks of dynamic biochemical systems supported by a macromolecular skeleton carrying informational RNA at the nodal intersections, allows for a component that is "genetic" in this over-extended sense.

Apart from the difficulties mentioned in the preceding paragraph, the existence of mechanisms that are "hereditary" but not "genetic" seems to me insufficiently substantiated. "Heredity" in this context depends primarily on cytoplasmic continuity in unicellular organisms multiplying by binary fission and secondarily on the mutual conformance of genetic patterns in the nucleus and biotonic patterns in the cytoplasm. On theoretical grounds discussed earlier, biotonic systems may be credited with specific pattern but it has not been demonstrated, to my knowledge, that these systems or any self-regulating metabolic systems or the like are stable unless they are linked with patterned macromolecules. More specifically, there is no decisive evidence that "cytoplasmic states" in the ciliates are stable in the absence of a macronucleus, whole or fragmented. "Cytoplasmic inheritance" in ciliates has more in common with embryonic induction than with heredity; the consequences depend as much or more on the "competence" of the responding tissue as on the quality of the inducing stimulus, which is conspicuously and specifically directive or selective, in a qualitative sense, only in the Group B mating types.

CHAPTER 12

The Organization and Development of the Metazoa

Studies of bacteria and various genetic fragments have disclosed indispensable information about genetic materials and genetic actions that could hardly be obtained from complex organisms but their limitations should be equally recognized. There is no compelling reason to believe that the relative simplicity of bacteria is primary or primitive in an evolutionary sense. Bacteria may have originated as fragments of true cells although they possess, as the transforming factors and other genetic fragments do not, the minimal machinery for independent existence as living organisms. For the study of developmental processes in the higher animals, which is our especial concern, the unicellular animals and plants are likely to be of greater evolutionary significance although they, too, have their limitations. The ciliated protozoa supply illuminating information about the selective utilization of the genome but contribute only minimally to the study of developmental processes. The developmental biology of multicellular organisms calls for different concepts, some of which will be considered in the present chapter.

I. The Cell Theory

The remarkable differentiation of specialized structures and "organs" in unicellular organisms seems to be possible only on a small scale; any substantial increase in size or complexity entails a subdivision of the organism into subunits, the cells. The natural unit, it has been proposed, is not the cell but the *energid*, by which is meant a nucleus and its "sphere of influence". The inability of a nucleus to control more than a limited amount of cytoplasm presents a basic problem in biology (Mazia, 1961). The division of multinucleated organisms into cells is a secondary phenomenon; it does not take place invariably and when it does it is not always complete (Russell, 1930). The division is usually effected in the higher animals, if not everywhere completely. Even when cells are apparently joined together by intercellular bridges, electron microscopy shows that there is no open communication between the interiors of contiguous cells (Fawcett, 1961). This does not necessarily mean that the cells are physiologically isolated from one another. Weiss (1958*a*) proposed that, dynamically, a cell does not end at its geometric outline and suggested that the cell surface is rarely naked but is covered with a coat of cell exudate which is usually of submicroscopic dimensions but sometimes microscopically visible as exemplified by slime cells and the surface coat of amphibian eggs. If this be true, all cell contacts are effected through enveloping films and Weiss considered a cell to be in "contact" with another body not only if the two surfaces were in direct opposition but also if they were separated by a narrow space occupied by a population of molecules whose free mobility was restrained. Not being subject to random dispersion, such molecules could form temporary links between the two surfaces and thus transmit effects that would be obliterated by a randomizing free diffusion. Grobstein and others have stressed the probable importance of the intercellular or *extra-cellular material* in mediating cell and tissue interactions. Some writers have credited it with wide, almost magical, powers as a complex continuum serving as a cell-binding framework, as well as an "information network", integrating the architecture and function of adult as well as embryonic tissues (Berrill, 1963; Grobstein, 1962; Moscona, 1960; Wilde, 1961*b*). Doubts about the wide powers attributed to the extra-cellular matrix have been expressed by several authors and discussed by Sigot (1965). Waddington (1962) questioned the relevance of the extra-cellular material to events in the closely-packed aggregates of cells such as are commonly found in early embryonic organs and quoted Grobstein's warning that, with certain *a posteriori* assumptions, the extra-cellular matrix can be made to explain almost anything. It is fairly obvious that, if interacting cells or groups of cells are separated by an extra-cellular matrix, that matrix will mediate the interactions. Nevertheless, the extra-cellular matrix is not an autonomous entity with fixed properties that imposes specific characters on the cells which it surrounds; it is a product of the cells embedded in it and its properties derive from those cells. It might be profitable to reconsider the extra-cellular matrix as a *consequence* of cell interaction and as a zone that allows cells to "come to terms" and interact with one another without losing their separate identities.

The cell theory of biological organization, as maintained by Virchow, holds that the organism is a "colony" or "republic" of independent cells and is built up of these cells and non-living intercellular materials much as a house may be built of bricks and mortar. There is no compulsive evidence from developmental or evolutionary biology to sustain this view or to indicate that colonial organisms hold the key to the organization of the metazoa. The cell theory, as held by Virchow, is vulnerable to destructive criticism from many sides (Cameron, 1952). An alternative "organismal" theory, advanced by Doncaster (1920) and supported by many others since, maintains that the organism is the individual and that the cells are the units into which it is divided to allow division of labour; the individual remains the complete unit and there is no final unit of space and time into which it can be divided.

As complexity increases, specialized cells interspersed with inter-cellular materials form tissues and organs but the fundamental unity of the whole organism persists (Radl, 1930; Russell, 1930).

The zygote itself is an "organism", the future organism in its simplest form and at every stage of development from the simple to the complex, in ontogeny and phylogeny the organism is an integrated whole (Russell, 1930). During development, two processes comprehended broadly by the terms *differentiation* and *integration* advance simultaneously. They are not antagonistic processes but different aspects of one process of evolution. Differentiation is not a primarily disruptive process which must be restrained or corrected by integration but a diversifying process which advances always in harmony with the integrated wholeness of the organism. If it were not so, the organism would die. The organismal theory of biological organization, like the kinetic theory of biotonic systems, has its shortcomings but the two theories concur in implying that biological organization at all levels is primarily and intrinsically integrative. Integration is a primary condition of life whereas differentiation is a secondary elaboration.

Weiss (1955) remarked that it is simply not true that nothing can be learned about "the organism as a whole" by studying its constituent parts and their interactions, adding, however, that it would be equally wrong to assume that mere preoccupation with the elements will tell the full story of their collective behaviour. More recently many biologists have been disturbed by the increasingly-reductionist trend of biological research. The basic difficulty is that the analytical procedures of chemistry and physics, when applied to living organisms, destroy or, at least, distort the object of study. It is not suggested that these procedures should not be used or that they are incapable of yielding informative results; the contention is that by themselves they are inadequate for studying the extreme intrinsic complexity of living organisms. Simpson (1962) maintained that reductionist analysis must be accompanied by its opposite, which he suggests might be called *composition*. He pointed out that biological explanation goes up as well as down the scale of levels of organization. The indispensable analytical experiments should be confronted whenever possible with the empirically-observed facts of life in the whole, and to be valid and useful the results must be accommodated in an integrative, holistic *biological* theory, which chemistry and physics, separately or together, cannot provide. For general biological discussions, reference may be made to essays by Commoner (1961); Elsasser (1958), Grobstein (1962), Ingle (1962), Sand (1965), Simpson (1962, 1963) and Weiss (1963).

II. Concepts of Development

Grobstein (1964) defined development as "the progression of the organism in its life history". It is not fully comprehensible without regard to the evolutionary history of the organism and to what is being developed, namely, an organism that lives, works, reproduces, grows old and dies. Development is a four-dimensional continuum in space and time and no wholly satisfactory way of describing or studying it without abstraction has yet been devised. The all-important time element is particularly difficult to handle.

In its narrower application to embryogenesis, development is a progression from the simpler to the more complex (Russell, 1930) and from the general to the special. Empirically, it is step-wise progressive and self-reinforcing. Like time, it goes in only one direction. Development is characterized by ongoing innovation which entails the progressive multiplication, diversification and specialization of cells, complex movements of cells and the organization of cells and their products into tissues and organs.

A. Preformation and Epigenesis

The evolution of ultimate complexity from apparent simplicity has always been a perplexing biological problem and two main theories, the preformationist and the epigenetic, have been advanced in various forms to account for it. Preformation theories hold that what things become exists from the beginning as potentialities or potencies, whereas epigenetic theories presume a stepwise development in which each step is a necessary cause or at least a precondition of the next and results in the emergence of new characters that were not preformed from the beginning. Modern embryology, as Oppenheimer notes, is based on epigenetic convictions and it stands on the concept that development proceeds from the general to the special or it falls (Oppenheimer, 1959a, 1963). The two theories do not now seem as sharply antagonistic as they were formerly thought to be, but they still reflect two different approaches to the study of development and heredity.

The preformation theory in the crude sense that the zygote contains a miniature organism or "homunculus" is no longer seriously maintained. In its present form, the theory is limited to the proposition that the inherited genetic material of the zygote predetermines the adult phenotype. Over thirty years ago Russell (1930) referred to "that stereotyped repetition of the course of development which we call heredity" and it is now apparent that the inheritance of phenotypic characters depends on the intervention of genetic materials in developmental processes. Geneticists maintain that development is controlled at every step by genes but the idea of a point-to-point correspondence between chromosomal genes and adult phenotypic characters is no longer tenable (Briggs and King, 1959). The concept of particulate determinant genes is now obsolete and obstructive. The zygote, as now believed, carries the whole of the genetic information needed for producing a complete

12. THE ORGANIZATION AND DEVELOPMENT OF THE METAZOA

adult organism but the information is *coded* in specifically and continuously patterned molecules of DNA. The *coded* information is preformed but the best available evidence indicates that the de-coding of the information is epigenetic and depends on progressively changing patterns of utilization of selected portions of the genetic material and on the linkage of genetic and biotonic patterns in mutual conformance.

The selective utilization of the total genome exemplifies a principle of great importance and wide application in biology described by Anfinssen (1959) as the principle of "functional adequacy of less than the whole". Anfinssen discusses this principle with especial reference to enzymes and hormones, whose biological effects are mediated by particular portions of their molecules, the remaining portions being inessential. Quastler (1959) deals with the same principle in a discussion of the application of information theory to biological integration and refers to it by the convenient term, *feature-sampling*. Briefly, Quastler means that in a particular biological process only a portion, or "sample", of the total available information of an input stimulus is actually used. In everyday life, different people often recognize the same object by observing different samples of its many features and ignoring others. As an extreme example of feature-sampling, it will suffice to mention the reliance of a blind man on touch and of a dog on smell for recognizing objects which most human beings identify by sight. In Quastler's terms, the effects of a compound input are mediated by only some of its features and not always by the same ones. It is proposed here that feature-sampling is an important key to the de-coding of the total genome during development; only a sample of the total genome is in effective use at one time and place and it is not always and everywhere the same sample.

Goldschmidt (1955) noted that the increasing complexity of organisms during evolution has not been accompanied by a parallel increase in the complexity of the genetic material, as judged by the appearances of the chromosomes. More recent observations have indicated that the mass of DNA increases with increasing complexity of the organism (Rich, 1962) but as Commoner (1961) has pointed out there is no consistent parallelism between the mass of DNA and the complexity of the organism. It has been suggested by Rich (1962) on the basis of DNA estimations and by Waddington (1962) on various other grounds that the genome has increased during evolution by duplication of segments of DNA so that in the higher animals a selectively advantageous redundancy of genetic information has been accumulated. Pontecorvo (1963) has emphasized another and perhaps more significant evolutionary trend, namely, an increase in the ratio of regulator to structural genes; the proportion of genes coding for polypeptides becomes smaller and smaller as the number of genes increases up the scale of organic complexity.

During ontogeny the *information* content of the organism is said to increase enormously and Elsasser (1958) maintained that this precludes a preformationist theory of development. Raven (1961), who estimated the number of "bits" of information in the zygote and in an adult organism, strongly disagreed with Elasser on many particulars. Apter and Wolpert (1965) have described such estimates of information content as highly arbitrary and of little value. The estimates take no account of spatial organization or of the fact that the increase in information content during development is attributable to the increase in the number of "units" making up the organism. Apter and Wolpert maintain that Information Theory is the wrong conceptual system for comparing stages of development and for analysing many of the processes involved; development is not equivalent to a communication system, to which information theory is applicable. They consider it more useful to think of the zygote as carrying *instructions for a programme of development*. Sand (1965) also stresses the problem of coding a programme of development and asserts that the translated genetic information is supplemented from the open system and is recoded with a novel component provided by tissue organization. Sand believes that development should be viewed as the history of an open system and that coding a programme for development seems to exceed the powers of DNA alone and requires the recognition of additional information carried in several higher levels of biological organization. The chemical concept of the nucleotide sequence code can specify only single-step relationships and has none of the integrative capacities required of a genetic code applicable to the wide scope of the problem of biological development.

Many other biologists have recently expressed misgiving about the adequacy of the DNA code to deal satisfactorily with all the phenomena of development and, more particularly, with stages beyond *translation* into protein structure; that all else follows automatically is a gratuitous assumption that leaves some of the most difficult problems of development untouched.

B. The Genetic Basis of Normal Development

For vertebrate development to advance to gastrulation and beyond, a complete and normally-balanced set of chromosomes must be present. The absence of one or more chromosomes arrests development and the defect cannot be made good by extra chromosomes that are not homologous with the missing ones. Polyploidy does not, in general, seriously impair development provided that the multiple sets of chromosomes are complete and normally balanced, but adding an incomplete set to a normally balanced complement usually causes abnormalities in early development and the severity of the effect is roughly proportional to the degree of imbalance; to ensure normal development, the sets must be complete and balanced. Haploidy does not necessarily preclude normal development but it often leads to a characteristic syndrome of abnormalities perhaps on account of "noise" in the developing system in the absence of the "redundancy" that ensures normal development of diploid or polyploid organisms.

The effects of genetic mutations, as observed especially in Drosophila, indicate that every step in development requires the participation of specific genetic loci. Mutations at specific loci produce specific patterns of abnormalities, which become apparent at various times spread through the whole period of development. Many loci are involved in each developmental process and different loci are active at different stages. Probably the whole of the genome is not in effective use at any one time but apparently all, or nearly all, of it is in effective use somewhere at one time or another (Briggs and King, 1959; Waddington, 1956).

The fertilized egg, or *zygote*, carries the *total genome* and is often described as *totipotent* or, even, as *omnipotent*, which certainly it is not. Grobstein (1959) points out that the zygote is formed by the fusion of two highly specialized cells neither of which would be expected, or has ever been observed, to transform into any other kind of cell. Moreover, the zygote itself has never been seen to transform into any kind of specialized cell much less into any or every kind. What the zygote can do is develop into an entire adult organism. Grobstein notes that *potency* and *potentiality* bear preformationist implications; development must entail loss of potencies, a gain being impossible because everything that comes about existed, potentially, before. Embryologists observe empirically that development is remarkable, above all, for increasing diversity and novelty and for gains of properties rather than losses. Grobstein maintains that in the study of development the idea of potentiality is of less immediate value than the idea of *capacity* which refers to "the particular developmental pathways immediately open to a region of an embryo" and to "the range and character of the demonstrable and immediate developmental alternatives". The genetic basis of capacity is the *facultative genome*, whereas the basis of potency is the inert *total genome*. Capacity, as defined by Grobstein, is similar to Waddington's *competence* but is a more generalized concept. It allows for the emergence of something "operationally new" at every stage of development.

C. The Programme of Development

The nature of the "instructions" for the programme of development is completely obscure but there is an abundance of empirical information about the programme itself. The programme includes not merely the production of an adult organism but also its reproductive activity, senescence and death. Comfort (1956) suggested that organisms become senescent because they "run out of programme". It seems at least as reasonable to suppose that senescence is part of the programme. The programme concept is especially relevant to the discussion of an important and distinctive feature of normal development that has been called *Directiveness*. Directiveness applies to the steady persistence of development towards a preordained goal, which is more constant than the means used to attain it. There is constant drive towards a normal and specific end or completion, which is usually related to one or another of the biological ends of self-maintenance, development or reproduction (Russell, 1930, 1945; Waddington, 1957, 1962). The phenomenon deserves attention here because it has been alleged that neoplastic development is sharply distinguishable from normal development in lacking directiveness towards a specific end-point. It is shown elsewhere that the proposed distinction is unreliable and that the concept of a programme of development is usefully applicable to neoplasia as well as to normal development.

Russell (1945) discussed many examples of directive activity in the maintenance and restoration of normal structure and function as well as in their development and drew some general conclusions about the characteristics of goal-directed activity. He noted that the goal is normally a terminus of action; when the goal is reached the activity stops, whereas if the goal is not reached activity usually persists. If the goal is not reached by one mechanism, other mechanisms may be brought into use. When the goal is normally reached by a combination of mechanisms, a deficiency in one mechanism may be compensated for by increased use of others. The same goal may be reached in different ways and from different starting points; the end-state is more constant than the method of attaining it. Goal-directed activity is limited by conditions but not determined by them.

The last two sentences of the preceding paragraph state inferences remarkably similar to those derived from the theoretical analysis of open-systems mentioned in Chapter 9, the penultimate sentence stating in effect the principle of *equifinality*. Many of the other characteristics mentioned by Russell are well exemplified by numerous homeostatic and compensating mechanisms and by the abilities of organisms to *regulate* during development and *adapt* physiologically during maturity. Barcroft's (1934) notable book on the architecture of physiological function presents from a different point of view and in different phraseology abundant evidence in the same general sense. It is worth noting that both Barcroft and Russell draw examples from adult organisms, including man, as well as from embryos; similar general principles apply throughout the life-cycle.

Gustafson and Wolpert (1963) strongly support Russell's contention that the end-state is more constant than the means of attaining it. Their studies of embryonic development of the sea urchin showed, as a striking feature, that although the end result was relatively precise and constant the events leading to it showed considerable variability. They suggested that the relative precision of the end-result might be due not to a single process but to several processes, each with a low degree of precision which, as a consequence of "overlap", might specify the end-result with considerable accuracy. Gustafson and Wolpert remarked that the whole problem of precision and imprecision in development has received insufficient attention and inferred from their own results that some cellular events might be specified with a much lower degree of precision than was previously supposed and did not require a complicated and precise regulating mechanism.

It is possible that the "normal" end-point is less precisely specified than has been thought; considerable variation in multiple constituent processes may be compatible with normality. Biochemical studies have disclosed a hitherto unsuspected variability in cell enzymes, the extent of which has not been throughly explored (Cohen, 1963; Villee, 1963). In more general terms, Weiss (1958c, 1962, 1963) has insisted that the total process has a greater degree of invariance than the individual component parts. Broadly speaking, the relative invariance of the end-point probably depends on the complexity of biological organization, on its net-like structure. It is the more remarkable that the programme of development can specify an effectively normal end-point without specifying in precise detail its constituent parts.

Russell maintained that the characteristic and remarkable feature of developmental processes is their anticipatory nature, their prospective significance, their building for the future. Barcroft expressed a concordant view in the words "the stage has been set before the play commences" and gives specific examples of *anticipation*. Waddington (1940) quoted the strange case of the callosities of ostriches as a memorable example. The callosities are apparently adaptive responses to the crouching position adopted by these birds but they develop in the embryo before the crouching stimulus can possibly operate. In a sense, *anticipation* is implied in the epigenetic theory of development but the mechanism is not thereby explained. The difficult conceptual problem is akin to that posed by the phenomenon of "decision in advance of performance" referred to elsewhere. Structure, as Russell insists, can develop completely independently of function. A sort of "morphogenetic memory" (Tomatis, 1967) is apparently built into the programme of development.

Barcroft commented, with special reference to *anticipation*, that the processes under discussion were of the kind to be expected if development had a teleological basis and were reasonable on any evolutionary basis. Russell said that directive action could rarely be called purposive but it is difficult to see why, unless it was to avoid the taint of teleology. As a consequence of evolutionary natural selection, most processes or structures that do not serve a useful purpose are soon eliminated and it is a fair inference that those that persist are useful and, in effect, purposive. Pittendrigh (1960–61) suggested that such processes should be called *teleonomic* to refer to "purpose" without the philosophical and theological implications about "design" that attach to "teleological". This term has been accepted by Davis (1961), Monod and Jacob (1961) and Simpson (1962) and seems to be a useful one.

The programme of development covers the whole life history of the organism, including its development from an ovum, its maintenance, its reproductive activities and its ultimate senescence and death. It does not specify in rigid detail the processes that are responsible for the sequence but it does, apparently, specify the architecture of spiders' webs and birds' nests (S. Wright, 1941). The ability of a linear DNA code to provide the instructions for so comprehensive a programme without substantial help has been questioned, but no other convincing identification of the instructions has yet been made. It may be suspected that the clue to the nature of the instructions is likely to be found in some of the phenomena that depend primarily on the *complexity* of biological organization, including the intrinsic properties of open systems, the organization of genetic and biotonic systems into a four-dimensional continuum and the factors responsible for the integrated "wholeness" of the organism.

The persistence and stability of the programme, even in unfavourable circumstances, is remarkable. It is possible to alter or distort the course of development by experimental interference or by changing the environment but it is not possible to produce an essentially different organism (Russell, 1930). It is difficult to alter the normal *order* of appearance of embryonic tissues and organs (Flickinger, 1962) and it is difficult, also, to tamper with the "biological clock" without killing the organism (Goodwin, 1963).

III. Early Vertebrate Development

A. THE GAMETES

The germ cells of sexually-reproducing animals are highly specialized cells, produced by a long and complex development in the gonads during which the number of chromosomes is halved. The spermatozoon is specialized exclusively, as it seems, for the preservation and transfer to the ovum of one copy of the male total genome carried in a haploid set of chromosomes. During spermatogenesis, interactions between nucleus and cytoplasm are minimal and in sperm heads the genetic DNA, accompanied by protamine instead of the usual histone, is in effectively crystalline form (Picken, 1960).

The specialization of the ovum is more complex; it not only provides a haploid set of chromosomes carrying the female total genome but is responsible also for cytoplasmic continuity from one generation to the next and for supplying the organized cytoplasmic pattern that serves as the basic pattern of future development (Raven, 1961, 1963). Nucleus and cytoplasm interact in a complex way during oogenesis. Intense nuclear activity seems to be directed to the synthesis of materials that pass into the cytoplasm. The cytoplasm of the ovum contains large quantities of RNA and a substantial amount of DNA. Embryologists are widely agreed that the initial pattern of development depends on the heterogeneity and anisotropy of the egg cytoplasm. In some eggs, visible heterogeneity of the cytoplasm corresponds fairly well with future morphogenetic events but in others, no correspondence is detectable and experimental redistribution of visible cytoplasmic materials does not disturb morphogenesis. The pattern of morphogenesis probably depends upon a more subtle heterogeneity than the light microscope reveals. The heterogeneity is evidently organized, not random, but the organization of the pre-fertilized egg, perhaps the most crucial event in development, is virtually uninvestigated (Oppenheimer, 1959*b*). It may depend, as Raven (1961) believes, on a relatively stable organization of the egg surface and the immediately underlying cytoplasm to form a "cortical field" having a stable pattern imposed, in part, by the genetic action of the egg nucleus and, in part, by an interaction between the egg and its ovarian environment.

B. Fertilization

Fertilization introduces a second haploid set of chromosomes into the ovum; normally it "activates" the ovum and sometimes it decides the plane of the first cleavage but artificial parthenogenesis shows that it is not indispensable for any of these effects. The known immediate effects of fertilization give no impressive clue to the nature of the "activation" which sets development on its way. Unless "activated", the ovum merely ages and dies; activation in some unknown way releases its ability to divide (Runnström *et al.*, 1959).

C. Cleavage

Enucleated eggs or eggs with abnormal nuclei can cleave to produce morulae and blastulae but these never begin to gastrulate. Normally, the nuclei multiply and the cytoplasm divides repeatedly into blastomeres but until the onset of gastrulation its substance does not notably increase. During cleavage, synthesis of DNA and associated protein is active but synthesis of RNA is not appreciable; the nucleus, it seems, is engaged almost exclusively in replication of the total genome, transcription being in abeyance. The nucleus exerts no important influence on the cytoplasm, which functions autonomously, as it can do for a while in enucleated eggs.

Formerly, sharp distinctions were drawn between *mosaic* and *regulative* eggs and between *determinate* and *indeterminate* cleavage but the distinctions have become much blurred (Grobstein, 1959). Mosaic eggs are found especially amongst the invertebrates; in amphibia and probably in all other vertebrates, to which the remainder of this chapter applies, the ova are of the regulative type and blastomeres are *not* irrevocably committed to their respective fates (Holtfreter and Hamburger, 1955).

The blastomeres are diversified as a consequence of the primary heterogeneity and anistropy of the egg cytoplasm and its cortex. The zygote thus becomes a multicellular system, the blastula, whose constituent units have similar nuclei but disparate cytoplasms.

D. Gastrulation

The functional isolation of the nuclei ends in the late blastula stage and the genome comes into effective use. The utilization of the genome is selective and, in accordance with the general principle of mutual conformance of genetic and biotonic patterns, it may be expected to differ in cells with disparate cytoplasms. Transcription and translation of selected effective genomes begin in the late blastula, and the consequences are manifested by biochemical and morphological differentiations, which become more conspicuous in the gastrula. Synthesis of new RNA becomes substantial and protein synthesis is evident, especially in the most actively developing regions. New antigenetic materials are synthesized between the blastula and gastrula stage in both frogs and newts. Metabolism changes considerably during gastrulation and neuralization. During gastrulation, the endoplasmic reticulum, previously very coarse and loose, becomes more complex, ribosomes increase in number and mitochondria increase in size and complexity, as well as in number.

Gastrulation marks the beginning of differentiation and morphogenesis. The surface cells of the dorsal half of the blastula begin to proliferate more rapidly and in one region they migrate inwards and then spread out to form a distinct inner layer of cells underlying the outer ectoderm. A permanent cavity, the archenteron, forms immediately below the inner layer and communicates with the exterior through an opening, the *blastopore*. Invagination of cells is most active at the dorsal lip of the blastopore and the invaginating cells produce important axial structures, including the notochord, and, lateral to it, the rudiments of the myotomes. The inner layer delaminates into the endoderm, forming the roof of the archenteron, which is the primitive gut, and the mesoderm between the endoderm and the surface ectoderm. The axis of the organism is now laid down and the three primary germ layers, ectoderm, endoderm and mesoderm, are established. Two parallel longitudinal folds of ectoderm, the neural folds, next project along the axis of the embryo, which by now is elongating, and fuse in the mid-line to form the

12. THE ORGANIZATION AND DEVELOPMENT OF THE METAZOA 317

neural canal, the rudiment of the brain and spinal cord. When the neural folds emerge the gastrula becomes a neurula.

The pioneer experiments of Spemann and his school gave convincing evidence of the importance of interactions between neighbouring tissues in morphogenesis. Spemann and his colleagues removed tissue from the dorsal lip of the blastopore of an early amphibian gastrula and implanted it in the blastocoele cavity or under the undifferentiated ectoderm of another early embryo. The graft continued to develop in its new site and, for the most part, it formed the same kinds of mesodermal and axial structures as it would have done in the intact embryo but, in addition, it induced the overlying ectoderm of the host embryo to form a neural plate, although without experimental interference it would have become epidermis. The graft also induced changes in the neighbouring mesoderm of the host. The graft and the modified host tissues tended to organize themselves so as to make a more or less complete embryo and Spemann called the material from the normal lip of the blastopore the *organizer*.

The observations of Spemann and others showed that dissimilar contiguous tissues could interact to form an organized whole and, further, that as a consequence of *induction* by a neighbouring tissue, a particular tissue may be constrained to differentiate into specialized structures that it never forms in normal development. Weiss (1953) described this latter phenomenon as *deflection*; in other contexts it is usually called *metaplasia*. After exposure to appropriate inductors, any part of the undifferentiated ectoderm of an early amphibian gastrula can develop into any kind of epidermal or mesodermal tissue or into some mesodermal structures. The *capacity* of the embryonic ectoderm for varied differentiations is far greater than could be inferred from its behaviour during normal development.

The induction of a neural plate by underlying axial mesoderm, referred to as *primary induction*, is followed by a chain of inductions, *secondary*, *tertiary* and so on, which stretches throughout embryonic development. Moreover, induction, or something like it, may be responsible for maintaining the integrity of structure and function in the mature normal animal.

In the frog *Rana esculenta*, an optic cup grafted under the ectoderm at an appropriate stage of development can induce the overlying ectoderm, which normally would differentiate into epidermis, to differentiate into an eye lens. This is another example of *deflection* of development. If an optic cup be removed from an embryo before a lens has developed, a lens will form, nevertheless, at the normal site, but it is never comparable in size or degree of differentiation with a normal lens; the presence of an optic cup is essential for complete and typical morphogenesis. In the intact animal, the optic cup does not select or deflect the path of differentiation; its chief action is morphogenetic, not differentiative. More generally, despite a measure of development in the absence of induction, an appropriate spatial and temporal coordination of inductors and responsive tissues seems to be necessary for the complete and typical development of many organs. It will be necessary to refer to certain aspects of induction and, more generally, of tissue interactions, at a later stage. Several comprehensive and detailed accounts of embryonic induction are available (Dalq, 1960; Grobstein, 1956b, 1959; Holtfreter and Hamburger, 1955; Saxen and Toivonen, 1962; Waddington, 1956; Yamada, 1961).

From gastrulation onwards cell *proliferation* and *growth*, *differentiation*, *morphogenesis* and *maintenance* are taking place simultaneously in a progressively changing integrated organism. These elements of development cannot be separated without abstraction, which must be tolerated, if only for purposes of exposition. The following chapters deal with certain aspects most relevant to problems of neoplastic development discussed in Part II and are, in that way, selective. As already mentioned, problems of growth, differentiation, morphogenesis and the integrated wholeness of the organism are posed by neoplastic development as well as by normal development. The impact of molecular biology has been greatest on problems of cell proliferation and of differentiation. Reference to cell proliferation was made in Chapter 5; differentiation will be discussed next.

IV. Differentiation

The term *differentiation*, in its broadest sense, applies to diversification in general. Grobstein (1959) held that it applies properly only to the diversity that is "in a broad sense, developmental", "stable beyond the initiating stimulus" and "relatively stable and progressive in terms of the life cycle of the organism". Further, "the fixity of ongoing innovation" is an essential characteristic; short term changes that are "repetitive without significant residue" and which correspond with what Weiss called *modulation* are not differentiative. Unfortunately, it is not universally agreed that differentiation and modulation are sharply separable and Weiss (1950) himself wrote that "Permanent differentiation can, therefore, be regarded as merely an advanced stage of an initially reversible modulation". In practice, *differentiation* is applied to a wide variety of processes and products contributing to the eventual diversification of organisms and especially to specialized structures and functions. Waddington (1962) and Grobstein (1964) have attempted to bring some order out of the semantic chaos but with little apparent effect on general usage. The term is applied most often to the specialization and maturation of cells and tissues, particularly when they are manifested by characteristic products or distinctive structural organization.

The most recent and narrowest definition of differentiation is based on protein synthesis. Jacob and Monod (1963) proposed that ". . . two cells are differentiated with respect to one another if, while they harbor the same genome, the pattern of proteins which they synthesize is different." This was based primarily on studies of bacteria but Grobstein (1963) accepted it, provisionally and with some modifications, for application to differentiation in vertebrate animals. The definition seems to me to be unduly restrictive and arbitrary; by making protein synthesis an exclusive criterion of differentiation, it seeks to define a whole phenomenon by one of its component processes. As Grobstein himself remarks, the production of defined amino acid sequences does not exhaust the developmentally-significant paths of macromolecular synthesis. Waddington (1962) notes that most of the conspicuous components of cells are probably not the direct products of "gene-protein systems" but are secondary elaborations which depend, in part, on organization of a *supramacromolecular* order of magnitude and on morphogenesis. Waddington distinguishes between a "gene-protein loop" and a subsequent "protein-phenotype loop". Goodwin (1963) also distinguishes between the primary and secondary consequences of translation. He supposes that the specific proteins travel from their site of origin to "some cellular locus" where they exert an influence on the metabolic state either by enzyme action or by some other means. Enzymatic proteins, it may be noted, may make a non-protein terminal product as in the parietal cells of the stomach, whose specialized function is to produce hydrochloric acid.

The changed state of a biological system implied by *differentiation* may manifest itself in detectable products, protein or non-protein, or in a change in reactivity detectable only by experiment (Weiss, 1958b). The "changed state" of, say, basal cells of the epidermis is not demonstrably attributable to, or accompanied by, specific protein synthesis; it is attributable to a changed developmental *capacity* or, as here proposed, to a changed *facultative genome* having a specific differential availability of potential effective genomes. Weiss now calls this *strain differentiation*. Probably many steps in differentiation involve no more than sequential changes in the facultative genome, detectable only by experimental tests. Too many studies of differentiation are centred almost exclusively on visible terminal products.

Weiss (1950, 1953, 1958b) has maintained for a long time that the crucial test of differentiation is "differential behaviour based on differential constitution" and has insisted on a distinction between "true differences in cell constitution" or "intrinsic cell character" on the one hand and, on the other hand, "the signs by which these differences become manifest." This distinction has been exposed to the criticism that "true" or "intrinsic" cell differences have not been demonstrated. Evidence that they do exist is accumulating and is implied by the recent opinion of Grobstein (1965) that "covert differentiation" may be the first basic step in differentiation.

Weiss described three successive steps in the specialization of cells: first, cells are selectively routed into one of several possible courses of development to become, for example, myoblasts; second, they acquire the production machinery for making specific differentiation products, such as myofibrils; and finally, the production of myofibrils actually takes place. Weiss pointed out that differentiation does not necessarily advance through all these stages, as becomes evident when reproducing cells or stem cells are segregated in special "germinal" layers or cords. The basal cells of the epidermis and of the neural epithelium are not "differentiated" in the sense of being visibly specialized but they are differentiated in the sense of having different *capacities* for further development; the basal cells of the epidermis yield epidermal cells whereas the neural epithelium yields only neural cells (Weiss, 1953, 1958*b*).

The three steps distinguished by Weiss can now be described somewhat more precisely in terms of selective utilization of the genome.

Step 1. The "selective routing" of Weiss corresponds with the *selection* of an effective genome from amongst those available in the facultative genome of the unspecialized parent cells.

Step 2. The elaboration of a "production machinery", can be subdivided into three sequential steps.

(a) *Transcription* of the selected effective genome yields a specific mRNA which leaves the nucleus and attaches to ribosomes in the cytoplasm.

(b) *Translation* of the mRNA on the ribosomes yields proteins, including enzymes.

(c) *Elaboration of the actual "production machinery"*, for which the term *operative cytosome* is suggested. This elaboration entails a supramacromolecular organization, including RNA, enzymes and other proteins and, perhaps, other macromolecules.

Step 3. The operative cytosome fabricates the specialized products.

The three subdivisions of Step 2 correspond well with Goodwin's scheme. The term *operative cytosome* is designed to apply comprehensively to the "production machinery" without narrowly specifying its nature or extent. It includes, at least, the endoplasmic reticulum and the mitochondrial membranes. At the present time, the steps of transcription and translation are being most actively studied. Many of the experiments include the use of actinomycin D to inhibit transcription and of puromycin or some other agent to inhibit translation. It may be well to note that some of these experiments are not easy to interpret. Although actinomycin D and puromycin do exert these actions, they can under certain circumstances, especially when used in high-dosage, exert other effects including lethal ones. Actinomycin seems to be the more specific in action but it is highly toxic *in vivo*. Puromycin activates

12. THE ORGANIZATION AND DEVELOPMENT OF THE METAZOA

glycogen breakdown in the liver and yields a class of abnormal peptides. Both substances can act as stressors, which stimulate the release of various hormones including adrenocorticosteroids (Hechter and Halkerston, 1965).

All investigators have found protein synthesis to be a necessary condition of differentiation but have found some differences in the requirement for sustained synthesis of mRNA by transcription of the nuclear genome. Observations on early frog gastrulae by Flickinger (1963) indicated that transcription of specific mRNA started shortly before the onset of differentiation. Exposure to actinomycin D interfered with the tissue differentiations due to occur soon afterwards but not with those expected at a later stage of development. Protein synthesis accompanied overt differentiation. Brachet and his colleagues also studied embryogenesis in frogs and found that protein synthesis increased sharply at the end of cleavage. The onset of differentiation was dependent on protein synthesis that could be inhibited at the stage of transcription by actinomycin D and at the stage of translation by puromycin, indicating, again, that new activities need new transcriptions of the genome (Brachet, 1963; Brachet and Denis, 1963; Brachet *et al.*, 1964; Denis, 1964*a*, *b*, *c*).

Laufer *et al.* (1964) showed that transcription of DNA was necessary for the pupation of larvae of the midge *Chironomus thummi*; partial inhibition of transcription by actinomycin D restricted development and was usually fatal. Furthermore, the salivary glands continued to secrete their characteristic product only so long as transcription was sustained and unimpaired. On the other hand, younger larvae survived exposure to actinomycin for several days, suggesting that the metabolic activities needed for *survival*, as contrasted with those needed for *new development* or specific performance, could continue for a relatively long time without sustained transcription of the genome. Several other investigations confirm the impression that *maintenance* processes are relatively resistant to inhibition by actinomycin probably, in part at least, because many or most of these essential processes take place in the mitochondria.

Grobstein and his colleagues used several methods in a comprehensive study of the differentiation of the exocrine acinar cells of the pancreas of eleven-day-old mouse embryos. They cultivated rudiments of the pancreas *in vitro* by the organ culture method and found that, on the second day of culture, electron microscopy revealed a poorly developed endoplasmic reticulum and abundant ribosomes disposed in formations resembling polysomes. By the third day, the endoplasmic reticulum was well developed and many cisternae were present; at this time the ultrastructure corresponded with that expected of cells that have begun to synthesize their characteristic end-product and in which protein is already accumulating in the cisternae of the endoplasmic reticulum and beginning to transform into secretion granules in the Golgi region. Prezymogen granules were small in size and few in number at this time but they were obvious in electron micrographs on the fourth day and discernible by light microscopy soon afterwards. The observations were consistent with the chemically-estimated amounts of amylase, which did not increase significantly before the end of the second day. Actinomycin almost completely suppressed the production of amylase if it was added to the cells during the first 24 hr of culture but if added at later times it had a diminishing effect until on the fourth day it had little effect on the synthesis of amylase. Grobstein suggested that by this time mRNA had been transferred to ribosomes. The observations imply that the synthesis of amylase continues thereafter without further transcription of the genome to yield labile mRNA. Amylase began to be synthesized in the absence of transcription at the same time as cell division stopped, an observation exemplifying the apparent antagonism between differentiation and cell proliferation, which will receive more attention later (Grobstein, 1964; Rutter et al., 1964; Wessells, 1964).

The evidence provided by Grobstein in favour of continued protein synthesis in the absence of sustained transcription of the genome is consistent with the evidence summarized elsewhere for a limited degree of autonomous functioning of "cytoplasm". Waddington (1962) expressed the opinion that the mechanisms that bring protein molecules into being are not necessarily the same as those responsible for their continued production. He thinks it conceivable that, in fully-functioning liver cells for example, the genes may have long-finished their work and handed over to the "productive machinery", which they were originally responsible for building up. The clearest indication of some such delegation of genetic function is supplied by the demonstration of protein synthesis in mammalian reticulocytes which have no nuclei. The current explanation is that the sustained protein synthesis depends on the translation of a relatively stable RNA transcript. Various messengers, it seems, either differ widely in their intrinsic stabilities or are variously stabilized by incorporation together with protein and perhaps other materials into organized structures such as the ribosomes and the mitochondrial membranes.

The foregoing discussion is relevant to the intepretation of some observations by Wilde (1961a) on non-nucleated fragments of disaggregated neuroepithelial cells of urodele embryos. Certain of these fragments underwent changes *in vitro* resembling those characteristic of normally differentiating pigment cells; they approached, to some extent, the dendritic form characteristic of melanophores and synthesized melanin in the absence of a nucleus. Wilde inferred that the contemporary presence of a nucleus is not necessary for certain differentiative processes in melanocytes and, more generally, that the cytoplasm can carry out differentiative activities in the absence of a nucleus. These inferences are compatible with the evidence for a durable operative cytosome already mentioned but Wilde maintained, further, that the "directional cues" for a specific differentiation can act directly on the cytoplasmic machinery and that his observations reduced the nucleus to "a somewhat secondary role in differentiation". These conclusions are unacceptable. The cells had received their "directional cues" before they were dissociated and apparently by that time they had acquired a stable operative cytosome. The role of the nucleus is indeed crucial. The "directional cues" act at the genetic level to select an effective genome which must be transcribed and translated before the synthetic machinery or operative cytosome can be built. Experiments that begin after the most critical steps in differentiation have been accomplished can be extremely misleading.

Laufer *et al*. (1964), whose own experiments were noticed earlier, maintained that morphogenesis and development in general are dependent on DNA-dependent synthesis of RNA and further that developmental processes may require *de novo* synthesis of different species of DNA-dependent RNA. It is indeed highly probable that development depends on a succession of differing mRNA's produced by transcription of differing effective genomes in the correct temporal order. In the closer analysis of differentiation, estimates of total RNA or total mRNA are not likely to be helpful; it will be desirable to differentiate between different effective genomes and their RNA transcripts. Ideally, the transcripts should be precisely specified in terms of base sequences. This is not yet possible but sufficient progress has been made to allow the demonstration of recognizably different *patterns of transcription* of the genome in diverse kinds of cells or in cells of one kind under different physiological conditions (Kidson and Kirby, 1964*a*, *b*; Paul, 1967).

The investigations discussed so far do not illuminate the critical first step in differentiation, the *strain differentiation*, which entails the preselection or selection of an effective genome, and they do not apply at all to supramacromolecular organization or morphogenesis. Much of the available information about these matters has emerged from studies of embryonic induction. The term has been applied indiscriminately to factors regulating the selection, transcription and translation of effective genomes and the subsequent steps of supramacromolecular organization and morphogenesis. Both the selective and morphogenetic aspects are illustrated in Spemann's studies of primary induction of the neural plate and secondary induction of an eye lens. In the main, the more recent studies of secondary inductions exclude the selective or directive aspect and deal only or chiefly with morphogenetic events.

Primary induction has been extensively studied by experimental embryologists and the voluminous literature has been comprehensively reviewed by Saxen and Toivonen (1962). Great energy and ingenuity have been devoted to the search for a specific inducing or neuralizing substance. None has been found. A great variety of tissues, dead or alive, from animals high or low in the evolutionary scale can neuralize amphibian ectoderm and so can a variety of chemical substances. The relevance of these experiments to primary induction in the intact embryo is questionable. An understanding of induction is more likely to be derived from the reacting cells than from the inducing materials, which will not be further considered here.

The importance of induction in vertebrate development is not questioned but it seems to be a secondary acquisition, essential perhaps for the evolution of vertebrates, that supplements without supplanting the basic mechanisms for controlling development as they are manifested in the mosaic eggs of invertebrates (Dalq, 1960; Goldschmidt, 1955). Goldschmidt interpreted induction as a technical advance for simplifying and integrating the "basic mechanisms", by which he clearly meant genetic mechanisms. The "basic mechanisms" hitherto have been discussed mainly in broad general terms of *competence* or *capacity*. These concepts will be discussed in more detail on account of their basic importance in the histology and histogenesis of tumours as discussed in Part II.

V. Competence and Capacity

Waddington (1932) introduced the term competence to refer to the ability of cells to differentiate into a particular tissue under certain conditions. Later writers have insisted that competence refers only to the specified response of a particular tissue, at a particular time to a specified stimulus (Holtfreter and Hamburger, 1955; Saxen and Toivonen, 1962). Grobstein (1959) pointed out that *capacity*, as defined by him, includes such competence without requiring specification of the stimuli eliciting particular differentiations; if competence be similarly freed from this requirement, as in practice it often is, then capacity is equivalent to the sum of all the competences of a tissue.

Competence is regarded as a state of physiological reactivity or responsiveness or, as Waddington (1940, 1956) suggested, as a state of unstable equilibrium in which a tissue is poised between two or more paths of development and may follow one or the other of them according to the prevailing environmental conditions. Competence has the essential quality of *choice* and more than two choices may be available. Extrinsic conditions select one of the available possibilities but genetic factors determine which possibilities are available. Inductors do not elicit structures that are not represented in the genetic repertoire of the tissue (Holtfreter and Hamburger, 1955); a change is not imposed on a cell but is "evoked from its multiple repertory" (Weiss and James, 1955). As an example, organs called balancers develop on the heads of larval newts but not on larval frogs. Balancer-inducing material is present in the head region of frogs as well as in newts but it does not induce a balancer except in newts (Holtfreter and Hamburger, 1955; Saxen and

Toivonen, 1962). In the terminology proposed here, the "multiple repertory" is encompassed by the facultative genome, which specifies the available developmental possibilities.

An outstandingly important property of competence is that it changes with time. The competence of the undifferentiated amphibian ectoderm to respond to induction by the normal organizer by forming a neural plate develops gradually, reaches a peak in the early gastrula and then declines until it is no longer detectable in the neurula. The change with time proceeds gradually and apparently autonomously; it is not a sudden event but *an autonomous process of aging* (Holtfreter and Hamburger, 1955; Waddington, 1956). Holtfreter and others have shown that primary competence declines in tissues cultivated *in vitro* just as it does *in vivo*; the response of the explanted tissue to an inductor is that proper to its actual age and not to the age when explanted. (Refs in Saxen and Toivonen, 1962).

Competence changes with time in a qualitative, as well as in a quantitative way. In *Ambystoma mexicanum* there are two successive periods of competence during which the ectoderm responds differently to the same kind of inductor; during the first period the inductor evokes fore-brain structures and in the second period it evokes hind-brain structures. The differences in response clearly must be attributed to alterations in the competence of the responding tissue. Similarly when the amphibian ectoderm loses its primary competence in the neurula stage, new competences for differentiation into such ectodermal derivatives as the lens of the eye, nasal epithelium and otocysts appear. This regional pattern of competence develops gradually and is demonstrable *in vitro* as well as *in vivo* (Holtfreter and Hamburger, 1955).

At its peak, competence may reach a state of *developmental imminence*; one of the possible differentiations is so imminent that it may occur in the absence of an identifiable inductor and in response to a great variety of non-specific stimuli (Grobstein, 1959; Grobstein and Parker, 1958). The study of primary induction has been greatly confused by the special case of developmental imminence known as *autoneuralization* of the early amphibian ectoderm. When competence for neuralization by the normal organizer is high, the process can be "triggered-off" by diverse non-specific stimuli bearing no relationship to the normal inductor. Many materials have been wrongly identified as inductors on the basis of a non-specific stimulation of imminent neuralization. Neuralization may become so *imminent* that, as now demonstrated beyond reasonable doubt, it takes place without any identifiable stimulus (Barth and Barth, 1959; Grobstein, 1959; Saxen and Toivonen, 1962). Waddington (1940) said that sometimes the function of the inductor might be so slight as to be replaceable by minor variations in conditions very difficult to identify. In the modern idiom, *noise* in the developing system may be a sufficient cause of "induction".

Developmental imminence implies that competence is more than a state of instability; it implies that, although a choice of developmental pathways is available, the choice is weighted, sometimes to such an extent that one of the choices will be preferred to all others, even in the absence of any identifiable extrinsic stimulus. The differential or preferential utilization of available developmental paths has been little considered or investigated. Comparable phenomena in the ciliated protozoa were discussed in Chapter 11 and interpreted in terms of selective utilization of the facultative genome. The phenomena now under discussion may be similarly interpreted.

There are one or two indications that preferential utilization of the facultative genome can be modified experimentally. When undifferentiated amphibian gastrula ectoderm is cultivated *in vitro* the preferential competence for neuralization is lost and replaced by a preferential competence for differentiation into mesodermal structures; autoneuralization gives place to autodermalization (Matsui, 1960 cited by Saxen and Toivonen, 1962). A similar phenomenon seems to have been revealed by experiments on the Hensen node of the chick blastoderm at the primitive streak stage. If the node is transplanted into another embryo, it develops ordinarily into ectoblast and, eventually, into neural tube but if the node is stained with Nile Blue before implantation, it develops into notochord and mesoblast or even into a whole new embryo (Dalq, 1960).

Saxen and Toivonen (1962) maintained that the investigations of autoneuralization show that the factors required for neuralization are already present in the ectoderm, at least in the form of their precursors; whether, during normal induction only an "unmasking" or release of factors occurs or whether there is an actual transfer of similar active factors from the inductor to the ectoderm remains to be decided. This statement is acceptable only with the important reservation that there is no evidence whatsoever that autoneuralization is due to factors, whether intrinsic or extrinsic, in the sense of specific inductive *substances*. The more likely interpretation is that autoneuralization is attributable to the imminence of a particular developmental pattern based on the preferential availability of a particular genetic pattern in the facultative genome. Preferential availability can be regarded as the *preselection* of a potential effective genome for later use, preselection being used in the sense discussed previously (p. 302).

Competence and inductive power are synchronized and Goldschmidt (1955) commented on the remarkable circumstance that inductors work only where they are needed, the competence being highest where inductor is available. Waddington (1962) stated that, in Drosophila, competence is determined by genetic constitution and, according to Goldschmidt genes control also the time-pattern of competence. Flickinger and his colleagues (1962, 1965) and Goodwin (1963, 1964) discussed the temporal changes of competence as the basis of sequential differentiation. Seemingly the changes are not controlled by environmental stimuli and Goodwin suggested that some kind of "biological clock" might be operating within the competent cells. Flickinger drew special attention to the extreme difficulty of altering the *order* of the sequential changes without killing the organism and he proposed that inductors merely provide the materials and energy for protein synthesis whereas competence is responsible for the sequential initiation of gene action. Both Flickinger and Goodwin imply that intrinsic properties of the genome are decisively important in the timing and qualitative specificity of differentiation and Flickinger, in particular, reduces the action of inductors to a permissive instead of a determinative one. This is plausibly true of the inductors that merely activate a preselected potential effective genome. Whether it applies to the experimental induction of a neural plate or, at a later stage of development, an eye lens in ectoderm that would become epidermis if left to itself, is more questionable. In these experimental inductions, the inductors exert a directive or selective action, which can hardly be considered to be merely permissive. It is desirable to distinguish between inductors which select or preselect a path of development or an effective genome and inductors which merely initiate or facilitate progression along preselected paths. Examples of the latter type of inductor will be given later.

It is doubtful if competence can be linked with genetic *action*. Its essential basis seems to be the differential availability or the preselection of one of a number of genetic patterns available in the facultative genome in advance of the actual utilization of any one of them as an effective genome. It depends more probably on genetic or chromosomal organization than on genetic *action*.

Weiss (1953, 1958b) discussed three important restrictions upon the effective utilization of the genome under the terms *discreteness*, *exclusivity* and *genetic limitation*. *Discreteness* means that the different cell types in an organism are rather sharply delimited from one another and do not intergrade; as Grobstein (1959) says, they are plainly one thing or another. *Exclusivity* means that a cell cannot, at one time, follow more than one of the discrete courses open to it; once it has embarked on one course, the other courses are automatically closed. Grobstein adds the comment that the properties of differentiated cells exist in sets; it is the *set* of characters that is exclusive. *Genetic limitation* means that the finite, though large, number of cell types is limited from the start by the constitution of the genome. The genome determines what a cell *can* do; extrinsic factors may decide what, within the genetic limitations, a cell *will* do.

Waddington's earlier concept of *canalization* covers discreteness, exclusivity and genetic limitation. More recently Waddington (1957, 1962) has elaborated a concept of stabilized, genetically-controlled pathways of development, which he calls *creodes*, to cover much the same range of empirical observations. Previously, Waddington had noted that the sharply contrasted alternatives depend on the presence of the normal genotype in which they are primarily ensured by natural selection. The constancy of the wild type is attributed to a "buffering" of the genotype against minor variations in both the developmental environment and the genetic make-up. Nevertheless the contrasts are blurred in many mutant types and intermediate types are found, especially in pathological conditions.

Grobstein has argued that the exclusivity of sets of characters does not mean that the properties of one set are universally linked nor does the absence of intergrading mean that differentiation is integral in the sense that it cannot be fractionated into independently variable characters. Nevertheless, the only example of *dissociation* of characters cited by Grobstein are, first, the loss of some of the characters with the retention of others during tissue culture *in vitro* (Evans *et al.*, 1952) and the dissociation of characters in neoplasia described by myself (Foulds, 1954). J. Needham (1942) discussed the *dissociation* or *disengagement* of characters in embryos and maintained that fundamental mechanisms can be thrown out of gear experimentally. He maintained that fundamental mechanisms are separable in fact and not merely in thought. More recently, A. E. Needham (1964) has discussed the relevance of dissociability to the principle of "multiple assurance" but no other comprehensive discussion has come to my notice. In the higher animals, dissociation of characters, if it occurs often except in neoplasia, has attracted remarkably little interest. It is reasonable to suppose that if cell characters are dissociable by experimental interference and in neoplastic disease, some unidentified regulatory mechanism ensures that normal development shall be discrete, exclusive and, in effect if not of necessity, integral. As recognized in Waddington's concept of *creodes*, the phenomena are not explicable without reference to the third restrictive element in development, genetic limitation. The frequency of the independent variability of characters, of the dissociation of characters and of the independent progression of characters in neoplastic development was emphasized in Part II of this book. The lack of integration, at all levels of analysis, is indeed one of the most consistent features of neoplastic disease.

An apparent break-down of exclusivity in certain non-neoplastic cells, described by Wilde (1961*a*) as *confused* cells, may not be a true exception to the general rule but is, nonetheless, instructive. Wilde's evidence for *confusion* came from some late-differentiating cells obtained by cultivating washed, disaggregated cells of *Ambystoma* embryos *in vitro*; he described melanogenesis together with myogenesis or with neurogenesis in single cells. The best-studied example of *confusion* was first observed by Fell and her colleagues whilst studying organ cultures of embryonic chick epidermis. By adding an excess of vitamin A to the culture medium, the epidermis could be changed into a mucus-secreting epithelium but the effect on individual cells depended on the stage of differentiation the cells had reached when the vitamin was added. Some of the superficial epidermal cells, although prevented from keratinizing, were too far "differentiated" in that direction to be converted into goblet cells. Goblet cells developed in the deeper, less differentiated layer. When a mucous epithelium was returned to a normal medium, the cells of the basal layer quickly resumed their normal keratinizing function but some of the more superficial cells contained mucus in their more superficial portions and fine filamentous material, characteristic of young keratinizing cells, in their deeper portions. Fell and her associates inferred that under certain conditions a cell might be differentiated for the simultaneous performance of two quite different specialized functions instead of being irreversibly committed to one type of specialization (Fell, 1956–57; Jackson and Fell, 1963; Pelc and Fell, 1960).

Strictly interpreted the most thoroughly studied examples of confusion show only that the visible products of two differentiative processes, which ordinarily are mutually exclusive, may co-exist in individual cells. They imply that in cells that have not advanced too far towards terminal specialization, transcription can be switched from one effective genome to an alternative one; transcription of one effective genome, for a while at least, does not suppress the availability of others present in the normal facultative genome. Observations on the serotype effective genomes in the macronuclei of Paramecium, as described in Chapter 11 led to the same conclusion. None of the observations on confusion seem to me to establish beyond question that two alternative effective genomes can be transcribed simultaneously. In view of the evidence already given for the appreciable stability of the operative cytosome in some kinds of differentiating cells, it is possible that in confused cells the operative cytosome established by the early transcription of one effective genome persists and continues to function at the same time as a new operative cytosome established by a later transcription of an alternative effective genome. At the most, this is a highly unusual situation.

The "switch" from one developmental pathway to another, aptly described by Weiss (1953) as *deflection* of development, corresponds with the phenomenon long known to pathologists as *metaplasia*. This phenomenon is so immediately relevant to problems of neoplasia, especially the histology and histogenesis of neoplasms that it warrants separate discussion here.

VI. Metaplasia

In its broadest sense *metaplasia* applies to the replacement of one differentiated cell type by another. Grobstein (1959) made an important distinction between (1) *cellular metaplasia*, which entails de-differentiation of cells followed by re-differentiation of the same cells along a different pathway and (2) *tissue metaplasia*, which is effected by the rejection of already differentiated cells and their replacement by differently differentiated cells generated from undifferentiated basal cells, stem cells or germinal cells, typified by the cells of the basal layer of the epidermis. Grobstein mentions many examples of cellular metaplasia, especially in the lower vertebrates; a great number of transformation of cells derived from mesenchyme have been described. Grobstein considers that changes of this kind are probably not uncommon but, in general, the transformations are between relatively unspecialized types of cell or from a relatively unspecialized to a specialized type. In the higher animals and in man, most cellular metaplasias are in the connective tissues, epithelial metaplasias being predominantly tissue metaplasias.

Willis (1958) gave a full account of pathological metaplasias in man. He defined metaplasia as the transformation in an adult of fully differentiated tissue of one kind into a differentiated tissue of another kind in response to abnormal circumstances, and he maintained that metaplasia occurs only in proliferating cells, being often a form of regeneration accompanied by atypical differentiation. Unfortunately, Willis did not distinguish between cellular metaplasia and tissue metaplasia; most of the regenerative metaplasias are probably tissue metaplasias. He claimed that the latent plasticity of each adult tissue, as manifested in its metaplasias, is a permanent residue of the much greater plasticity of the early embryo and he wrote ". . . Literally then, regeneration is resumed embryonic development and metaplasias are the evidence of resumed embryonic plasticity." I cannot see what is gained by evoking "resumed embryonic plasticity"; the "plasticity" is there all the time in the facultative genome of the transforming stem cells as an adult character. Some important general principles will be illustrated here by reference to observations on the embryonic chick epidermis. Following the original observations by Fell and Mellanby (1953) on the effect of an excess of vitamin A on embryo chick epidermis cultivated *in vitro* by Fell's organ-culture technique, Fell and her colleagues have extended the observations in several important respects (Fell, 1964; Jackson and Fell, 1963).

Mucous metaplasia of the embryo chick epidermis is a particularly instructive example of *tissue metaplasia*. When an excess of vitamin A is added to the culture medium, mucoid differentiation takes place below the normal keratinized layer which, eventually, is sloughed away. Similarly, when the excess of vitamin A is withdrawn, the cells newly differentiating from the basal layer are of the keratinizing type and they develop below the mucoid cells which are cast off. Attention is directed to two important facts: first, there is no transformation of one type of differentiated cell into another; metaplasia is effected by deflecting the differentiation of cells newly emerging from the undifferentiated basal layer of stem cells into a new abnormal pathway; second, the mucoid differentiation persists only so long as the excess of vitamin A is maintained; when it is withdrawn, the newly differentiating cells resume their normal keratinizing function. Weiss and James (1955) described a persistence of mucoid differentiation after withdrawal of vitamin A but Lasnitski (1958) gave good reasons for believing that it was attributable to vitamin A retained in the tissue through the period of observation.

McLoughlin (1961a, b, c) notably extended the observations by studying *in vitro* the reactions of epidermis separated from its underlying mesenchyme. The separated epidermis keratinized almost completely and abnormally; the basal layer keratinized and did not grow, mitoses being remarkably rare. McLoughlin inferred from various observations that the epidermal cells had a basic tendency to immediate keratinization and did keratinize unless prevented by some extrinsic influence which, normally, was apparently supplied by the subjacent mesenchyme. The isolated epidermis behaved variously when implanted on different kinds of mesenchyme in organ cultures. On gizzard mesenchyme it did not keratinize but secreted mucin and sometimes became ciliated. On proventriculus mesenchyme, it secreted mucin at first but after a few days it reverted to its normal habit and keratinized. When implanted on myoblasts from head mesenchyme, the epidermis spread out in a single layer of squamous cells, whereas on fibroblasts from the same source it keratinized more densely than in the control cultures.

Contact with certain types of mesenchyme evidently deflected the epithelium from its normal course of development but several observations indicated that the effect was reversible and that the mesenchymal influences must act continuously to maintain the unusual type of differentiation. McLoughlin found also that an excess of vitamin A inhibited the keratinization of isolated epidermis, which instead secreted mucin and sometimes became ciliated. This observation indicated that vitamin A acted directly on the epidermal cells and not through a primary action on mesenchyme.

Several inferences from these findings deserve emphasis. McLoughlin noted that the response of isolated epidermis to gizzard mesenchyme was remarkably similar to its response to vitamin A although there was no evidence that the mesenchyme produced substantial amounts of the vitamin. She inferred that the epidermal cells had only a limited repertory of types of differentiation, their reactions to any one of a wide range of extrinsic conditions being restricted to one of a limited number of possible responses. Varied extrinsic stimuli could select the same type of differentiation. McLoughlin's further inference that the epidermal cells of the five-day-old chick embryo possess the complete equipment for keratinizing and also for secreting mucus needs qualification. It is most plausibly applicable to some exceptional *confused* cells, but there is no evidence that normal epidermal cells carry operative cytosomes for fabricating both keratin and mucin as they should in order to be credited with the "complete equipment" for carrying out these syntheses. What they do possess is a facultative genome carrying potential effective genomes for keratinization and for producing mucin; these are available for alternative use but not, normally, for simultaneous use.

The demonstration of a strong intrinsic tendency to keratinization is unexpected since many experiments to be mentioned later indicate that, in general, isolation of an epithelium from its normal surroundings impedes differentiation. According to McLoughlin the positive controlling mechanism in normal intact skin is the repression of keratinization in the basal layer of the epidermis by contact with the basement membrane; cells that have lost contact with the basement membrane keratinize because they have escaped from repression. More recently Sengel (1965), in agreement with McLoughlin, remarked that contrary to what has been thought, the essential role of the dermis is not to provoke the differentiation of basal cells but to prevent their anarchic evolution towards keratinization and so preserve normal stratification of the epidermal layers. It is noteworthy that *living* dermal tissue is not essential; the necessary substratum can be supplied by a collagen gel (Dodson, 1963). Fell (1964) suggested that the presence of a substratum that allowed a normal polarization of epidermal cells was an important factor in epidermoid differentiation of epidermal cells. A similar conclusion had been reached earlier from a study of the epidermoid differentiation of thymic epithelium under experimental conditions (Foulds, 1934*b*). Nevertheless, polarization of epithelial cells is probably not the only factor in epidermoid differentiation; mitotic activity may be another important one.

Many, perhaps all, types of epithelial cells are capable of epidermoid differentiation under abnormal conditions *in vivo* or *in vitro* but they lack the *developmental bias* towards epidermoid differentiation that is so conspicuous in the epidermal cells of the chick embryo. The basic innate tendency of epidermal cells to keratinize unless actively prevented by extrinsic conditions from doing so is not shared by other epithelia that do not normally keratinize. As Weiss (1953) implied, special environmental conditions are needed to deflect differentiation into an abnormal pathway but *normal* development needs no special conditions, being achieved through an "intrinsic pattern" of "maturation". The basis of this "intrinsic pattern" is our immediate concern. The *developmental bias*, which in the embryonic chick epidermis *approaches developmental imminence*, reflects a high probability that a particular one of the several alternative potential effective genomes available in the facultative genome will in fact be transcribed as an effective genome. The essence of developmental bias, it is proposed, is that one of the available potential effective genomes has been *preselected* for future use as an effective genome and will in fact be transcribed and translated unless differentiation is actively deflected into an abnormal path by unusual extrinsic conditions.

12. THE ORGANIZATION AND DEVELOPMENT OF THE METAZOA

The *preselection* of an effective genome corresponds with what embryologists have called *crypto-differentiation, chemo-differentiation* or, most often, *determination*. When a tissue is said to be *determined* to develop in a certain way, it will, in fact, develop in that way under all circumstances in which it can develop at all (Waddington, 1940). Grobstein (1959), Dalq (1960) and others have criticized this usage on the ground that no certainty of fate has been demonstrated; cells that are supposed to be fully *determined* and that will, in fact, develop in the predicted way if left undisturbed, may develop in an entirely different way if transferred to a new environment. Moreover, Grobstein (1959) formerly disputed the reality of chemo- or crypto-differentiation of cells, arguing that, although states of differentiative bias can exist in the absence of other currently detectable signs, there is no evidence that this relates to the properties of individual cells rather than to groups of cells. More recently, Grobstein seems to have modified this opinion, although to an extent that is not entirely clear. He now admits, as one possible interpretation of recent experimental findings, that differentiation may begin with a "covert" phase, which is, in fact, the fundamental phase, not only differing from the "overt" phase but entirely separate from it and able to persist for long periods without overt expression. This interpretation, according to Grobstein, requires the assumption that, *inter alia*, the covert phase of differentiation begins with the essential event of differentiation that determines the differentiated type and also that "true differentiation" is "covert with respect to the criteria applicable to the final state", stable from the outset and presumably propagable because cells often continue to divide during the covert phase. Moreover, this interpretation requires a different definition of differentiation, namely one in terms of operations that determine what cells will do later on (Grobstein, 1964).

The substance is in danger of getting lost in the polemics. The distinction between the new "covert" differentiation and the old "crypto" differentiation is not evident. The valid objection to the term "determination" is that it implies irreversibility. A different term, not carrying that implication, would be preferable but the phenomenon to which the term has been applied is a real and fundamental one. The first step in differentiation, as here proposed, is a change in the facultative genome that makes one of a limited number of utilizable genetic patterns preferentially available for later use as an effective genome; the first step, that is to say, is the *preselection* for future use of one of the potential effective genomes present in the facultative genome. The preselection is not irrevocable, as a consideration of *deflection* and *metaplasia* abundantly shows, but it depends, nevertheless, on a change in cells that is stable and replicable in the absence of transcription and translation of the preselected effective genome. It is presumed, tentatively, as a working hypothesis, that *preselection* effects a stable, replicable modification of the facultative genome without altering the total genome. The physico-chemical basis of the modification which probably depends on chromosomal rather than purely "genetic" organization will be discussed later. It may be remarked here that *preselection* exemplifies or perhaps, even underlies, the important, more general phenomena of *decision in advance of performance* and the directive and anticipatory characteristics of epigenetic development.

Many studies of embryonic "induction" have been concerned with the "activation" and "expression" of an effective genome that has been preselected before the observations began. They are relevant to the transcription and translation of effective genomes but not to their selection; they are often relevant also to the later stages of supramacromolecular organization and morphogenesis and to stages in which the interactions of cells and tissues importantly influence development.

VII. Embryonic Induction and Tissue Interactions

Grobstein applied the term *induction* to "all developmentally-significant tissue interactions". Some embryologists have extended it even further to include the action of chemical substances on developing tissues so that the term has become, as Weiss (1959a) has said, ". . . merely a property of the observer, who links together all kinds of phenomena which are intrinsically different and have different mechanisms. . . ." The resulting confusion is extreme and is directly related to the diverse usages of the term *differentiation*. It is used almost indiscriminately by some writers to apply to the preselection or selection of effective genomes, to the transcription and translation of those genomes and to the subsequent supra-macromolecular organization and morphogenesis. The position is further confused by the diversity of the materials investigated; varied tissues from diverse regions of embryos of different species examined at various stages of development cannot be expected to yield concordant results and, in fact, they do not.

A. Secondary Inductions and Tissue Interactions

The secondary inductions described at later stages of development comprise a medley of phenomena implicated in the selection, transcription and translation of effective genomes, in supramacromolecular organization and in morphogenesis; no single description can cover all of them. A few examples will be described here to illustrate some of the more important principles and mechanisms that have been disclosed. For more comprehensive and detailed information reference may be made to Grobstein (1956b, 1959), Holtfreter and Hamburger (1955) and Waddington (1956, 1962).

Long chains of developmental processes lead up to the manifestation of phenotypic characters and development can be enhanced, inhibited or deflected by a great variety of mechanisms, by no means limited to inductive tissue interactions, acting at any link of the chain. Varied extrinsic circumstances described by Grobstein (1959) as "general systemic factors" and by Gaillard (1957) as "outside factors" and which include nutritional, hormonal and nervous factors, are not properly called "inductive" although some of them are often so described. Nevertheless some of these factors may be limiting or decisive for the expression of specific characters that have already undergone an adequate "determination" (Gaillard, 1957). Many of the systemic factors merely accelerate or retard development along a course that has already been selected and some of them are clearly unspecific in their action. Moscona and Hubby (1963) substantially enhanced the glutamotransferase activity of the neural retina of chick embryos simply by isolating the tissue and cultivating it *in vitro*. They inferred that the activity of enzymes in embryonic development might be controlled by inhibitory or repressor mechanisms or by the availability of substrates. Cells may interact metabolically when one type of cell produces a metabolite that another types needs. For example, protein synthesis in one tissue may be controlled by the supply of amino acids produced in a neighbouring tissue (Herrmann, 1960).

Wilde (1959) has discussed the action of substances of low molecular weight with special reference to the development of pigment cells in urodeles. He found that the differentiation of explanted neuroepithelium into pigment cells and ectomesenchyme could be inhibited specifically by adding structural analogues of phenylalanine to the culture medium and, further, that added phenylalanine converted ventral ectoderm into pigment cells. Wilde inferred that phenylalanine was the causal inducing agent leading to the differentiation of pigment cells and ectomesenchyme from the neuroepithelium. It is doubtfully justifiable or useful to describe the action as either causal or inductive. The experiments indicate only that phenylalanine is an essential amino acid in the synthesis of pigment granules and that under the conditions of the experiment it is available in adequate amounts in neuroepithelium but not in ventral ectoderm. No other comparable example of dependence of a particular kind of differentiation on a single amino acid seems to have been described and it may be inferred, tentatively, that the availability of essential amino acids is not ordinarily a limiting factor in normal development.

B. Induction of the Lens in the Amphibian Eye

In his pioneer experimental studies of inductive tissue interactions in amphibia, Spemann first showed that the optic cup induced overlying ectoderm to form a lens. Many later investigators have experimented with this system and have confirmed and extended Spemann's work.

The inductive effect requires contact or, at least, close apposition of the interacting tissues. Weiss (1959a) distinguishes two phases in the induction of a lens. During the first phase, a lens area, more extensive than the definitive lens rudiment, is roughly blocked out; during the second phase, one portion of this area becomes the definitive lens rudiment. According to Weiss, "contact" of the interacting tissue is not necessary during the first phase but it is essential during the second phase and the definitive lens area is co-extensive with the area of contact. The contact must last for a substantial period of time, much longer than is required for primary induction. The induction, indeed, is a "continuing effect" (Grobstein, 1956b, 1959).

The optic cup induces a lens to develop only if the apposed ectoderm is in an appropriate state of competence. The competence of ectoderm to respond to the inductive stimulus provided by an eye-cup by forming a lens is not permanent; after a while it gives place to a different competence and the optic cup then induces not lens but cornea.

In some species of amphibia the competent ectoderm corresponds fairly closely in extent with the definitive lens rudiment but in other species it is much more extensive and the optic cup is responsible for localizing the development of a lens to its own immediate vicinity. To a degree that varies considerably from species to species a lens can develop without an identifiable stimulus, as a result presumably of developmental imminence, at the site where a lens normally develops. At other sites, where the development of a lens is not imminent, an optic cup may deflect the normal course of development into an abnormal course leading to lens. In normal development, by contrast, the optic cup does not deflect its course. The inductive stimulus is not essential for starting the differentiation of ectoderm into lens tissue, the quality of the differentiation being decided by the competence of the ectoderm, but it is essential for ensuring the complete and normal morphogenesis of a lens.

C. Inductive Tissue Interactions in Mice

Grobstein (1954, 1956a, 1959) introduced some new techniques for studying the tissue interactions involved in the morphogenesis of certain organs in mice. The three important procedures used by Grobstein and, subsequently, by other investigators comprised (1) cultivation of embryonic rudiments *in vitro* by the organ culture method of Fell. (2) Dissociation of the rudiments into their epithelial and mesenchymal components by trypsinization and cultivation of the two components *in vitro*, separately and after recombination. (3) Cultivation of the two components together but with a filter membrane ("millipore" filter) interposed between them. Grobstein himself used mouse embryonic tissues and, in particular, rudiments of salivary gland, metanephros and pancreas from eleven-day-old embryos. At the time of use, the paths of development of the tissues and their effective genomes had already been *preselected* or *determined* and the experimental procedures did not alter their developmental fates qualitatively.

Intact rudiments of the tissues mentioned grew and differentiated normally *in vitro* but neither of the two components separated by trypsinization grew or differentiated when cultivated separately; after recombination, they grew and differentiated typically. The epithelial components of salivary gland and of ureteric bud continued to develop only if they were recombined with homologous mesenchyme. By contrast, both salivary gland epithelium and epithelium of the ureteric bud stimulated metanephrogenic mesenchyme to differentiate and form kidney tubules and so also did dorsal spinal cord of mouse or duck embryos. The interaction is clearly not species- or tissue-specific but killed tissues are *not* effective inductors in these systems as they are in primary induction (Grobstein, 1954, 1955a, b, 1956a, 1959).

There is substantial evidence that the apposition of ureteric bud and metanephrogenic mesenchyme is necessary for a mouse kidney to develop normally *in vivo*. Certain lethal mutants of the house mouse lack kidneys and have highly abnormal urogenital systems. In viable heterozygotes, kidneys do not develop unless the tip of the ureteric bud reaches the metanephrogenic mesenchyme. Investigations by the organ culture method show that spinal cord from either normal or mutant mouse embryos can induce tubules to develop in metanephrogenic mesenchyme from either normal or mutant mice. The abnormalities of development in the mutant mice, therefore, are not attributable to a lack of competence of the mesenchyme to respond to an inducing stimulus but to a failure of the ureteric bud to make contact with the mesenchyme at a critical time (Gluecksohn-Schoenheimer and Rota, 1949; Gluecksohn-Waelsh, 1963).

Mouse metanephrogenic mesenchyme can form tubules without being exposed to the known inductors, ureteric bud or dorsal spinal cord, if it is implanted in the anterior chamber of the mouse eye and also, less consistently, in various other sites in mice and in the coelomic cavity of the chick embryo. The addition of iris, cornea or intra-ocular fluid to the mesenchyme *in vitro* does not induce tubules to develop. The "*in vivo* effect" does not depend on cell contact or on tissue contact and it is not species- or tissue-specific. Grobstein and Parker, after discussing possible interpretations of the phenomenon, favour the view that the formation of tubules in the metanephrogenic mesenchyme is *imminent* in a probability sense. In normal development *in situ* the ureteric bud is determinative but under the conditions of culture or transplantation other factors may act as determinants by raising or lowering the probability of tubules forming and the important factor may be a general one favouring differentiation rather than a specific one for metanephrogenic mesenchyme (Grobstein, 1959; Grobstein and Parker, 1958).

In general secondary inductions require contact or close apposition of the interacting tissues and the effects are limited closely to the area of apposition. This does not apply to the induction of kidney tubules by spinal cord but it applies to the other systems used by Grobstein, who investigated the nature of the necessary "contact" by interposing filter membranes between the interacting tissues. Grobstein found that two tissues could interact when separated by a membrane up to 80μ thick with an average pore diameter of $0 \cdot 1\mu$. Light and electronmicroscopy showed that membranes of this specification prevented cytoplasmic contact between the two tissues. When the inductor tissue was labelled with a radioactive isotope, radioactivity extended through filters to approximately the same extent as inducing activity. Small molecules, including amino acids that could pass through an intact cellophane membrane were not inductive; if the tissues were separated by a cellophane membrane with a hole in it, inductive action was limited to the area of the hole. Grobstein recognized that it was hazardous to deduce the size of the effective molecules from the physical properties of filters but he inferred that the effective inducing agents are macromolecules of such low mobility in their native and active form that normally they are present only in close association with their originating tissue. His general conclusion is that induction depends on "contact" in the physiological rather than the physical sense of the term and that induction is mediated by the microenvironment of cells, which allows restricted movements of inductive agents but not the free mobility of materials in solution. Grobstein emphasizes the probable importance of the inter-cellular or extra-cellular matrix in mediating tissue interactions (Grobstein, 1954, 1956a, 1959, 1961a, b; Koch and Grobstein, 1963). The nature of cell "contact" and the significance of the extra-cellular matrix were discussed in a previous section with some reservations about the importance of the matrix.

The more recent observations of Grobstein and his colleagues on the differentiation of embryonic mouse pancreas, to which some reference has been made already, correspond, in general, with the earlier ones on kidney. Ordinarily epithelial tissue, by itself, did not differentiate *in vitro* but varied mesenchymal tissues induced it to do so; tissue and species-specificity was notably lacking. Moreover, epithelial tissue, by itself, could grow and differentiate if the culture medium contained 10–20% of embryo juice instead of the usual 3% and the addition to the standard culture medium of certain particles sedimented by centrifugation from embryo juice could similarly induce the epithelium to grow and differentiate. Trypsin inactivated the particles, suggesting that the active component might be a protein, but the chemical constitution of the particles was not clearly determined.

The interposition of a Millipore filter between epithelium and mesenchyme allowed the two tissues to be separated at any desired time thereafter; the epithelium could then be cultured alone. It transpired that the separated epithelium synthesized zymogen granules consistently if it had been exposed to mesenchyme for 48 hr and to a less extent after exposure for a minimum of 30 hr. Grobstein inferred that after exposure to mesenchyme for 48 hr, at a time preceding by 18 hr the detection of enzymatic or structural signs of differentiation, some at least of the epithelial cells were so changed that they could differentiate without a continuing inductive stimulus. As summarized by Grobstein (1964), during the period between approximately the 30th and 48th hr of culture the epithelium shows none of the signs of definitive pancreatic differentiation but it undergoes "a critical change" and only time seems to be needed for the change to be manifested. Grobstein remarked that the change was of the kind variously referred to as determination, chemodifferentiation or covert differentiation. It is difficult to accept this opinion because Rutter et al., (1964) expressly disclaimed having demonstrated that either mesenchyme or the material obtained from embryo juice conferred "an essential pancreatic character" on the epithelial cells or that the experiments were particularly relevant to "classical embryonic induction". They recognized, indeed, that the epithelium might already be "biased" in the direction of pancreas formation at the beginning of the experiment. This developmental "bias" is the essence of determination or crypto differentiation, a precursor of the "critical change" not a consequence of it.

It is noteworthy that in the experiments under discussion, the cells adjacent to mesenchyme increased in number thirty-fold during a culture period of five days. Without exposure to mesenchyme, the epithelial tissue by itself spread without acquiring a morphological pattern and the rate of cell proliferation was judged to be significantly lower than in epithelium exposed to mesenchyme. The interacting system, therefore, involved the regulation of cell duplication as well as specific morphological and, presumably, metabolic events (Grobstein, 1946; Rutter et al., 1964; Wessells, 1964). It is possible that the "critical change", which was not an abrupt one but the consequence of sustained exposure to the inductive materials, consists in the attainment of a minimal cell mass, which as will be mentioned again elsewhere, is important in many kinds of "differentiation". The ability of the epithelium to go on differentiating after withdrawal of the inductive stimulus was abolished by dividing the mass into eight fragments of equal size, none of which differentiated when cultured separately although differentiation advanced again in the mass formed by re-aggregating them. It was concluded that differentiation was stable at the tissue level but not stable at the cell level.

Auerbach (1960) noted that, consistently with a previous suggestion by Grobstein, the differentiation of kidney tubules becomes "cell-stable" at later stages of development and quoted Moscona's observations showing that the re-aggregated cells of dissociated kidney tubules re-organized immediately into tubules. Differentiation has been observed also in isolated myoblasts (Moscona, 1964; Konigsberg, 1963) and Elsdale and Jones (1963) reported that isolated dissociated cells of amphibian neurulae differentiated *in vitro* into chorda cells, nerve cells and mesenchyme. Elsdale and Jones inferred that at the neurula stage, the timing and specificity of differentiation are vested in individual cells acting as cell-sufficient units and not on a "field" imposing control from outside the cells. Apart from an observation of Wilde (1961*a*, *b*) on confusion in isolated cells, the available evidence indicates that cell-mass, entailing interactions between cells of one kind does not direct or deflect the course of differentiation but only facilitates its progression along a preselected course.

The more recent observations of Wessells and Cohen (1967) on the morphogenesis of mouse pancreas showed that preselection occurred at an unexpectedly early stage of development. Endodermal tissue, located in the dorsal wall of the gut of eight-day mouse embryos, was "committed" towards developing into pancreas before condensed pancreatic mesoderm or even loose mesoderm was associated with the dorsal gut tissue and before pancreas morphogenesis began. This is a more positive recognition that the preselection of the path of development had taken place before the "inductive" action of mesoderm had begun to operate than the earlier statement that the epithelium may have been "biased" in the direction of forming pancreas at the beginning of the experiments. Wessels and Cohen's experiments also cast doubt on the minimal mass requirement. They found that large tissue mass *per se* was not a prerequisite of overt differentiation. The failure of fragmented rudiments to differentiate, as observed in the earlier experiments was correlated with the continuation of mitotic activity in the fragments. Their observations suggested that, in chronologically older tissues, non-dividing cells could accumulate after fewer cell cycles than would be required in the ancestral population with the result that overt differentiation could occur in a small tissue mass.

The observations discussed so far of Grobstein and his colleagues can be summarized in the form of a provisional working-hypothesis as follows:—

The first step in differentiation, as here proposed, is the *preselection* of an effective genome. This step entails a modification of the facultative genome making one of a number of potential effective genomes preferentially available for later use. The physical basis of the modification of the facultative genome is not known but empricically it is stable and replicable; it can endure through successive cell divisions without being transcribed or translated. With the important reservation that the preselection is not irrevocable, it corresponds closely with the determination, crypto-differentiation and chemodifferentiation of earlier writers, with the "latent" differentiation of Dettlaff (1964) and with the "covert" differentiation of Grobstein (1964).

The second step of *activation* brings the preselected genome into active use as an effective genome by linking it with biotonic systems so that it can be transcribed. The principle of mutual conformance of genetic and biotonic patterns now comes into play. Both may need readjustment. The effective genome is, at first, apparently intrinsically unstable and must be extrinsically stabilized and maintained by mutually conforming biotonic systems which in turn seem to depend on the interaction between the differentiating tissue and another tissue of a different kind. If the extrinsic conditions, and presumably the biotonic systems, are sufficiently abnormal, the preselection may be annulled and a new effective genome selected; if this happens, differentiation is deflected into a new path. Under normal or near normal circumstances, the preselected effective genome is the one actually used and the "inductive" tissue interaction has no selective or directive action. At this stage, cell proliferation seems to be an important factor in the differentiative process.

3. After a limited but substantial period of tissue interaction, the differentiating cells undergo a "critical change" which enables them to dispense with inductive tissue interaction on condition that they are associated together in groups or masses. The nature of the critical change is not known. It is a gradual change and possibly consists of a progressive stabilization of the mutual conformance of genetic and biotonic systems.

4. Eventually the dependence on "group" factors wanes and disappears; the effective genome is then intrinsically stabilized and replicable like the effective genome in the ciliate macronucleus which persists through repeated vegetative fissions and which can be transcribed and translated without the help of extrinsic factors emanating from contiguous tissues of like or unlike kind.

CHAPTER 13

The Relationship between Cell Proliferation and Cytodifferentiation

The main effect of many of the secondary and later inductions is more a morphogenetic than a differentiative one. Even in primary induction the initial observed effect is not differentiative. As described by Dalq (1960), initial processes of cell division, cell movements and general transformations of cells and tissues precede cyto-differentiation. Cell proliferation is implicated also in secondary and later inductions as exemplified by the observation of Grobstein and his colleagues on the embryogenesis of the mouse pancreas. This preparatory role of cell proliferation has been overshadowed hitherto by the contentious inverse relationship or antagonism between cell proliferation and cyto-differentiation. Both aspects of the inter-relationship between the two processes will be discussed in this chapter.

I. The Preparatory Role of Cell Proliferation

Holtzer (1963) maintained that in both embryonic and adult tissues cell division is seen predominantly in connective tissue cells and in unspecialized "basal", "reserve" or "stem" cells; the usual sequence *in vivo* is that the unspecialized cells divide and then one or both of their daughter cells begin to synthesize a specialized product. Holtzer inferred that cell division is a common antecedent and, perhaps, a necessary precursor of cytodifferentiation and suggested that mitosis might release or activate information present in the nucleus. Dettlaff (1964) reviewed in detail the significance of the number of cell generations and the duration of interkinetic states as factors in differentiation and the observations of Wessells and Cohen mentioned previously seem to confirm the importance of these factors. Stockdale and Topper (1966) reported that epithelial cells of the mouse mammary gland explanted *in vitro* must first divide before they can synthesize casein in in response to stimulation by insulin, hydrocortisone and prolactin. Their observations suggested that some of the environmental stimuli that elicit differentiative processes may act on cells only when they are proliferating. Most recently Flickinger *et al.* (1967) have suggested that embryonic cells that have divided less frequently (endoderm cells) may have different patterns of transcription of DNA from cells that have divided more often (dorsal ectoderm and mesoderm cells). The outcome of experiments to test this hypothesis should be illuminating.

It may be recalled that, in *Paramecium*, a new mating-type effective genome can be selected and established only whilst a new macronucleus is being formed as a consequence of conjugation or autogamy. Apparently it is only during this period that the facultative genome is fully "available" for the selection of a new effective genome. It is possible, similarly, that mitosis is an important prelude to cytodifferentiation in vertebrate animals because it makes the facultative genome freely and immediately available for the selection of effective genomes. This interpretation is consistent with the suggestion by Stockdale and Topper (1966) that the stability of the differentiated state may be attributable to the infrequency with which cells that are synthesizing specialized products divide, the non-dividing cells seldom being vulnerable to the environmental factors that are capable of altering their differentiated state. Previously, Mazia (1961) suggested that mitosis might erase the organization of the parent cell so that the daughter cells could start with a "clean slate" upon which they could write their own destinies. The still earlier statement by Willis (1958) that cell proliferation invariably accompanies metaplasia is also relevant. It is proposed that cell proliferation establishes a "clean" facultative genome available for a new selection of effective genomes as well as a relatively "clean" cytoplasm resulting from the drastic disorganization of the operative cytosome that takes place during mitosis.

A. E. Needham (1964) proposed as a reasonable generalization that the processes of growth and differentiation are directly rather than inversely correlated, complementary rather than reciprocal or antagonistic. Highly differentiated cells can grow and a positive antagonism between the state of differentiation and growth has not been proved. There is more evidence that growth controls differentiation than *vice versa*. Ebert and Kaighn (1966) have reviewed further evidence to the effect that, in some cells at least, new transcription may depend upon an immediately preceding replication.

II. The Inverse Relationship between Cell Proliferation and Differentiation

When cells multiply at their maximum rates, they do not differentiate; they seem to be incapable of fabricating the organized ultrastructure or operative cystosome that is necessary for producing specialized structures or for sustaining specialized functions. When signs of specialized structures or functions become apparent in cells that have been dividing at less than the maximum rate cell division usually stops (Berrill, 1963; Grobstein, 1959).

Nevertheless the failure of a cell to divide does not inevitably result in differentiation and, conversely many undifferentiated cells do not divide (Bertalanffy, 1960; Ducoff and Ehret, 1959). There is indeed abundant uncontroverted evidence of a certain degree of inverse relationship between cell division and cytodifferentiation in a variety of circumstances but the proposition that there is a fundamental antagonism or inverse relationship between them is subject to so many qualifications that it has little operational value. Much of the controversy that has encircled it is traceable to confusion between different levels of analysis and between different stages and components of the two processes and especially perhaps, as Weiss maintained, to the diverse usages of the term differentiation. Some of the confusion can be mitigated by breaking down the unanswerable question into a number of answerable ones.

Weiss has distinguished between two main kinds or stages of "differentiation". One is the "selective routing" of cells into diverse pathways of development, which he now (1967a, b) calls *strain differentiation* and the other concerns the overt manifestation of specialized products and functions. In the terminology proposed in this book, strain differentiation corresponds with the perselection of one of the potential effective genomes available in the facultative genome of the differentiating cells; the second type comprises the" expression" of the selected genetic pattern. The "expression" entails a chain of processes including the transcription and translation of the chosen effective genome and the subsequent processes of supramacromolecular organization and morphogenesis. It should be noted that in the medical literature "differentiation" applies ordinarily to the stage of "expression" and rarely if ever, to strain differentiation. This usage has been in operation so widely and so long that it seemed hopeless to attempt to modify it in the earlier part of this book but, from here on, strain differentiation will receive much attention.

A distinction should be made at the outset between the behaviour of *tissues* and of the individual *cells* of which they are composed. The cells of a tissue divide asynchronously and, ordinarily, only a small fraction of them are dividing at any one time. Consequently it is possible for the tissue as a whole to be structurally differentiated and functionally active although it is growing at a substantial rate. This happens conspicuously in neoplasms; epidermal tumours can grow and at the same time produce abundant keratin. There is no consistent inverse relationship between rate of growth and degree of differentiation although, in a general way, the faster growing tumours tend to be the less well differentiated. Rapidly regenerating liver, following partial hepatectomy in adult mammals, seems to be able to carry out its specialized functions (Le Breton and Moulé, 1961).

The more important questions concern the behaviour of individual cells. Broadly speaking, differentiated cells tend not to divide and dividing cells tend not to be differentiated and the more clearly a cell is specialized the less likely it is to divide (Swann, 1958). Nevertheless, viable differentiated cells rarely lose irrevocably their capacity to divide. The proposition that a differentiated cell cannot divide is untenable (Weiss, 1959b). The contrary assertion by Wilde (1961a) that every cell with a viable nucleus should be deemed capable of division is probably too sweeping without substantial qualification. It does not apply to the static tissues in adult mammals; although mitosis has been seen even in neurones, it is extremely rare. The fact that most kinds of differentiated cells *can* divide, although often only under abnormal conditions, should not be allowed to obscure the other fact that under normal conditions many of them do not.

The failure to divide many cells that are not highly specialized in structure is unexplained (Swann, 1958). The accumulation of dead material, so prominent in keratinizing epidermal cells, is probably not a common cause of inhibition of mitosis in most other types of tissues and it is not likely that competition for nutrients or energy is an important factor. The inhibitory effect of ongoing production or renewal of intracellular materials is probably more significant especially if it be taken to apply more generally to sustained specialized function. So extended it may be applicable to the almost complete inhibition of mitosis in some *expanding* tissues notably the mammalian liver whose functional activity is continuous and intense. The failure of the cells to divide is certainly not due to a lack of capacity to do so; after partial hepatectomy, liver cells multiply at a great rate.

Profitably to carry this discussion further requires attention to recent studies of the mitotic cycle. In the first place a distinction should be made between mitosis proper and the much longer-lasting interphase during which the essential preparations for mitosis are made. It is reasonably sure that during mitosis proper, the genome is not available for transcription; the synthesis of RNA is suppressed except in the earliest and latest phases. Moreover, during mitosis the operative cytosome is so severely disorganized that it is unlikely that translation and supramacromolecular organization can be performed. Stable products already present at the onset of mitosis may persist, although often they do not, but during the short time occupied by mitosis it is highly probable that the selection, transcription and translation of effective genomes are suppressed. In this sense, mitosis is antagonistic to differentiation but current interest has shifted to the interphase.

Several authors have maintained that DNA cannot be replicated and transcribed at the same time (Brachet, 1964; Lark, 1963; Prescott and Kimball, 1961). The details of replication and transcription of DNA in the chromosomes of the higher animals are not well understood. It is thought that replication requires the unwinding and separation of the double helix, both strands of which are replicated. Replication may begin at one end of the

13. CELL PROLIFERATION AND CYTODIFFERENTIATION

strand and spread like a wave to the other end or, as some believe, it may be initiated at multiple points along the strand. On either view, replication does not take place simultaneously throughout the whole length of the strand. It seems that *in vivo*, although not *in vitro*, only one of the two DNA strands is transcribed and that this can be effected either without separation of the strands or with separation only of the segments that are to be transcribed. It is possible that, at a given time, one portion of the DNA may be engaged in replication whilst at the same time another portion is engaged in transcription (Goldstein and Brown, 1961; Mazia, 1965). Studies by Rusch *et al.* (1964) on the myxomycete *Physarum polycephalum* indicated that the messenger RNA concerned in mitosis was synthesized early in interphase, simultaneously with the replication of DNA. Technical difficulties present similar observations being made on tissues of higher animals but it is known that in them the synthesis of RNA continues through the S phase, during which DNA is replicated. At the present time it does not seem possible categorically to affirm or deny that simultaneous replication and transcription of the genome in a single cell is impossible. As yet, there is no decisive evidence for an intrinsic antagonism between proliferation and differentiation *at this level of analysis* (Herrmann *et al.*, 1967).

Swann (1958) argued that cell differentiation is best regarded simply as another form of differentiation. He believed that differentiated cells have ceased to make mitotic protein and that their various division mechanisms have been brought to a halt at some point in the mitotic cycle well in advance of the doubling of DNA. The resumption of cell division, in his view, must involve a major switch in every way comparable with the switch involved in embryonic induction whereby differentiation is brought about in the course of development.

Bullough (1964, 1965, 1967) has greatly extended and elaborated Swann's concept with special reference to tissue hemeostasis in adult mammals, which he has sought to explain in terms of a mutual antagonism between mitosis and differentiation. Bullough's hypothesis has two distinctive features. First, it presumes that cell division and cell specialization depend on two mutually-exclusive patterns of transcription of the genome based on a "mitosis operon" and on "ageing-tissue operons" respectively and that the mutual-exclusion is effected by regulatory circuits of the Jacob-Monod type. Second, it attributes *selection* of one or other of these alternatives primarily to the action of tissue-specific substances challed *chalones*.

Bullough defines a chalone as an internal secretion produced by a tissue for the purpose of controlling, by inhibition, the mitotic activity of that same tissue. As a consequence of the action of chalones, the cells that are prevented for undergoing mitosis tend to embark on an ageing pathway controlled by an *ageing operon*, which, during the course of evolution has become linked with *tissue operons* that control specific tissue functions; once embarked on the ageing pathway the high concentration of chalones in the tissue inhibits the ageing operon so that progress along the pathway is slow and, also, the tissue operons are activated by what may be a simple trigger mechanism.

Bullough considers it reasonable to regard all the genes controlling mitosis as forming a single mitosis operon controlled by an operator gene and considers it reasonable, also, following Jacob and Monod, to suggest that the operator is controlled by a repressor molecule produced by a regulator gene and that this repressor is inactive in the absence of the tissue chalone. The chalone is regarded as a double molecule produced in part by an effector gene and in part by a tissue gene. The complete chalone is tissue specific and therefore able to pass outwards through the cell walls which are also tissue specific. The normal balance between mitoses and function, it is proposed, is achieved in terms of the intracellular concentration of tissue-specific chalone, this concentration being determined by the balance between chalone synthesis and chalone loss from the cells.

It is impossible adequately to discuss here Bullough's highly detailed argument for which reference should be made to his recent book (Bullough, 1967). It is inevitably to a great extent speculative and entails many subsidiary assumptions which are difficult to evaluate but it represents a serious attempt to analyse the important phenomenon of tissue homeostasis in terms of modern concepts of regulation.

In several discussions, the inverse relationship between cell proliferation and cytodifferentiation has been minimized or denied (Davidson, 1964; Fell, 1957; Grobstein, 1959; Konigsberg, 1963). Fell thought, as did Grobstein, that the important factor in limiting cytodifferentiation was not proliferation *per se* but the migration and dispersal of cells which often follows proliferation. She remarked that in organ cultures, which do not encourage cell dispersal, the more numerous the mitoses, the greater, as a rule, was the degree of differentiation. This observation conforms with the proposition that mitosis is important in preparing for differentiation but it does not illuminate the basic problem, which concerns the ability of individual cells to divide and differentiate *at the same time*.

Grobstein emphasizes the favourable effect on differentiation of the aggregation of cells into compact masses exceeding a certain minimal size. He notes that in all animals from the protozoa to the vertebrates a certain minimal mass of cells seems to be necessary for continued differentiation. With the agreement of other authors, Grobstein concludes that close apposition and crowding of cells within a confined space as well as increasing total mass favour differentiation whereas disposal of cells, a low ratio of cell mass to volume of medium and decreasing total mass repress it (Elsdale and Jones, 1963; Grobstein, 1959; Holtfreter and Hamburger, 1955; Konigsberg, 1963; Wilde, 1961a).

The arguments against an inverse relationship between cell differentiation and cytodifferentiation lean heavily on observations of cells and tissues cultured *in vitro*. The relevance of such observations to the general problems of proliferation and differentiation *in vivo* will be discussed in the next section.

III. Cell Proliferation, Differentiation and De-differentiation in vitro

The three main kinds of culture in vitro are: (1) *organ culture* as developed especially by Fell (2) conventional *tissue culture* and (3) the more recent *cell culture* in monolayers. These methods, in the order given, entail increasing loss of visible signs of differentiation, increasing abstraction and difficulties of interpretation and increasing risk of technical errors and artefacts.

Organ culture preserves the intimate relationship of unlike cells and tissues and encourages compact growth. Differentiation persists and may advance so that, for example, an appropriate axial skeleton develops *in vitro* in a chick limb-bud.

Conventional tissue culture annuls the normal relationship between cells and tissues of unlike type and encourages the dispersion of cells of one type without abolishing cell contacts. Specialized structure and function may persist for a considerable time. Compact growth helps to maintain the differentiated state but the overt signs of differentiation commonly decrease. Specialized products such as pigment granules often disappear but the loss is reversible, at least for a time and the specialized products or activities may reappear if the conditions of culture are appropriately adjusted. Differentiative characters tend to diminish as culture is prolonged and it becomes progressively more difficult to reverse the de-differentiation although some properties such as those distinguishing epithelial cells from fibroblast-like cells may persist for a long time (Grobstein, 1959).

In the general experience, monolayer cultures quickly lose many of their overt specialized characters and serial propagation becomes increasingly difficult until, after a few weeks or a few months, either death of the cultures puts an end to serial propagation or surviving cultures undergo *transformation* to yield long-term or established cell lines, which are capable of rapid proliferation and indefinitely-sustained serial propagation *in vitro*. The "short-term" and "long-term" cultures differ in important ways and need separate consideration as does the obscure phenomenon of transformation. The following account deals mainly with a few matters of especial relevance to general problems of proliferation, differentiation and de-differentiation *in vitro*. Much additional information may be found in reviews by Eagle and Levintov (1965), Levintow and Eagle (1961), Davidson (1964), Sanford (1965, 1967), in the report of the Second Decennial Tissue Culture Conference

(*Nat. Cancer Inst. Monogr.*, **26**, 1967) and in the comprehensive treatise on tissue culture edited by Willmer (1965).

A. Short-term Cultures

The loss of differentiative characters begins almost immediately after explantation and may precede the onset of cell division. Different types of cells de-differentiate to varied extents which depend, in considerable measure, on the technical details of culture. Monolayer cultures tend to lose the majority of their specialized features but there are some notable exceptions; myoblasts, even in dispersed monolayers, may differentiate into contractile striated muscle fibres (Konigsberg, 1963; Moscona, 1964). Cells of connective tissue origin often retain tissue-specific biochemical functions; various cell lines of diverse connective tissue origins secrete acid mucopolysaccharides and some produce collagen. Certain tissue-specific complements of enzymes may be retained for a long time and some neoplastic endocrine tissue cells continue to secrete their characteristic hormones.

Chromosome abnormalities develop frequently but erratically in cultured cells. The normal diploid state may persist for many months; on the other hand, highly abnormal complements of chromosomes many of which are individually abnormal, may appear. Commonly the "chromosomal disorder" begins simultaneously in a substantial proportion of the cells soon after explantation and at nearly the same time as the loss of differentiative characteristics. The chromosome abnormalities differ from cell to cell and are not present in all cells so they are not likely to be responsible for the uniform de-differentiation seen in short-term cultures.

There is substantial evidence to indicate that much, or perhaps all, of the apparent de-differentiation in short-term cultures is reversible. Specialized structures and functions have been restored to de-differentiated cultures in several ways, notably by implanting the cells into animals or on to the chick chorioallantoic membrane. The reasonable inference from this evidence is that the tissue-specific potential effective genome persists in a stable replicable form in the facultative genome of the de-differentiated cells but that transcription and translation of the preselected genome do not take place under the standard conditions of culture; overt differentiation can be restored by appropriate manipulations of the cells' environment which allow transcription and translation to proceed.

B. Transformation

When transformation takes place, the short-term cultures are overgrown by cells of a new type distinguishable from those formerly present by their morphological features, by their rapid proliferation and by their capacity for sustained serial propagation.

Some years ago, several investigators showed that many alleged "transformations" were due to contamination of the cultures, by widely-used established cell lines that were being maintained concurrently in the same laboratories, the most notorious contaminants being Earle's "L" strain of mouse fibroblasts and Gey's "HeLa" strain of human carcinoma cells. It is scarcely questionable that such contaminations account for many reputed examples of transformation but subsequent experience indicates that it does not account for all of them. It is fairly sure, too, that many established cell lines harbour viruses or pleuropneumonia-like organisms but it has not been proved that they are responsible for the transformations. No "cause" being found for the transformations they are described as "spontaneous", without prejudice to the future discovery of an inciting agency. Sanford has given good evidence that the opinion expressed by some workers that all established cell lines are neoplastic is incorrect but it is difficult to tell from the literature which are and which are not and this has to be borne in mind in interpreting the results. Our concern in this chapter is not with the neoplastic properties but with the differentiative ones.

C. Established Cell Lines

Characteristically there is a general similarity or, as Eagle says, a "disappointing uniformity" amongst established cell lines irrespective of their origins. Even cell lines from normal and from neoplastic tissues respectively are indistinguishable. Almost all lines are hypotetraploid and all proliferate rapidly. In general, they have similar nutritional requirements and similar enzymatic equipments. Enzyme induction, repression and end-point inhibition do not seem to be important control mechanisms in the metabolism of these cells. With a few important exceptions they do not carry out the specialized functions that characterize their parent tissues but they retain certain properties that are common to all types of cells and are little affected by environmental control mechanisms. Species-specific and strain-specific antigens persist in cells grown for many years *in vitro* even when grown in heterospecific or chemically-defined media. Tissue-specific antigens also persist. It has been reported also that human cells, after many years of cultivation, still retain esterases with specific electrophoretic patterns corresponding with those of fresh human tissues (Davidson, 1964; Waymouth, 1967).

Eagle (1963) maintained that the only established cell lines that clearly perform organotypic functions originated from neoplastic tissues. The best examples include a cell line, derived from a mast-cell tumour, that produces serotonin and histamine (Schindler *et al.*, 1959) and a line originated from a malignant melanoma that continues to form melanin (Moore *et al.*, 1962). From about the time of Eagle's paper, the situation has altered and examples of tissue-specific function in cell lines derived from normal tissues have been accumulating. Davidson and his colleagues, for example, reported that a line derived from normal connective tissue of the eye of a weanling rat continued to produce acid mucopolysaccharides, mostly hyaluronic acid, and made the further observation that this specific function was inhibited by actinomycin-D, implying that it depended on sustained transcription of a specific effective genome (Davidson, 1963; Davidson *et al.*, 1963).

It should be noted that, for many years, workers in tissue culture laboratories were preoccupied with a search for culture conditions permitting maximum *growth* and in the main were less concerned with differentiation and function. It is now being increasingly recognized that the requirements for differentiation may be more exacting than those for growth. In one respect the requirements for growth and differentiation are similar; a certain "minimal mass" of cells seems to be needed for each and differentiation in particular is favoured by compact growth and hindered by dispersion of cells. Many years ago Fischer affirmed that single normal cells would not multiply under the usual conditions of tissue culture. Later investigations in Earle's laboratory showed that single cells of Earle's well-established strain L would not multiply under the conditions that were being used satisfactorily for maintaining the strain but would divide if confined within a capillary tube in about 0·04 ml. of culture medium. The general conclusions from the investigations were that the culture medium and conditions of culture ordinarily used were not adequate for the growth and division of single cells unless the medium had already been "conditioned" by the metabolic activity of growing cells. These conclusions were not universally applicable; single cells of certain kinds would divide in ordinary media that had not been conditioned (Earle *et al.*, 1951; Likely *et al.*, 1952; Sanford *et al.*, 1948–49). The value of Puck's "feeder-layer" method for obtaining clones from single cells by cultivating them on a layer of irradiated living but non-dividing cells probably depends on a similar "conditioning" of the medium. Eagle, who has studied the nutritional requirements of cultured cells extensively, suggested that essential metabolic intermediates, cofactors and, perhaps, macromolecules, which are retained in the cells of organized tissue, may escape from cells dispersed in a relatively large volume of fluid so quickly that the biosynthetic capacities of the cells are not equal to replacing them fast enough to sustain an adequate intracellular concentration. The critical factor in the "minimal mass" requirement may not be the absolute amount of differentiating tissue but the ratio of cell mass to the amount of surrounding medium with which it is equilibrated (Eagle, 1963; Eagle and Piez, 1962).

Konigsberg (1963), who obtained clones from single myoblasts in "conditioned" culture medium, came to essentially the same conclusion as Eagle and proposed that the minimum tissue-mass phenomenon depends on factors that limit diffusion from cells or replace types of molecules that are lost by diffusion. It should be emphasized nevertheless that satisfying the minimal mass requirement is not sufficient to ensure differentiation of particular kinds of cells that need special conditions to reveal their capacity for tissue-specific differentiation. A propensity for synthesizing and storing glycogen persists for a long time in certain cell lines derived from tissues specialized for carrying out those activities *in vivo* and the extent to which the activities are manifest varies widely with the conditions of culture and notably with the amount of glucose in the medium. Characteristic functions of the tissue of origin, although not evident under the standard conditions of culture are disclosed in many cell lines by special procedures as, for example, by administering hormones to which the parent tissue is specifically responsive.

From his review of varied phenomena of the types referred to in the preceding pragraph, Davidson (1964) concluded that they supported the general proposition that cells of long-term cultures often retain "latent capabilities for histotypic response". Writing with particular reference to strain L cells, which originated from mouse fibroblasts but ordinarily produce no collagen unless peptone is added to the culture medium, Davidson remarked that the cells of strain L and of certain other cell-lines of similar provenance "preserve in latent form the ancestral differentiation-acquired ability to make extra-cellular collagen fibres though the conditions inspiring this behaviour remain obscure in detail". This conclusion can be otherwise expressed in terms of the persistence of a stable and replicable tissue-specific potential effective genome in the facultive genome of cells of established cell lines; overt differentiation can be restored by appropriate manipulations of the conditions of culture as it can in short-term cultures. More recently Herrmann *et al.* (1967) inferred from their experiments that the change from differentiation to proliferation as well as from proliferation to differentiation requires DNA-dependent RNA synthesis, and in this sense interconversion of proliferation and differentiation is similar to the induction of specific differentiation itself. One difference should be pointed out; the interconversion does not entail the selection of one of a number of histotypic effective genomes but it does entail a switch from histotypic differentiation to proliferation. Herrmann and his colleagues noted that while the production of proteins needed for proliferation on the one hand and for specialization on the other hand could be easily reversed, attempts to change production of protein for one type of specialization to another had failed. It seems from this that the preselection of a histotypic effective genome is extremely stable and, perhaps, even irreversible in established cell lines. Contemporaneously, Waymouth (1967) discussed the effects of extrinsic factors on the "expression of differentiated functions *in vitro*" and called attention to the need to pay more heed to the use of specific keys to unlock specific capacities of cultured cells.

To summarize the foregoing discussion, it can be said that established cell lines in general do not retain the overt tissue-specific differentiative characters of their parent tissues and, so far as the evidence goes, they do not retain the "specialized equipment", or operative cytosome, for establishing those characters; the evidence shows only that some lines retain, to a greater or less extent, not characters but *capacities* for histotypic differention based on stable, replicable preselected effective genomes available in their facultative genomes. The "de-differentiation" is reversible insofar as the circumstances of cultivation can be arranged to bring about transcription and translation of the preselected genome. It cannot be taken for granted that this is true of all established cell lines or that the facultative genome is retained in its entirety in any of them. It is noteworthy that species-specific characters and other properties common to all the tissues of an organism are long retained and that all established lines tend to converge towards a common simplified type. It is probable that the genetic controls of species-specific characters and of the basic mechanisms that are essential for maintaining and propagating cells reduced to the simplest generalized form are different from those implicated in specific differentiation.

The analysis of differentiation in established cell lines is complicated by two kinds of alteration of the total genome; the karyotypes are notably unstable and various karyotypes are present even within single clones of cells and, furthermore, the genomes are liable to point-mutation. The karyotypes continue to alter during serial propagation but the nature of the phenotypic changes that are attributable to gross abnormalities of the karyotype is not known. It is unjustifiable to suppose at present that visible abnormalities of the chromosomes reflect specific changes in the genome at least insofar as it affects the production of a substantial number of enzymes that have been studied (Davidson, 1964). Judging from the limited information available, alterations of chromosomes and malignant transformation in cultures seem to be independent, unrelated phenomena (Sanford, 1965). The distinctive characteristics of certain cell lines are probably due to point mutations. The characteristics, which have no evident relationship to the tissues from which the strains originated, include resistance to various toxic agents such as amethopterin and 6-mercaptopurine in strains that were formerly sensitive to them. Changes of this kind are often attributable to changes in a single enzyme and, according to all the available evidence, result from random mutations of low frequency. The same mechanism seems to operate in originating sublines that are independent of an extrinsic source of certain materials required by the parent line and lines that are able to utilize various sugars in place of glucose. Point-mutation and selective processes are often recognizable in the adaption of cell lines to new media; the adapted lines stem from only a fraction of the former cell population, which they overgrow. Cell lines have characteristic patterns of amino acid uptake and release; no two lines have exactly the same pattern. Cell lines are distinguishable also by the distribution of various enzymes, by sensitivity or resistance of varied degrees to virus infection and by various other properties including morphology and growth rate (Davidson, 1964).

The significance of the karyotypic abnormalities in cultured cells is obscure and their relevance to normal differentiation is extremely doubtful. Briggs et al. (1964) pointed out that similar abnormalities of karyotype complicate the interpretation of the results of transplantation of amphibian cell nuclei into enucleated eggs and remarked that these abnormalities must be understood or controlled before either cell culture or nuclear transplantation can be fully exploited in their different ways in the analysis of cell differentiation. Similar considerations apply to the transplantation of tumours *in vivo* and are especially serious in the serial propagation of ascites tumours. It seems most likely that in all three systems the abnormalities of karyotype are artefacts consequent on the experimental procedure and irrelevant to the basic mechanisms of differentiation or neoplasia.

CHAPTER 14

Differential Utilization of the Genome

All recent discussions of differentiation presume that its basis is to be sought in differential utilization of the genome although the mechanisms involved in this utilization are still controversial (e.g. Abercrombie, 1967a; Bullough, 1967; Markert, 1965). General acceptance of this hypothesis is due in considerable measure to visible evidence of chemical differentiation and differential gene function during development in chromosomes of two kinds that provide exceptionally favourable opportunities for study; these are the polytene chromosomes of Diptera and the "lamp-brush" chromosomes of vertebrate oocytes.

I. The Polytene Chromosomes of Diptera

The polytene chromosomes of *Diptera* larvae, of which the "giant" salivary gland chromosomes of *Drosophila* are the best known, consist of hundreds or thousands of fine fibrils aligned side by side. The conspicuous cross-banding results from the lateral apposition of expansions, called chromomeres in which the fibrils are uncoiled. The positions of the bands and the characteristics of individual bands are highly specific and constant in insects of the same species and are the same in various differentiated tissues of an individual insect. This is the best morphological evidence so far available that the *total genome* is the same in all differentiated tissues and it adds to the abundant, if indirect, evidence from other sources that differentiation is not attributable to mutation, deletion or segregation of particulate, determinant genes.

Balbiani, the discoverer of the giant chromosomes of *Drosophila*, noticed localized swellings, now called *Balbiani rings*, which, at first sight, look like annular masses of material surrounding an otherwise normal chromosome. The rings are found only in salivary glands but they are now recognized as extreme examples of localized expansions on *puffs* that are present, less conspicuously, on the chromosomes of other larval tissues. The rings and puffs are formed by the breaking-up of sections of the chromosomes into a large number of uncoiled, fine threads which project laterally from the chromosomes in loops to form a diffuse swelling in which ribonucleoprotein accumulates. Each puff develops from a single band or interband. On the supposition that each thread is an uncoiled chromosome, it is calculated that each genetic unit, as determined by Mendelian analysis, corresponds with a stretch of chromonema at least 5μ long. The puffs are much less numerous than the bands and develop only at certain specific sites along the chromosomes to make a characteristic pattern, which differs from one tissue to another. Moreover, the pattern changes during larval development; one set of puffs regresses and another set develops until, at pupation, all regress. The puffs seem to correspond with reversible functional changes at specific genetic loci and to be related to cell function and to specificity of cell type. The Balbiani rings, apparently, are puffs of a special kind, larger and more stable than ordinary puffs and perhaps concerned especially with secretion, which is extremely active in larval salivary glands.

The cytological observations are reasonably interpreted as showing that synthetic activity is concentrated at specific loci on the chromosomes and that different loci are active at different stages of development and in different tissues at the same stage. Beerman describes puffing as a manifestation of "gene activation in a qualitative sense" and suggests that puffs represent changes in the functional state of genetic loci that switch on a new type of steady-state or nuclear-cytoplasmic feed-back cycle. Consistently with Beerman's interpretation it may be supposed that puffs are sites of selective utilization of the genome as effective genomes linked with biotonic systems (Beerman, 1959; Breuer and Pavan, 1955; Briggs and King, 1959; Callan, 1963; Callan and Lloyd, 1960; Fischberg and Blackler, 1963; Gall, 1958; Pavan, 1959).

II. Lamp-brush Chromosomes of Oocytes

The unfertilized eggs of many vertebrates and invertebrates contain remarkable "lamp-brush" chromosomes, which have been most studied in amphibia, those in salamanders being especially large. These chromosomes attain lengths of 1 mm or more during meiotic prophase. The "lamp-brush" appearance is due to an abundance of fine loops protruding at right angles from the chromosomes. The loops consist of a fine axial thread of DNA surrounded by a matrix containing RNA and protein. The axial thread is probably an extended and projecting portion of a chromonema that extends continuously through the whole length of the chromosome. At each end the loop is attached to a chromomere and there is no other connexion between the chromomeres at the two ends. The axial thread seems to be a permanent structural part of the chromosome that uncoils and projects laterally during synthesis and folds and retracts into the main axis when functionally inert.

14. DIFFERENTIAL UTILIZATION OF THE GENOME

The loops, numbering hundreds, are remarkably varied in form from one locus to another but have specific differential characteristics at each locus. As the oocyte develops, loops emerge at certain loci, acquire a heavy coating of nucleoprotein and then wane, to be succeeded by a new set of loops at different loci. Loops never detach from the chromosomes; apparently they regress by discharging their ribonucleoprotein and folding back into the chromosome, possibly into the chromomeres which seem to be larger where there are no loops.

Most probably the loops are sites of RNA synthesis and liberate RNA or RN-protein in granular form into the cytoplasm. Particles containing RNA and ranging in size from the lower limit of visibility up to $4-5\mu$ in diameter accumulate in the cytoplasm as the egg develops together with larger particles, usually called nucleoli, and the evidence indicates that they are produced at specific sites on the chromosomes. It suggests, too, that each locus has a specific activity and, perhaps, makes a specific RNA by transcription of a segment of the DNA thread of considerable length. Loops come and go, different loci being active at different times. Many hundreds of loops regress during the latter part of oogenesis and typical meiotic chromosomes appear at the first metaphase. Since these chromosomes furnish half of the genetic material of the zygote they must carry the total genome and the loops must be interpreted as fully-reversible visible expressions of physiological changes in genetic action (Allfrey and Mirsky, 1963; Callan, 1963; Callan and Lloyd, 1960; Fischberg and Blackler, 1963; Gall, 1958).

The observations on lamp-brush chromosomes agree remarkably well with those on Balbiani rings and puffs. Both materials provide visible evidence of selective transcription of disconnected segments of the DNA, these segments being much larger than geneticist's "genes" and they indicate that the "primary gene product" is RNA. They provide no evidence of transcription of the total genome in continuity. Beerman credits the Balbiani rings with considerable stability and Breuer and Pavan (1955) and Pavan (1959) described a residual excess of DNA after the regression of some puffs. With these possible exceptions, which deserve further study, the reversibility of the puffs and of the lamp-brush loops seems to be complete, "without significant residue". Otherwise expressed, the utilization of one segment of the DNA as an effective genome entails no permanent change in the total genome.

Observations on the polytene chromosomes of the dipteran fly *Chironomus thummi* have yielded interesting results. The pattern of puffing of chromosomes of late-instar larvae can be changed by injecting the insect hormone ecdysone and the changes correspond with those that occur during normal development. During the transition from larva to pupa, puffs appear and disappear in a specific sequence. Characteristically, puffs appear only in certain combinations; some appear more than once and some are present all the time. It is remarkable that the typical combinations of puffs and the sequence of patterns of puffing are maintained despite extreme variations in experimental conditions and this seems to be an outstanding feature of the behaviour of puffs. The pattern of puffing normally seems to be dependent on the hormone ecdysone but it is possible to activate a graded sequence of genetic loci experimentally by effecting stepwise changes of the Na^+/K^+ ratio within the cells. Each Na^+/K^+ ratio incites a particular group of genetic loci to form puffs and the sequence of patterns produced by stepwise changes in the ratio corresponds with that seen in the salivary glands during the transition from the larval stage to the pupa (Clever and Karlson, 1960; Karlson and Sekeris, 1964; Kroeger, 1963a, b).

It has been suggested that the Na^+/K^+ ratio in the nuclear sap actually controls the pattern of puffing and probably, therefore, controls the pattern of transcription of the genome at a particular time but the ratio may merely reflect some other progressive change in the biotonic systems of the nuclei, which seems to advance autonomously or, at least, with little regard to extrinsic conditions. The exceptional interest of the phenomenon is that the *sequential changes* occur in a particular order that seems to be specified by a rather rigid programme, which experimental interference does not alter.

III. Some General Principles of Differential Utilization of the Genome

It is desirable to preface a consideration of mechanisms by a survey of the pheonomena which they are expected to bring about. Some of the more important empirical observations and theoretical problems will be summarized in the following paragraphs; most of them have been mentioned already in previous chapters but some of the most important of them have been insufficiently regarded in general discussions of the subject. Much more is at issue than turning genes "on" or "off".

(1) A basic, and as it seems to me incontrovertible, presumption of this discussion is that "genes", and more specifically DNA, have no intrinsic "activity" at all. DNA supplies genetic *patterns*; dynamic biotonic systems provide all the *activity*. According to this, Abercrombie's (1967a) attribution of differentiation to the differential repression or release of the activities of DNA is unfortunate and inaccurate.

(2) Differentiation is not adequately described in terms of the activation of individual genes; it entails the activation of integrated genetic patterns. The activated genetic patterns constitute *effective genomes*.

(3) Differential utilization of the genome presupposes a choice between multiple integrated genetic patterns *available* for use as effective genomes. The available choices constitute the *facultative genome*.

(4) Strain differentiation establishes a facultative genome which is characteristic for each tissue and which survives repeated cell divisions as shown most conspicuously in the stem cells of renewing tissues. It is inferred that the facultative genome is stable and replicable. The facultative genome supplies the genetic basis for *competence* or *capacity* in embryos, for metaplasia or divergent differentiation in the tissues of adult animals and for the multiplicity of the differentiative types of neoplasms derived from a single tissue.

One remarkable feature of the facultative genome is the preferential or differential availability of the diverse alternative genetic patterns. This is particularly well illustrated by "autoneuralization" of the embryonic amphibian ectoderm; the probability that the neural effective genome will be "expressed" in preference to all available alternatives becomes so high that the ectoderm reaches a state of *developmental imminence* whereupon neuralization takes place in response to trivial non-specific extrinsic stimuli or even, as it seems, as a result of random perturbations of the biotonic systems of the cells. Preferential availability characterizes also the states variously referred to as determination, crypto-differentiation or covert differentiation; one of the available potentialities effective genomes is *preselected* for effective use at a later time. The preselection is stable and replicable but it is *not* irreversible (Waddington, 1962).

During embryonic development, competence and, inferentially, the facultative genome, changes, apparently autonomously, with time. It is not clear whether this depends on a qualitative change in the pattern of potential effective genomes or in their differential availabilities. Be that as it may, there is during embryonic development a sequential activation of diverse genetic patterns in a strictly specified order, which is of the essence of epigenetic development. Reference was made to the evidence of sequential activation of "puffs" of the polytene chromosomes of *Chironomus* earlier in the chapter. Markert (1965) referred to the "programmed production of molecular specificity" and Flickinger (1962) discussed more specifically, "sequential gene action" but the subject has received much less attention than its importance warrants.

(5) In an earlier chapter the successive stages of cytodifferentiation were described in terms of the *selection, transcription* and *translation* of effective genomes and the subsequent processes of *supramacromolecular organization* and *morphogenesis*. This sequence can be interrupted at any stage.

Selection, in this context, ordinarily means the confirmation of an earlier preselection; linkage of preselected genetic patterns with conforming biotonic patterns establishes an effective genome. In some circumstances, nevertheless, the biotonic systems are truly selective in that they annul the preselection and bring about *metaplasia* or *deflection* of development from the preselected path. The selected genome is now "activated" and the immediate consequence is transcription of the effective genome followed normally by translation and protein synthesis. Markert (1965) pointed out that protein synthesis requires the integration of many events in a complex pattern which can be disrupted at various stages in many different ways. It may be misleading therefore to infer from the absence of a particular kind of protein synthesis that the apppropriate gene is not functioning. Ideally the identification of specific mRNA by its base sequence should be used as the index of "gene function". This is not yet practicable but substantial progress is being made towards the recognition of specific *patterns of transcription*, which are at the root of cytodifferentiation. It should be recalled here that, in conformity with the mutual exclusion principle, the selection of one effective genome automatically prevents the "activation" of any alternative one.

(6) Markert believed that although the reciprocal interaction between the genome and its environment may not account for all the relevant events in differentiation, it certainly includes the central phenomena. This is accepted in the present discussion which emphasizes perhaps even more strongly the complete dependence of genetic action on the interaction of genetic and biotonic systems. This terminology is more specific and, as I think, more illuminating than the more usual discussion of gene-cytoplasmic interactions. The biotonic systems, as described in Chapter 9, provide the complexity, the integration and the pattern of events that Markert presumed necessary for protein synthesis. Moreover, they provide a multiplicity of feedback mechanisms and an interpretation of the action of the "environment" on the genome without recourse to "trigger" mechanisms or "signals". Extrinsic stimuli it is presumed act only on and through the biotonic systems by deforming their pattern. It is noteworthy that industrious search for specific inducing *substances* in embryonic development has been notably unrewarding and that there has been a drift of interest from the identity of inductors to the mechanisms of response of the developing tissues.

(7) There is some evidence that genetic patterns that are not in effective use nevertheless still persist in the genome (Markert, 1965). Evidence that unused genetic patterns in the macronucleus of *Paramecium* are not expunged from the genome was given in Chapter 11. The integrity of the total genome has been convincingly demonstrated in differentiated cells of plants; whole plants can be regenerated from cuttings of differentiated tissue (Mather, 1948; Steward, 1963). The most satisfactory evidence that the total genome is retained intact in differentiated cells of metazoa comes from some nuclear transplantation experiments of Gurdon to which reference will be made later in this chapter. Cell metaplasia demonstrates the preservation of alternative genetic patterns in the facultative genome of differentiated cells

of vertebrate animals; tissue metaplasia demonstrates only the preservation of the facultative genome in the unspecialized stem cells of renewing tissues. Neither kind of metaplasia yields any evidence about the integrity of the total genome. The variety of differentiative types of neoplasm derived from cells of one kind provides additional evidence to the same effect and with similar limitations. The "inappropriate" endocrine functions of certain neoplasms referred to in Chapter 6 imply that certain genetic patterns not normally demonstrable in the facultative genome of the normal parent cells become "available" in their neoplastic descendants, and that more of the total genome is present than can ordinarily be demonstrated. It is likely enough that differentiation is consistent with an invariant total genome, but with the exception of Gurdon's experiments there is little evidence one way or the other. It is probable that the integrity of the total and facultative genomes is *not* preserved under certain abnormal conditions notably in cells repeatedly subcultured *in vivo* or *in vitro* and in nuclei transplanted into alien cytoplasm. It is questionable if either the total genome or the facultative genome persists in highly specialized cells that are incapable of division. It seems justifiable to attribute differentiation to the differential utilization of an invariant total genome with the proviso that the invariance can rarely be rigorously proved and may well be inessential.

(8) Abercrombie (1965, 1967a) has emphasized the stability, and indeed the irreversibility as well as the heritability of the results of differentiation as general characteristics of differentiation but in the discussion of the 1967 paper, Paul pointed out that the exceptions to the generalization were numerous and P. Weiss urged that the term differentiation should not be restricted to recognizable products of differentiation but should embrace *strain differentiation*. Without the extension proposed by Weiss and without specification of the type and stage of differentiation under consideration, generalizations about "differentiation" have little relevance or validity.

According to the best available evidence, strain differentiation, being based on a facultative genome that is stable and replicable, may be credited with high stability and with heritability but with the reservation that the normal *developmental bias* inherent in the facultative genome, corresponding with the preselection of one of the available integrated genetic patterns available for future use as an effective genome, can be overcome by extrinsic factors impinging on the biotonic environment of the genome.

The next question concerns the stability and heritability of an effective genome that is engaged in transcription. The effective genomes of cells of tissue that are liable to *cell metaplasia* are apparently unstable; environmental conditions can inactivate one effective genome and establish a different one in its place. In tissues that are liable to *tissue metaplasia* the position is not entirely clear. Ordinarily, once it has been established an effective genome persists until the cell dies, but the uncommon "confused" cells indicate that alternative effective genomes are available in differentiating cells and can be activated by extrinsic conditions provided cell specialization has not already advanced too far.

The persistence of products of differentiation through successive generations of cells without dilution, from which Abercrombie inferred some kind of cell or tissue heredity, justifies that interpretation only if it can be shown that new synthesis of the product continues without sustained transcription and translation of the appropriate effective genome; if that can be shown it implies that the operative cytosome of a differentiated cell is stable and replicable. Continued function in the absence of sustained transcription has indeed been reported and attributed to translation of stable, long-lived mRNA (e.g. Stewart and Papaconstantinou, 1967). It has not been demonstrated, to my knowledge, that long-lived mRNA can replicate but since virus RNA can replicate, the possibility that it may do so cannot be dismissed out of hand.

According to the evidence already given, dedifferentiation is usually due to the inadequacy of the cell environment for sustaining one or more of the processes of transcription, translation and supramacromolecular organization. The biotonic systems contribute to a critical degree in maintaining the stability and activity of effective genomes, whose intrinsic stability and replicability are doubtful.

The general inference from this discussion is that the most biologically-important, intrinsically stable and replicable element in differential utilization of the genome is the facultative genome. The integrity of the total genome is presumed but as an entity it is biologically inert. The replicability of the effective genome is doubtful and its stability probably depends on the linkage between conforming genetic and biotonic systems.

(9) Markert posed a fundamental question that remains unanswered: Is the programmed production of molecular specificity during development the whole story?; is all the information required to make an organism encoded in DNA? As mentioned in Chapter 12, several biologists have maintained that it is not and that development should be studied in terms of instructions for a programme of development, which is beyond the capacity of the DNA code alone. As Markert pointed out life is a continuum in time and each generation passes on to the next a highly complex cellular organization of which DNA is only the central part. The interaction between DNA and the rest of the cell provides the motive power for differentiation and cell change. In the terminology used here, the linkage of biotonic systems with DNA is essential for differentiation and the biotonic organization of open systems carries information and leads to coding at a higher level of organization than the nucleotide triplets of DNA, and to an important degree of chromosomal

organization (Apter and Wolpert, 1965; Flickinger, 1962; Goodwin, 1964; Sand, 1965).

IV. Mechanisms of Differential Utilization of the Genome

A. THE HISTONE HYPOTHESIS

Many attempts have been made to account for differential utilization of the genome by the selective "masking", "screening-off" or "blocking" of certain segments of the genome by ancillary materials of the chromosomes. Following the original suggestion of Stedman and Stedman (1950), the most popular hypothesis has been that histones, which are the most characteristic proteins of metazoan chromosomes but absent from bacterial "chromosomes", act as gene inhibitors or modifiers and suppress the activity of those segments of the DNA to which they are attached. It was shown by Huang and Bonner (1962) and others that histones do indeed inhibit transcription of DNA and much evidence in favour of the histone hypothesis was advanced especially by Allfrey, Mirsky and their colleagues (Allfrey et al., 1963; Allfrey and Mirsky, 1963; Frenster et al., 1963; Mirsky and Osawa, 1961). Nevertheless, the histone hypothesis have been severely criticized on several grounds and it has been suspected that the main function of histones in the chromosomes is to serve as a mechanical support for the long, folded DNA molecules (Bullough, 1967; Jacob, 1964).

A particular difficulty in applying the histone hypothesis has been to account for the specific affinity of histones for strictly defined segments of the genome that are to be "turned off". It is presumed that particular histones must be able to recognize particular segments of the DNA and no chemical basis for this specific recognition has been found. Barr and Butler (1963) noted that the histones from a variety of mammalian tissues resembled one another to a remarkable degree. Their own attempts at fractionation disclosed no more than ten to fifteen separate types of histones, a number grossly inadequate to account for the presumed number of inactive genes. They inferred that the various histones could distinguish at most between different classes of genes and not between individual genes and, further, that even if there is a correspondence between the sequences of bases in the DNA and the sequence of amino acids in the inactivating histones, which is an open question, some additional mechanism must be presumed to ensure the presence and removal of particular histone molecules from the DNA at the appropriate times.

Other investigations have led to similar conclusions and to a consequent modification of the histone hypothesis. Histone is now credited with a general repressive action on transcription of the whole genome but with no specific affinity for any particular segments of it; the specific activation of particular segments is now ascribed to selective *de-repression* by other components of the chromosomes, possibly acidic proteins (Frenster, 1965; Paul, 1967). According to Frenster's interpretation, nuclear polyanions displace histone from certain regions of the DNA and so allow opening of the double helix and give an opportunity for a specific repressor RNA to hybridize with one of the single DNA strands and set the other strand free for transcription.

It is not clear that the elaborations of the histone hypothesis overcome the basic difficulties. In Frenster's hypothesis the onus of selecting the appropriate segment is shifted to the nuclear polyanions which, it seems, must be capable of "recognizing" the histones to be displaced. Once the double helix has been opened, the proposed action of a specific repressor RNA is plausible but the general belief is that the repressors are proteins (Bretscher, 1968).

Possibly the wrong questions are being asked about the selective activation or repression of genes. Markert (1965) pointed out that the regulatory mechanisms can scarcely be as numerous as the genes themselves; if they were, the informational potentialities of the genome would be consumed in self-regulation with nothing left over for heterosynthetic functions. Markert suggested, therefore, that the specificity of regulation must depend on the specific arrangements of molecules and not on the matching of specific molecular structure with corresponding genetic structure. This approaches close to the concept of the mutual conformance of genetic and biotonic *patterns* repeatedly advanced in this book.

Many discussions of the selective utilization, including those concerning the role of histones imply that all genes would be "active" if not forcibly repressed. This is contrary to the argument of Item 1 in the preceding section. Markert (1958, 1960), also believing that this concept is of dubious validity, started from the contrary assumption that all genes would be quiescent if not activated. He proposed that as cleavage of the ovum proceeds, the chromosomes in various parts of the embryo are subjected to different chemical environments and that, by mechanisms that are still obscure, specific metabolic patterns in the cell impose correspondingly specific proteins on the DNA of the chromosomes. Markert presumed that the coupling of these proteins to specific regions of the genome produces functional nucleoproteins or "activated genes"; in the absence of the protein the same regions of the DNA are presumed non-functional. The activated gene, following Markerts argument, initiates a specific synthesis, of an enzyme for example, and thereby alters the metabolic pattern of the cell, whereupon the new metabolic pattern activates additional genes. The cyclic interaction between genes and metabolic pattern activates additional genes. The cyclic interaction between genes and metabolic patterns continues until stability is reached by the chromosomes becoming completely "differentiated" or by the metabolic patterns ceasing to change. In a later paper, Markert (1964) maintained that the genome is in a dependent responding position rather than an autonomous director of the cell's activities and that an acceptable explanation of differentiation must involve some mechanism whereby the function of the genome is, in effect, regulated by the surrounding protoplasmic environment. The Jacob-Monod hypothesis to be discussed next takes account of this requirement.

B. THE JACOB-MONOD HYPOTHESIS

Jacob and Monod's interpretation of selective utilization of an invariant total genome in bacteria is based on three main presumptions: first, the total genome is functionally discontinuous, being segmented into separately-controlled "units of transcription", the *operons*; second, the transcription of particular operons is regulated by "cytoplasmic" factors linking spatially distinct *regulator* genes with the operons; third, the control is essentially a negative one effected by *repressors* produced by the regulator genes. The active system comprises two genetic sites on the chromosome linked by biotonic systems and incorporates a negative feed-back control of transcription. It is possible to devise model circuits based on these presumptions which, under prescribed circumstances, would be intrinsically stable and persistent; so that a selected effective genome, once established, would continue to operate indefinitely. Monod and Jacob (1961) proposed that comparable mechanisms, being "directed, highly specific and efficient" might provide for the orderly differentiation of higher animals.

Bullough (1967) has recently maintained that there is no essential difference between the labile differentiation of bacteria and the stable differentiation of the metazoans. It seems to me that the stability constitutes an "essential difference" and to account for it Bullough suggests that during the evolution of metazoa a series of repressors, which are able to bind particularly strongly with their relevant genes or operators, has been developed. His contention that, wherever it is found, differentiation involves the selection of some pattern of genetic activity that is relevant to the needs of the organism is fully agreeable to the argument of this chapter; the further statement that in all cases this is evidently achieved as a response to the presence of specific chemical messengers is more controversial.

In Bullough's opinion it is possible to conclude only in the most general terms that, as in the bacterial cell, so also in the more complex metazoan cell, a gene remains inactive while it or its operator is combined with a specific repressor substance the nature of which is unknown, that the histone of the chromosome acts as a skeletal support for the very long and highly folded DNA molecules but that it may also play some role in gene repression, that in metazoan cells as in bacterial cells the genes are repressed or derepressed in response to the presence of certain chemical messengers which may be metabolities or specific effectors or inducing agents and that in a metazoan cell gene activation is accompanied by the breaking of the DNA-histone complex and the uncoiling of the DNA molecule.

In general, I question the adequacy or validity of the bacterial model for the study of the progressive differentiation of metazoan organisms. The essence of epigenetic *development*, which is our primary concern, is completely lacking from bacteria and so also is the incorporation of the genetic DNA into elaborately organized chromosomes. Admittedly, selective utilization of the genome is common to bacteria and to the metazoa and it is reasonable to suppose that regulatory circuits comparable with those proposed by Jacob and Monod are implicated in the selection of effective genomes but evidence from the ciliated protozoa negates the proposition that selection necessarily depends on specific chemical messengers. One of the drawbacks of the Jacob-Monod regulatory circuits, which their originators were amongst the first to point out, is that they can be arranged to explain almost anything. They can be arranged to account for the selection and maintenance of extrinsically-stabilized effective genomes. They can be applied to the mucous metaplasia of chick epidermis that evoked and maintained by excess of vitamin A and perhaps, although less easily, to the intrinsic developmental bias of the epidermis towards squamous differentiation. One the other hand, I have seen no plausible explanation in these terms of the stability and replicability of *strain differentiation*. More generally, the bacterial models do not illuminate some of the most important and urgent problems in normal and neoplastic development such as the phenomena of competence or capacity, developmental bias and developmental imminence, determination and decision in advance of performance or, as Nanney described it "differentiation" in the absence of "expression". The models are plausibly applicable to effective genomes in operation but not easily applicable to facultative genomes; most importantly perhaps, they are not evidently helpful in accounting for the sequential activation of genetic patterns in a precisely specified order that is at the basis of epigenetic development.

The interpretation of genetic *action* in terms of a linkage between genetic and biotonic patterns provides for regulatory circuits of the Jacob-Monod type and for feed-back control; it does not entirely exclude but it certainly does not require or presume the operation of specific signals or inducers. The question at issue is not so much the existence of these circuits as their adequacy for accounting for *all* the phenomena of differentiation and in particular for the facultative genome, the concept of which Bullough accepts and uses, and all the phenomena which are associated with it. The sequential changes in the facultative genome with time and its high stability and replicability probably require coding at a higher level of organization than the

DNA double helix and call for accessory chromosomal organization as maintained in Item 9 in a preceding section. The facultative genome, being apparently deeply implicated in the instructions for the programme of development, is not likely to yield easily to analysis. Some aspects of nuclear differentiation and chromosomal organization which may be pertinent to this problem are discussed next.

C. Nuclear Transplantation in Amphibia

Briggs and King (1952) developed an elegant method of transplanting nuclei embryonic frog cells into enucleated frogs of the same species (*Rana pipiens*). Eggs which received nuclei from blastulae comprising 8–16,000 cells developed into completely normal embryos and larvae; apparently the nuclei were functionally equivalent to the zygote nucleus being "totipotent" and completely undifferentiated. It may be noted that the transplanted nuclei were taken at a stage of development before the genome has come into effective use. When nuclei were taken from various portions of early gastrulae, again most of the eggs developed into normal embryos but, with increasing age of the donor animals, the nuclei were progressively less able to sustain development beyond the blastula stage, at which time the genome normally comes into effective use and, of the embryos that survived longer, most were defective. Briggs and King (1959), interpreted their experiments in terms of nuclear differentiation which they thought might be ascribed to "stabilized changes in the condition of specific genetic loci" but they did not exclude the possibility that some non-chromosomal nuclear component or perinuclear organelle might be implicated. They concluded from a variety of experiments that in some cases at least the nucleus or some nucleus-associated organelle might in the course of differentiation undergo genetic changes of a type needed to confer stability on differentiated cells.

Other investigators have repeated and extended the experiments with some variations in transplantation technique and in the species of frog used. In a general way they have confirmed the observations of Briggs and King by showing, in particular, that the ability of transplanted nuclei to sustain normal development of enucleated eggs decreases progressively with increasing age of the donors of the nuclei, that the "nuclear differentiation" is stable and "heritable" in the sense that it persists through an indefinite series of back transfers of nuclei to enucleated eggs and that "totipotent" nuclei remain "totipotent" through similar series of back-transfers. The newer observations provide no fresh evidence of a specific relationship between the tissue-source of the nuclei and the type of the developmental defect (Elsdale *et al.*, 1960; Fischberg and Blackler, 1963; Fischberg *et al.*, 1958; Gurdon, 1960).

The limitations of transplanted nuclei become apparent much later in *Xenopus* frogs than in *Rana* not being demonstrable until the donor embryos have reached the neurula or tail-bud stage. Even then and at much later stages of development, the limitations of nuclei seem to be less frequently and less consistently manifested (Gurdon, 1962a, b). In Gurdon's experiments some eggs implanted with nuclei from *Xenopus laevis* donors ranging from late blastulae to free-swimming tadpoles developed into adult frogs. Gurdon (1962b) transplanted nuclei obtained from gut epithelial cells from feeding tadpoles; unlike most donor cells, these were "differentiated" in the sense of being visibly specialized cells with a striated border. Making allowance for certain technical limitations, Gurdon estimated that 24% of the nuclei of these specialized cells were able to promote the development of enucleated eggs into normal, feeding tadpoles. Most recently Gurdon and Uehlinger (1966) transplanted nuclei of intestinal cells obtained from *Xenopus* tadpoles which had just started feeding into ova and obtained adult frogs some of which were fertile. They inferred that least some differentiated intestinal cells had not lost the genetic factors necessary for the development of a fertile, adult frog. Less decisive evidence in the same sense has been reported by Picheral (1962) and Simnett (1964).

Gurdon's argument implies that cell differentiation does not necessarily entail any irreversible alteration in the total genome; it does not exclude stabilized but reversible alterations in the pattern of utilization of the genome. In all experiments, the results of transplantation are extremely varied; some nuclei seem to be "totipotent" whereas others of similar provenance are completely ineffectual. Gurdon maintained that the stable and replicable changes in the genome of some transplanted nuclei are not essential for cell differentiation and have not been proved to originate during normal development; until proof to the contrary is forthcoming, it is possible to attribute the "nuclear differentiations" as artefacts produced by the experimented procedure and irrelevant to normal development (Gurdon, 1963, 1964, 1967). Gurdon's suggestion that the artefacts might be attributable to the action of recipient cytoplasm on the implanted nuclei gains support from observations on the transplantation of nuclei into frogs of a different species. As might be expected, transplanted nuclei have only a limited ability to sustain the development of ova from different species. During the limited period of development that is possible, the nuclei seem to undergo irreversible changes and gross changes of karyotype are found and are attributed to faulty mitosis of the nuclei in the foreign cytoplasm (Gurdon, 1964; Hennen, 1963, 1965; Moore, 1958, 1962). There is evidence that similar disturbances of mitoses occur also as a result of transplantation within the same species (Briggs *et al.*, 1964; Di Berardino and King, 1965; Gallien *et al.*, 1963; Subtelny, 1965a, b).

Flickinger (1962) has argued that if transplanted nuclei are specifically "differentiated" the recipient eggs would be expected to show a definite type of developmental failure so that only that type of tissue corresponding with that from which the nuclei are derived would develop; this behaviour is not observed. The argument implies that the nuclei of differentiated cells carry intrinsically-stable and replicable effective genomes, which is not established, and that these effective genomes have the ability to convert enucleated eggs into specifically differentiated adult cells. The argument overlooks the fact that an egg with its own nucleus, or any other nucleus, cannot transform directly into any kind of differentiated cell (Grobstein, 1959). The differentiated cell is the end-result of an ordered sequence of changes some of which are not manifested by overt signs; it is a product of epigenetic development which requires that diverse genetic patterns should be available for use in a precisely ordered sequence and requires, no less, a mutual conformance of genetic and biotonic patterns at successive steps of the sequence. During development, genetic and biotonic patterns must change progressively, concurrently and in mutual conformance. When the nucleus of a differentiated cell is implanted in an enucleated ovum the genetic and biotonic patterns are "out of phase" and normal development can proceed only if the full range of genetic patterns is not only present but available in the transplanted nucleus. Gurdons observations indicate that this can happen but only occasionally and for completely unknown reasons; much more commonly the full range of genetic patterns is not available and the result of transplantation is either complete failure of development beyond the stage of development at which the genome comes into effective use or abnormal development depending on the availability of some genetic patterns allowing limited development. As Briggs *et al.* (1964) have noted the occurrence of karyotypic abnormalities due to faulty mitosis severely complicates the interpretation of the results but it seems reasonable to infer that although the total genome is not materially diminished during differentiation, the availability of its component alternative genetic patterns is progressively restricted, although not irreversibly. The results are consistent with a substantial degree of reversible chromosomal differentiation.

D. Paramutation and Treption

Two closely allied hypotheses proposed by Brink (1960, 1964) and by Serra (1964) are relevant to the differential utilization of the genome although they throw little light on mechanisms. They were advanced primarily to account for the stability and replicability of differentive changes by presuming alterations at the genetic level that do not entail material changes in DNA and which, therefore, are not "mutations" in the strict sense of the word. Brink calls them *paramutations* whereas Serra calls them *treptions* (*Gr.*

treptos, turned, changed) but it is evident that both authors are dealing with similar phenomena, often in similar terms. The changes differ from mutations in several important ways: they occur naturally and regularly in certain somatic cells but not in germ cells; they are gradual not abrupt; they are metastable and may be easily reversible or virtually irreversible, apparently with all grades of stability in between; they are adaptive and directed, being determined by physiological processes and occurring as responses to stimuli provided by the internal milieu of the cell or, indirectly, by extrinsic agencies altering that milieu.

Brink's hypothesis of paramutation is based primarily on studies of paramutation in maize but he mentions comparable phenomena in the protista. His hypothesis embodied the presumption that chromosomes include two kinds of genetic material called, respectively, orthochromatin and parachromatin which, Brink suspected, might correspond with euchromatin and heterochromatin if those two materials could be better defined. More recently, Brink (1964), admitting that the material basis of paramutation remained a matter of speculation, suggested that it reflected the action of heterochromatin in interfering with the uncoiling of chromosomes, which is an essential event in the processes leading to the "activation" of genes; tight coiling entails partial or complete local repression of gene action. Brink described "repression of gene action by pycnosis" as a common and basic process that does not involve material alteration of the genes but affects their expression.

According to Allfrey and Mirsky heterochromatin or "pycnosis" results from tight coiling and the accretion of histone protein. If this be true, Brink's hypothesis seems to be a modification of the histone hypothesis, orthochromatin being identifiable as DNA, parachromatin as histone and heterochromatin as tightly-coiled DNA having an accretion of histone. Brink presumed that parachromatin was coextensive with the orthochromatin and credited it with the general function of "activating" the orthochromatin; the parachromatin was believed to be the product of a special class of chromosomal elements and to be capable of replicating in phase with the orthochromatin so that it functioned as "quasi-genetic material", receiving, recording and transmitting from cell to cell during mitoses, information from outside the chromosomes and eliciting specific responses in the orthochromatin. Brink emphasized, in particular, the sensitivity and specific responsiveness of parachromatin and of paramutation to changes in the internal milieu of the cell and to extrinsic factors that modify it during differentiation and development.

Serra's hypothesis of *treption* refers more specifically to the role of treption in determining mating-types and serotypes in the ciliated protozoa and he maintained that it must be involved also in embryonic and tissue differentiation in metazoa. Serra did not discuss the mechanism of treption in detail but suggested that irreversible or long-lasting treption might be established by "cellular enzyme-RNA systems" inducing a permanent change in the chromosomal DNA and in his respect his interpretation is at variance with that of Brink.

The terms paramutation and treption, of which paramutation seems the more likely to come into general use, are applicable in a descriptive sense to many empirically-observed phenomena described in this chapter and in earlier ones and their greatest value is probably to focus attention on these important but widely neglected phenomena; their explanatory value is probably small.

Mather (1961) described phenomena in the fly *Sciara coprophila* which resemble Brink's paramutation in three respects: they are "directed", they are "determinate" and, in some circumstances, even inevitable and they are "metastable" being reversible under certain conditions. Mather noted other possible examples of paramutation and was tempted to ascribe to paramutation the stability that cells achieve at their end-points of differentiation. Nevertheless, he pointed out that paramutation, by itself, does not provide a satisfying explanation of differentiation, which evidently, in this context, applies to strain differentiation. For an understanding of differentiation it is still necessary to turn to the cytoplasm although paramutation may be involved as a secondary response to differences in the cytoplasm. In Mather's view the probable significance of paramutation is to stabilize changes that are initially cytoplasmic and to give to cell heredity a precision that purely cytoplasmic transmission could not guarantee. Mather further noted that if cytoplasmic changes can bring about paramutational alterations of the chromosomes, external interventions, through their effects on cytoplasm, might bring about directed changes of the hereditary materials; he did not refer specifically to intrinsic changes in the DNA.

V. Conclusions

Unfortunately, the chief conclusion seems to be that none of the proposed mechanisms is convincingly adequate; none covers more than a fraction of the phenomena and problems itemized in a previous section. In Mather's discussion, "cytoplasmic" seems to be an inappropriate word; the interactions are, presumably, intranuclear. Substituting biotonic systems or metabolic patterns for "cytoplasmic" mechanisms, Mather's exposition has much in common with Markert's; each embodies important and valid considerations but the illuminating synthesis has yet to be made. The *selective* action of biotonic systems is recognized by Markert and by Mather but, so far as I can recall none of the authors mentioned in this chapter, with the exception of Bullough, who accepts and uses the concept of a facultative genome, pays any attention to what "genes" are *available* for selective "activation". Differently expressed, the fundamental problem of *competence* is ignored.

CHAPTER 15

Applications of Recent Concepts of Biological Organization to some Problems of Normal and Neoplastic Development

I. Normal Development

A. GENETIC MATERIALS AND GENETIC ACTIONS

A few years ago, Waddington (1962) suggested that biologists should carry molecular biology as far as it will go along its own lines but try, also, to find "the irritant facts" with which it is unable to deal. I have adopted a similar attitude in this book and it has seemed to be a profitable one.

1. Differentiation

The most immediate impact of molecular biology upon developmental biology has been on the study of differentiation, which is now emerging from the confusion and obscurity, which have long encompassed it. Differentiation can now be rationally studied, if not yet fully "explained", in terms of differential utilization of the genome or differential *patterns of transcription*. Differentiation entails the preselection and selection of effective genomes and the transcription and translation of these genomes by linkage with conforming dynamic biochemical systems. The various steps are differently controlled.

The problem of *preselection* is inextricably related to that of the nature of *competence* which, as here proposed, depends on the differential availability of the utilizable genetic patterns constituting the facultative genome. The physical basis of the differential availability has not been elucidated but the present indications are that it probably depends on chromosomal organization implicating ancillary materials of the chromosomes without necessarily compromising the integrity of the DNA of the total genome. The facultative genome was invented (Foulds, 1963) to take account of the empirical evidence for the differential availability of alternative genetic patterns for use as effective genomes under a variety of circumstances; it provides a genetic basis for pathological *metaplasia* as well as for normal developmental *competence* and capacity and is closely relevant to many problems of neoplastic development as will be described later.

Preselection of an effective genome corresponds closely with what has been called *determination* but without the implication of irrevocability; it corresponds also with the early stage of differentiation that is now called *strain differentiation*. As well exemplified by the basal stem cells of the epidermis and other renewing tissues, the preselection and, inferentially, the facultative genome is extremely stable and indefinitely replicable in the absence of phenotypic expression of the preselected effective genome. Competence, determination and developmental imminence all imply a comparable stable and replicable preselection of an effective genome in advance of the engagement with biotonic systems that is needed for its transcription and translation or, more generally, for its effective use. According to the best available evidence, competence is an intrinsic quality of the genome that changes autonomously with time and that is not subject to environmental control. Similarly, preselection seems to imply an intrinsic change in the facultative genome that is independant of extrinsic controlling factors and of engagement with dynamic biotonic systems. It should be emphasized that transcription and translation and all subsequent steps in the phenotypic *expression* of the preselected effective genome are strictly dependent on linkage with biotonic systems and are liable to modification or even suppression by extrinsic factors impinging on those systems.

In some earlier sections the term *selection* of an effective genome has been applied to the first step in the activation of a potential effective genome present in the facultative genome as a result of the linkage of conforming genetic and biotonic patterns. Ordinarily, in normal development, this "selection" seems to entail no more than the confirmation and expression of an earlier preselection but under certain circumstances the biotonic systems may override the preselection so that their action is truly selective. This terminology is not ideal but the distinction to be emphasized is between a preselection inherent in the facultative genome and a choice between alternative genetic patterns, available in the facultative genome, that is effected by linkage with biotonic systems and subject to extrinsic controlling factors operating through those systems.

Strain differentiation based on an intrinsically stable and replicable preselection in the facultative genome must be separated sharply from all subsequent steps of "differentiation", starting with transcription and translation of the selected effective genome, that lead to overt phenotypic expression. Failure to appreciate this distinction has been responsible for much confusion and disputation about the "stability" and "reversibility" of "differentiation". In short-term cell cultures *in vitro* de-differentiation often results from the inability of the culture medium to sustain the biotonic dynamic systems that are needed to ensure continued expression of the strain differentiation of the cultured cells; by adding non-specific nutrients or specific hormones to the medium or by transferring the cells to an *in vivo* environment, the cells can be induced to "redifferentiate" as a consequence of renewed expression of the "strain differentiation" which has persisted and replicated during the phase of de-differentiation. The de-differentiation implies only the suppression of the transcription and translation of an effective genome; it does not entail any suppression or reversibility of the strain differentiation which depends on an intrinsically stable and replicable facultative genome.

Under certain pathological and experimental conditions, extrinsic factors may bring about deflection of re-differentiation or metaplasia by expression of an effective genome other than the preselected one. In general, the extrinsically selected genome must also be extrinsically maintained; if "normal" external conditions are restored, the "normal" preselection is again expressed as is well shown in Fell's experiments on the effect of vitamin A on chick epidermis. Extrinsic conditions impinging on the biotonic systems of undifferentiated cells can override the preselection inherent in the facultative genome and select an alternative effective genome but they do not expunge the original preselection which is expressed again when the external conditions are appropriate. In short, the abnormal extrinsic conditions do not modify the intrinsic qualities of the facultative genome although they may modify or suppress their expression as overt phenotypic characters. The "minimal mass requirement" for differentiation is now under re-investigation but it is apparent that it does not imply a lack of capacity of individual cells for specific differentiation but only the favourable effects of cell associations for the expression of those capacities by facilitating the transcription and translation of preselected effective genomes or the subsequent steps culminating in overt phenotypic characters.

B. Extra-Nuclear Genetic Mechanisms

The study of the *differentiation* of specialized structure and function has overshadowed the importance of those basic mechanisms of *maintenance* that are common to all unspecialized cells and bacteria and that allow them to keep alive, grow and multiply. It seems that all cells that can be indefinitely cultivated *in vitro* are degraded to a common state in which little if any trace of their origin from one or another normal tissue or even from normal or neoplastic tissue can be distinguished. A few indications have been mentioned earlier to the effect that *maintenance* as contrasted with differentiation is not dependent on sustained transcription of nuclear DNA, being unaffected by actinomycin D. One possible explanation is that genetic functions of the nucleus are in some way delegated to cytoplasmic organelles, carrying either genetic DNA or to stable, replicable, informational RNA.

The highly complex and specialized infraciliature of ciliated protozoa may be interpreted as an extreme example of delegation or decentralization of genetic function allowing a substantial degree of local developmental autonomy as illustrated in particular by Tartar's observations on *Stentor*. It is doubtful if any comparable specialization is to be found in the higher animals but the relative autonomy of mitochondria present in all organisms from the protista to the primates and of the chloroplasts in plants is possibly an important factor in the process of maintenance. In the protista, the mitochondria and chloroplasts, which carry the respiratory enzymes, are probably responsible for the substantial powers of survival of the "cytoplasm" of certain organisms, notably *Acetabularia*, in the absence of a nucleus. The discussion of the relative "autonomy" of mitochondria in higher animals and of its genetic basis need not be repeated here. The relevant observations are comparatively new and their interpretation not always clear but it is evident that they may become important for the understanding of maintenance processes and of extranuclear genetic mechanisms.

The demonstration of DNA in mitochondria does not necessarily mean that it behaves in all or most ways like the DNA of chromosomes or that it acts as "genetic material" although it is a reasonable working-hypothesis to suppose that it does. If this hypothesis be substantiated, it remains possible that other mechanisms of de-centralized genetic control also operate, most probably, according to present ideas, through informational RNA that is either intrinsically stable at the time of its transcription or that begins as unstable mRNA and becomes stabilized and replicable by incorporation in ribosomes or other components of the ultrastructure of cells. The resistance of haemoglobin-production in anucleate mammalian reticulocytes was mentioned earlier as a possible example of transcription and translation of a stable informational RNA.

C. The Dynamic Component of Biological Organization

The dramatic change in the comprehension of the stable macromolecular components of biological organization has not been accompanied by a similar increase in the understanding of the dynamic biochemical flux. Pitot (1963) wrote that, for the biochemist, the multicellular organism is "a nightmare of complexity". Two basic features of this complexity are being increasingly, if tardily, recognized; complexity *per se* is an essential condition of life and it is an organized complexity. The concept of a cell as "a bag of enzymes" was never congenial to biologists and is now being renounced by biochemists, but no comprehensive and operationally-useful biochemical concept of organized complexity has yet replaced it. The basic problem is to conceive an organized pattern of materials in continual flux linked, in mutual conformance, with the pattern of macromolecular and supramacromolecular organization disclosed by molecular biologists and electron microscopists.

It was argued in Chapter 9 that equations could be re-written, without violence to accepted chemical principles as "biological" equations which could be used to construct representations of complex open-systems so interconnected and interdependent as to form *organized patterns of materials in motion*. It should be emphasized that the diagrams presented in Chapter 9, in particular Fig. 70, represent an organized flux of molecules and not of chemical "substances" that can be isolated and characterized. The substances isolated by biochemical procedures, apart from the stable macromolecular ones are, in the main, artefacts produced by disrupting the organized systems.

The idealized systems represented in the diagrams are abstractions; their stability or even their existence in isolation may be doubted. The same may be true of the more stable macromolecular structural organization of cells. In the living organisms, the effective system must comprise a linkage of mutually conformable and interdependent dynamic and structural patterns. It was suggested in Chapter 9 that the "reaction regions" shown in the diagrams might represent the sites of linkage of the dynamic systems with the organized ultrastructure. In the previous section of the present chapter it was further implied that the mitochondrial membranes and, perhaps the ribosomes of the endoplasmic reticulum might act as sites of decentralized or delegated genetic action by virtue of their content of either extra-nuclear DNA or stable informational RNA. It may be recalled that the structure of "cytoplasm" can be severely disrupted mechanically as well as during mitosis without permanent disorganization of its structure or function; it is credible that the dynamic pattern is in some measure responsible for restoring the structural pattern.

The proposed organization fulfills one of the requirements of biological organization that several biologists have been urging for a number of years namely the interrelatedness and interdependence of all biochemical processes within cells (Weiss, 1958c); the output of one system for example serves as the input of another. It can also easily incorporate control circuits of the Jacob-Monod type as specialized components. These circuits are of outstanding significance in that they provide the first persuasive interpretation, based on specific examples in bacteria, of the integration of the stable and mobile components of biological organization in single self-regulating systems. It seems likely that they embody general principles of regulation that will prove to be of wide validity. Nevertheless, it is my impression that some contemporary writers are pressing too hard the *detailed* application of these circuits, based primarily on observations on bacteria, to problems of regulation in vertebrate animals. It should be remembered that the components of the circuits are not joined together by wires like those of electronic circuits but by organized systems of molecules in chemical flux, which cannot be completely specified in terms of stable chemical substances and which probably cannot be fully elucidated by reductionist analysis.

D. The "Network" Concept of Biological Organization

An essential feature of Fig. 70 is that it depicts a network of chemical activity such that every portion of the chemical net is connected directly or indirectly and closely or remotely with every other portion of the net. Judah and his colleagues, who studied the effects of toxic drugs on mammalian tissues, found that a considerable variety of drugs produced much the same biochemical injury of cells and interpreted their findings as evidence for a network organization of biochemical activity. They remarked that a force applied to any part of a network will be transmitted in varied degrees to all parts of the net and that the location of the conspicuous deformation or break may be independent of the point of action of the disturbing force, being determined more by local variations in the strength of the net (McLean et al., 1965).

Paul Weiss, who has repeatedly emphasized the interrelatedness and interdependence of the biochemical processes within cells, has recently re-stated his views in terms of the "network" hypothesis of biological organization. His remarks were directed primarily to the control of growth but are much more widely applicable. Weiss maintains that no single method of growth-control or feedback system can be conceded an explanatory monopoly. As in development in general, we are faced not with the old time linear, mechanistic cause-effect chain where, if one link is cut, all the rest falls out, but with network systems where, if one part is destroyed, the remainder will still produce something that looks to be a good likeness of what would have been seen if that particular thread of the fabric had not been severed. We are dealing with multiple-mesh fabrics with numerous pathways connecting any two points so that, when one connection fails, many by-passes remain to preserve uninterrupted communication. Weiss suggests that simplification of the problem of growth control will presumably depend on finding general rules of network behaviour rather than on finding a linear cause-effect relation chain that could be considered to be *the* control mechanism (Weiss, 1967*b*).

The network organization is closely relevant to the conspicuous manifestation at all stages of development of multiple mechanisms achieving, maintaining and restoring the "normal" structure and function of the organism. Biologists are concerned with *regulation* and *adaptation*, physiologists with homeostatic mechanisms and pathologists with *compensatory devices* for mitigating the effect of disturbances exceeding those that can be corrected by the normal homeostatic controls. All these mechanisms may be viewed as factors in the *directiveness* of biological development. In the present context it is important to emphasize that, as in Judah's description, varied stimuli applied at diverse portions of the network may produce similar or apparently identical results and that deformations at one place may be more or less successfully corrected by compensatory modifications at other places so that the total function of the net, although not its minute organization, may be maintained. The network organization is consistent with the principle of "multiple assurance" discussed by A. E. Needham (1964) and with B. E. Wright's (1966) views on biological organization.

II. Neoplastic Development

No general discussion of theories of "cancer" or of neoplastic development will be attempted at this stage, the present section being restricted to a few matters closely related to the preceding consideration of biological organization.

In "An Attack on Cytologism", Smithers (1962) assailed a number of what he considered to be widely-held misconceptions about the cellular basis of neoplasia. In particular he maintained that no specific, irreversible, heritable change had been demonstrated in individual cancer cells. This composite statement raises several important questions which call for serious but separate consideration. It seems to me scarcely questionable that neoplastic cells are abnormal cells; the facts of metastasis in man and transplantation in animals are hardly explicable on any other basis. I maintain, therefore, that some change has been demonstrated in neoplastic cells. The nature of that change is arguable.

If Smithers means what I think he means by "specific" change, I am in complete agreement with him and deprecate as much as he does, descriptions of "The Cancer Cell", which imply that all malignant neoplastic cells possess, in common, the same distinctive, definable abnormality. It is desirable nevertheless to state more explicitly the basis for this opinion. So far as I am aware no change attributable to the loss or abnormality of any particular *structural gene* or group of structural genes has been *consistently* demonstrated in neoplastic cells. Furthermore no diminution or abnormality of the total genome has been *consistently* demonstrated; abnormalities have been found in abundance in some neoplasms but they are much more probably incidental accompaniments or consequences of neoplasia than its "cause". These statements are tantamount to a rejection of the somatic mutation theory of neoplasia unless the term *mutation* be extended far beyond its historical and customary meaning. Some of the readers who have welcomed Smithers' criticisms most warmly, seem to have been actuated mainly by a not unreasonable dislike of that theory, but it should be emphasized that several of the other assertions made by Smithers do not follow automatically from the denial of somatic mutation.

In the discussion of normal differentiation, the importance of stable replicable changes located tentatively in the facultative genome was emphasized as a basic feature of *strain differentiation*. The material basis of the facultative genome is not known but is coming within the range of experimental study. The conclusion to be emphasized here is that mechanisms of stable replicable change, not dependent on mutation in the sense of irreversible change in the total genome, are now being envisaged, if not yet far elucidated. It is no longer admissible to presume that stable replicable change *must* be due to mutation; it can now be ascribed plausibly to stable, replicable change in the pattern of utilization of an intact total genome. It was further emphasized earlier that the changes in the facultative genome are stable and replicable in the absence of phenotypic expression; in other words, they are *intrinsically* stable and replicable whereas the phenotypic expression consequent on transcription and translation of a selected effective genome is subject to modification or control by extrinsic, environmental factors. In my view, it is desirable to reconsider the "irreversibility" and "heritability" of all changes in neoplasia in similar terms. "Irreversibility" is almost impossible to prove and "heritability" in somatic cells has not the same significance as the Mendelian inheritance of adult characters of intact animals through the germ cells.

There are at least two well-substantiated analogies with *strain differentiation* in neoplasia, namely, *incipient neoplasia* and *residual neoplasia*; both are characterized by stable and replicable cell changes that persist in the absence of phenotypic expression of neoplastic properties. According to the available evidence, both incipient neoplasia and residual neoplasia, once established are permanent; their irreversibility can scarcely be proved but their reversibility if it existed could be, but has not been, proved. It seems legitimate to me to adopt the working hypothesis that the two neoplastic states differ from normal strain differentiation in being irreversible. Residual neoplasia is of especially important in the consideration of regressions of hormone-dependent tumours; the *capacity* for neoplastic growth persists through periods of regression but the *expression* of that capacity depends on extrinsic stimuli; the *expression* of the capacity is therefore reversible, but the neoplastic capacity itself vested as here proposed in an abnormal facultative genome, according to the evidence now available is *not* reversible.

The evidence for regression and maturation of tumours is not disputed and, indeed, was extended beyond that usually presented in the chapter on

15. PROBLEMS OF NORMAL AND NEOPLASTIC DEVELOPMENT 383

Retrogressive Neoplasia. Reversibility of neoplasia is not disputed here on theoretical grounds but because I can find no unequivocal empirical evidence that reversibility, in the sense of reversion of neoplastic cells to normal cells has ever been observed. At the most the "expression" of an irreversible neoplastic capacity can be reversed but not by the reversion of individual cells to normal cells. Alternative and more plausible interpretations of regression are discussed in Chapter 5 and need not be repeated here.

It is convenient to comment here on the alleged "reversion to embryonic type" which is still being extolled as a characteristic feature of malignant tumours. It is true that many malignant tumours are, like many embryonic tissues, "undifferentiated" but the evidence for any real "reversion" to embryonic type is non-existent. The relationship between embryonic cells and "anaplastic" neoplastic cells is much the same as that between a child and a senile man in second childhood; in the one potentialities are undeveloped and in the other they are gone forever (Foulds, 1963). It is noteworthy that the relatively few neoplasms that credibly originate in embryonic tissues show little tendency to maintain the embryonic state of their parent tissues, much less to "revert" to an earlier state. Often, indeed, they are apt to advance to more mature states; some are unexpectedly responsive to therapy and others are liable to spontaneous regression. On this showing, the embryonic state is not particularly conducive to "malignant" behaviour.

The mass of evidence provided by the study of metastasis of tumours in man and the transplantation of tumours in animals indicates that neoplastic behaviour depends on a stable, probably irreversible, replicable intracellular change. It suggests further that the change is present in all the neoplastic cells. The decisive test of cloning, which is required to substantiate this belief is beset by technical difficulties which were first encountered in growing isolated normal cells *in vitro*. Until recently the only successful cloning of neoplastic cells was achieved by Furth and Kahn (1937) using a transplantable mouse leukaemia whose parent normal cells are adapted to a solitary existence. The cloning of a transplantable teratocarcinoma provides convincing evidence of a stable replicable mechanism within individual cells which is responsible for all details of their neoplastic structure and behaviour and which carries, more over the "morphogenetic memory" or "programme of development" of extremely complex tumours.

In the light of the discussion of "differentiation" in tumours presented in Part II of this volume it is difficult to accept without disabling qualification the hypothesis advanced by Paul (1967) amongst others that cancer is due to faulty cytodifferentiation or the reference by Bullough (1967) to "that collapse of differentiation which leads to cancer". There is no conspicuous abnormality of structural or functional differentiation in a great number of tumours including many that are accepted as being "cancers". These tumours may be called "minimal deviation tumours", using that term in a wider and different sense from that proposed by Potter. Highly various degrees of "deviation" are discernible in the capacity of mammary tumours to carry out the normal function of their parent tissue of secreting a milky fluid. The "minimal deviation" from the normal gland is the cystadenoma, a benign tumour that secretes in response to the same hormonal stimuli as the normal gland from which it differs, so far as can be seen, in its lack of morphological integration into the excretory system of the gland. Milky secretion, similarly responsible to normal hormonal stimuli, is seen also, in transplanted, supposedly malignant, mammary tumours in mice and rats and it has been shown that the secretion in rat tumours has the biochemical characteristics of milk. In general, tumours can differentiate in any one of the ways in which the stem cells of the parent tissues can differentiate under normal or pathological conditions. The stem cells of transplanted malignant teratomas can differentiate into a wide variety of tissues; the fault is not in differentiation but in integration.

The anomalies of differentiation in tumours are of a more subtle kind than is revealed by histological studies of cytodifferentiation. Of the three characteristics of normal differentiation called by Weiss *discreteness*, *exclusivity* and *genetic limitation*, genetic limitation is conspicuously operative in neoplastic development but *exclusivity* and *discreteness* are impaired and often severely disturbed (Foulds, 1963). This is well shown by the individuality of the different tumours induced by the same procedure in the same tissue. In the series of Morris hepatomas for example, no two are exactly alike in every respect. This individuality of tumours can be viewed as a special case of the widely prevalent dissociation of independently variable characters to which studies of tumour progression first drew wide attention. Another special case is the "inappropriate secretion" of hormones by neoplastic tissues. These "inappropriate" secretions imply that genetic patterns not normally available in the facultative genome persist in the total genome and can be activated in neoplasia but not, so far as is known at present, under other circumstances.

In the continuing search for some common basis for neoplastic behaviour, attention is now being increasingly directed towards some defect or abnormality in the regulatory systems of cells (Bullough, 1967; Pitot, 1966). The prevalance of the dissociation of independently variable characters in neoplasia, the individuality of tumours and the "inappropriate" biochemical activities of some tumours, suggest that dissociation *per se* may be an important or even fundamental characteristic of neoplasia (Foulds, 1967). The dissociation might be due to alterations in regulator genes or in regulatory circuits of the Jacob-Monod type. According to the "network" hypothesis of biological organization, dissociation might occur in many ways and at many points in the net with similar results; there need be no single or specific "cause" of the

15. PROBLEMS OF NORMAL AND NEOPLASTIC DEVELOPMENT

disruption of the net and no single specific site or type of injury of the net. For the time being, perhaps as good a way as any to sum up the situation is by slightly misquoting one of my countrymen whose copyright has expired:

> "The cell is out of joint: O cursed spite
> That ever I was born to set it right".

References

Abercrombie, M. (1957). *Symp. Soc. exp. Biol.* **11**, 235–254. Localized formation of new tissue in an adult mammal.
Abercrombie, M. (1965). *In* "Ideas in Modern Biology" (J. A. Moore, ed.), pp. 259–280. The Natural History Press, Garden City, N.Y. Cellular interactions in development.
Abercrombie, M. (1967*a*). *In* "Cell Differentiation" (A. V. S. De Reuk and J. Knight, eds.), pp. 3–17. Ciba Foundation Symp. J. & A. Churchill, London. General Review of the nature of differentiation.
Abercrombie, M. (1967*b*). *Natn. Cancer Inst. Monogr.* **26**, 249–273. Contact inhibition: the phenomenon and its biological implications.
Abercrombie, M. and Ambrose, E. J. (1962). *Cancer Res.* **22**, 525–548. The surface properties of cancer cells: a review.
Adams, W. R. and Prince, A. M. (1959). *Ann. N.Y. Acad. Sci.* **81**, 89–100. Cellular changes associated with infection of the Ehrlich ascites tumor with Newcastle disease virus.
Alexander, S. (1968). *Proc. R. Soc. Med.* **61**, 464–466. The effects of carcinoma on the skin.
Algire, G. H., Chalkley, H. W., Legallais, F. Y. and Park, H. D. (1945). *J. natn. Cancer Inst.* **6**, 73–85. Vascular reactions of normal and malignant tissues *in vivo*. I. Vascular reactions of mice to wounds and to normal and neoplastic transplants.
Allen, A. C. (1940). *Archs Path.* **29**, 589–624. So-called mixed tumors of the mammary gland of dog and man. With special reference to the general problem of cartilage and bone formation.
Allfrey, V. G., Littau, V. C. and Mirsky, A. E. (1963). *Proc. natn. Acad. Sci. U.S.A.* **49**, 414–421. On the role of histones in regulating ribonucleic acid synthesis in the cell nucleus.
Allfrey, V. G. and Mirsky, A. E. (1963). *Cold Spring Harb. Symp. quant. Biol.* **28**, 247–262. Mechanisms of synthesis and control of protein and ribonucleic acid synthesis in the cell nucleus.
Andervont, H. B. (1950). *J. natn. Cancer Inst.* **11**, 73–81. Attempt to detect a mammary tumor-agent in strain C mice by estrogenic stimulation.
Andervont, H. B., Shimkin, M. B. and Canter, H. Y. (1957). *J. natn. Cancer Inst.* **18**, 1–39. Effect of discontinued estrogenic stimulation upon the development and growth of testicular tumors in mice.
Anfinsen, C. B. (1959). "The Molecular Basis of Evolution." John Wiley & Sons Inc. New York.
Apolant, H. (1906). *Arb. K. Inst. exp. Ther. Frankf. a.M.* **1**, 7–62. Die epithelialen Geschwülste der Maus.
Apter, M. J. and Wolpert, L. (1965). *J. theoret. Biol.* **8**, 244–257. Cybernetics and development. I. Information theory.
Archer, F. L. and Orlando, R. A. (1968). *Cancer Res.* **28**, 217–224. Morphology, natural history and enzyme patterns in mammary tumors of the rat induced by 7, 12-dimethylbenz(a)anthracene.
Askanazy, M. (1936). *Z. Krebsforsch.* **43**, 405–433. Funktionen des Geschwulstgewebes.

Attardi, G. and Smith, J. (1962). *Cold Spring Harb. Symp. quant. Biol.* 27, 271–292. Virus specific protein and ribonucleic acid associated with ribsosomes in poliovirus-infected He La cells.

Auerbach, C. (1967). *Science, N.Y.* 158, 1141–1147. The chemical production of mutations.

Auerbach, R. (1960). *In* "Self-Organizing Systems" (M. C. Yovits and S. Cameron, eds.), pp. 101–107. Pergamon Press, New York. The organization and reorganization of embryonic cells.

Austin, M. L., Widmayer, D. and Walker, L. M. (1956). *Physiol. Zoöl.* 29, 261–287. Antigenic transformation as adaptive response of *Paramecium aurelia* to patulin; relation to cell division.

Axhausen, G. (1909). *Virchows Arch. path. Anat. Physiol.* 195, 358–461. Histologische Studien über die Ursachen und den Ablauf des Knochenumbaus im osteoplastischen Karcinom.

Azzopardi, J. G. and Williams, E. D. (1968). *Cancer* 22, 274–286. Pathology of nonendocrine tumors associated with Cushing's syndrome.

Bagg, H. J. (1936). *Am. J. Cancer.* 26, 69–84. Experimental production of teratoma testis in fowl.

Baltus, E. and Brachet, J. (1963). *Biochim. biophys. Acta* 76, 490–492. Presence of deoxyribonucleic acid in the chloroplasts of *Acetabularia mediterranea*.

Barcroft, J. (1934). "Features in the Architecture of Physiological Function." Cambridge Univ. Press, Cambridge.

Bardawil, W. A. and Toy, B. L. (1959). *Ann. N.Y. Acad. Sci.* 80, 197–261. The natural history of choriocarcinoma: problems of immunity and spontaneous regression.

Barr, G. C. and Butler, J. A. V. (1963). *Nature, Lond.* 199, 1170–1172. Histones and gene function.

Barrett, M. K. (1958). *J. chron. Dis.* 8, 136–157. A critical analysis of "tumor immunity".

Barth, L. G. and Barth, L. J. (1959). *J. Embryol. exp. Morph.* 7, 210–222. Differentiation of cells of the *Rana pipiens* gastrula in unconditioned medium.

Baserga, R. (1965). *Cancer Res.* 25, 581–595. The relationship of the cell cycle to tumor growth and control of cell division: a review.

Bashford, E. F. (1908). *Scient. Rep. Invest. imp. Cancer Res. Fund.* 3, pp. 441–447. Appendix. Draft scheme for enquiry into the nature, cause, prevention and treatment of cancer.

Bashford, E. F. (1911). *Scient. Rep. Invest. imp. Cancer Res. Fund.* 4, 131–214. The behaviour of tumour cells during propagation.

Bashford, E. F. and Murray, J. A. (1904). *Scient. Rep. Invest. imp. Cancer Res. Fund.* 1, pp. 3–36. The transmissibility of malignant new growths from one animal to another.

Bashford, E. F. Murray, J. A. and Cramer, W. (1905). *Scient. Rep. Invest. imp. Cancer Res. Fund.* 2, Part 2, pp. 1–96. The growth of cancer under natural and experimental conditions.

Bashford, E. F. Murray, J. A. and Haaland, M. (1908). *Scient. Rep. Invest. imp. Cancer Res. Fund.* 3, 359–397. Resistance and susceptibility to inoculated cancer.

Bashford, E. F., Murray, J. A., Haaland, M. and Bowen, W. H. (1908). *Scient. Rep. Invest. imp. Cancer Res. Fund.* 3, 262–283. General results of propagation of malignant new growths.

Bauman, R. (1935). *Z. Krebsforsch.* 42, 178–191. Wachstum und Differenzierung papillärer Drusencarcinome.

Beale, G. H. (1954). "The Genetics of *Paramecium aurelia*". Cambridge Univ. Press, Cambridge.

REFERENCES

Beale, G. H. (1957). *Int. Rev. Cytol.* 6, 1–23. The antigen system of *Paramecium aurelia*.
Beale, G. H. (1964). *In* "Cellular Control Mechanisms and Cancer" (P. Emmelot and O. Mühlbock, eds.), pp. 8–18. Elsevier Publishing Co., Amsterdam. Genes and cytoplasmic particles in *Paramecium*.
Beard, J. W. and Rous, P. (1934). *J. exp. Med.* 60, 723–740. A virus-induced mammalian growth with the characters of a tumor. (The Shope rabbit papilloma.) II. Experimental alterations of the growth on the skin: morphological considerations: the phenomena of retrogression.
Beerman, W. (1959). *In* "Biological Organization" (C. H. Waddington, ed.), Chapter 3: Pergamon Press, London. Organization of the Chromosome.
Bell, W. B. (1925). *Lancet ii*, 1003. The specific character of malignant neoplasia.
Bell, W. B. (1926). *Br. med. J. i*, 687–690. Theory and practice in relation to the treatment of cancer with lead.
Belling, J. (1933). *Genetics* 18, 389–413. Crossing over and gene rearrangement in flowering plants.
Benson, C. D., Mustard, W. T., Ravitch, M. M., Snyder, W. H. Jr., and Welch, J. K. (1962). Cited by Everson, T. C. and Cole, W. H. (1966).
Benzer, S. (1957). *In* "The Chemical Basis of Development" (W. D. McElroy and B. Glass, eds.), pp. 70–94. Johns Hopkins Press, Baltimore. The elementary units of heredity.
Berenblum, I. (1944). *Arch. Path.* 38, 233–244. Irritation and carcinogenesis.
Berenblum, I. (1949). *J. natn. Cancer Inst.* 10, 167–174. The carcinogenic action of 9-10-dimethyl-1, 2-benzanthracene on the skin and subcutaneous tissues of the mouse, rabbit, rat and guinea pig.
Berenblum, I. and Shubik, P. (1947a). *Br. J. Cancer.* 1, 379–382. The role of croton oil applications, associated with a single painting of a carcinogen, in tumour induction of the mouse's skin.
Berenblum, I. and Shubik, P. (1947b). *Br. J. Cancer.* 1, 383–391. A new quantitative approach to the study of chemical carcinogenesis in the mouse's skin.
Berenblum, I. and Shubik, P. (1949a). *Br. J. Cancer.* 3, 109–118. An experimental study of the initiating stage of carcinogenesis and a re-examination of the somatic mutation theory of cancer.
Berenblum, I. and Shubik, P. (1949b). *Br. J. Cancer.* 3, 384–386. The persistence of latent tumour cells induced in the mouse's skin by a single application of 9:10-dimethyl-1:2-benzanthracene.
Bernal, J. D. (1962). *In* "Horizons in Biochemistry" (M. Kasha and B. Pullman, eds.), pp. 11–22. Academic Press, New York. Biochemical Evolution.
Berrill, N. J. (1963). *Devl Biol.* 7, 342–347. Morphogenetic fields, their growth and development.
Bertalanffy, F. D. and Lau, C. (1962). *Int. Rev. Cytol.* 13, 357–366. Cell renewal.
von Bertalanffy, L. (1960). *In* "Fundamental Aspects of Normal and Malignant Growth" (W. W. Nowinski, ed.), pp. 137–259. Elsevier Publishing Co. Amsterdam. Principles and theory of growth.
Bertani, G. (1958). *Adv. Virus Res.* 5, 151–193. Lysogeny.
Bielschowsky, F. (1955). *Br. J. Cancer.* 9, 80–116. Neoplasia and internal environment.
Bielschowsky, M., Bielschowsky, F. and Lindsay, D. (1956). *Br. J. Cancer.* 10, 688–699. A new strain of mice with a high incidence of mammary tumours and enlargement of the pituitary.
Biggs, R. (1947). *J. Path. Bact.* 59, 437–444. The myothelium in certain tumours of the breast.

Billingham, R. E. (1961). In "Transplantation of Tissues and Cells" (R. E. Billingham and W. K. Silvers, eds.), pp. 65–67. The Wistar Press, Philadelphia. Transplantation of tissues to the cheek pouch of the Syrian hamster.
Bittner, J. J. (1936). *Science*, *N.Y.* 84, 162. Some possible effects of nursing on the mammary gland tumor incidence in mice.
Bittner, J. J. (1937). *Am. J. Cancer* 30, 530–538. Mammary tumors in mice in relation to nursing.
Bittner, J. J. (1952). *Cancer Res.* 12, 387–398. Transfer of the agent for mammary cancer in mice by the male.
Black, J. W. (1964). *Br. J. Cancer*. 18, 143–145. The localization of metastatic Brown-Pearce carcinoma in granulation tissue.
Bloom, H. J. G. (1964). *Ann. N.Y. Acad. Sci.* 114, Article 2, 747–754. The natural history of untreated breast cancer.
Böhmig, R. (1937). *Verh. dt. path. Ges.* 30, 329–337. Die morphologisch fassbaren Wachtumsgesetze drüsenbilder Karzinome und ihrer Metastasen.
Böhmig, R. (1950). "Form. und Wachstumgesetze drusenbilder Karzinome." Grune, Stuttgart, Thieme.
Bonner, J. T. (1958). "The Evolution of Development." Cambridge Univ. Press, Cambridge.
Bonser, G. M. (1945). *J. Path. Bact.* 57, 413–422. A microscopical study of the evolution of mouse mammary cancer. The effect of the milk factor and a comparison with the human disease.
Bonser, G. M. (1949). *Acta. Unio. Intern. Contra Cancrum.* 6, 595–601. A comparison of the evolution of spontaneous and chemically induced mammary cancer in mice.
Bonser, G. M. (1954). *J. Path. Bact.* 68, 531–546. The evolution of mammary cancer induced in virgin female IF mice with minimal doses of locally-acting methylcholanthrene.
Bonser, G. M. (1958). In "International Symposium on Mammary Cancer" (L. Severi, ed.), pp. 575–584. Division of Cancer Research, Perugia. The mammary changes in IF female mice following a limited dose of four carcinogenic chemicals.
Borst, M. (1902). "Die Lehre von den Geschwülsten", Bd 1. J. F. Bergman, Wiesbaden.
Boutwell, R. K. (1964). *Prog. exp. Tumor Res.* 4, 207–250. Some biological aspects of skin carcinogenesis.
Böving, B. G. (1959). *Ann. N.Y. Acad. Sci.* 80, 21–43. The biology of trophoblast.
Boyd, J. S. K. (1956). *Biol. Rev.* 31, 71–107. Bacteriophage.
Boyd, W. L. (1966). "The Spontaneous Regression of Cancer". Charles C. Thomas, Springfield, Illinois.
Brachet, J. (1961). In "The Cell" (J. Brachet and A. E. Mirsky, eds.), pp. 771–841. Academic Press, New York. Nucleocytoplasmic interactions in unicellular organisms.
Brachet, J. (1963). In "Biological Organization at the Cellular and Supercellular Level" (R. J. C. Harris, ed.), pp. 167–182. Academic Press, London. The role of the nucleic acids in the process of induction, regulation and differentiation in the amphibian embryo and the unicellular alga, *Acetabularia mediterranea*.
Brachet, J. (1964). *Adv. Morphogen.* 3, 247–300. The role of nucleic acids and sulphydril groups in morphogenesis (amphibian egg development, regeneration in *Acetabularia*).
Brachet, J. and Denis, H. (1963). *Nature, Lond.* 198, 205–206. Effects of actinomycin D on morphogenesis.
Brachet, J., Denis, H. and de Vitry, F. (1964). *Devl Biol.* 9, 398–434. The effects of actinomycin D and puromycin on morphogenesis in amphibian eggs and *Acetabularia mediterranea*.
Braun, A. C. (1961). *Can. Cancer Conf.* 4, 89–98. The plant tumor cell as an experimental tool for studies on the nature of autonomous growth.

Braun, A. C. (1963). *In* "General Physiology of Cell Specialization" (D. Mazia and A. Tyler, eds.), pp. 73–79. McGraw-Hill Book Co., New York. Biochemical changes of a heritable type that result in autonomy.
Brenner, S. (1959). *In* "Biological Organization" (C. H. Waddington, ed.), p. 33. Pergamon Press, London. Discussion.
Brenner, S. (1961). *Cold Spring Harb. Symp. quant. Biol.* 26, 101–110. RNA, ribosomes and protein synthesis.
Bresler, V. M. (1959). *Probl. Oncol.* 5, No. 12, 24–30. Experimental teratoids of white mouse testis induced by testosterone and copper sulphate.
Bresler, V. M. (1964). *Acta. Unio. Intern. Contra Cancrum.* 20, 1501–1503. On the dynamics of blastomogenesis in the testis.
Bretscher, M. S. (1968). *Nature, Lond.* 217, 509–511. How repressor molecules function.
Breuer, M. E. and Pavan, C. (1955). *Chromosoma* 7, 371–386. Behaviour of polytene chromosomes of *Rhynchosciara angelae* at different stages of larval development.
Briggs, R. and King, T. J. (1952). *Proc. natn. Acad. Sci. U.S.A.* 38, 455–463. Transplantation of living nuclei from blastula cells into enucleated frogs' eggs.
Briggs, R. and King, T. J. (1959). *In* "The Cell" (J. Brachet and A. E. Mirsky, eds.), Vol. 1, pp. 537–617. Academic Press, New York. Nucleocytoplasmic interactions in eggs and embryos.
Briggs, R., Signoret, J. and Humphrey, R. R. (1964). *Devl Biol.* 10, 233–246. Transplantation of nuclei of various cell types from neurulae of the Mexican axolotl (*Ambystoma mexicanum*).
Brink, R. A. (1960). *Q.Rev. Biol.* 35, 120–137. Paramutation and chromosome organization.
Brink, R. A. (1964). *In* "The Role of Chromosomes in Development" (M. Locke, ed.), pp. 183–230. Academic Press, New York. Genetic repression of R action in maize.
Brookes, P. (1966). *Cancer Res.* 26, 1994–2003. Quantitative aspects of the reaction of some carcinogens with nucleic acids and the possible significance of such reactions in the process of carcinogenesis.
Brunschwig, A. and Tschetter, D. (1938). *Surgery Gynec. Obstet.* 67, 715–721. The mode of inception and lateral spread of certain squamous cell carcinomas. A histopathologic and experimental study.
Bullock, F. D. and Curtis, M. R. (1920). *Proc. N.Y. path. Soc.* N.S. 20, 149–175. The experimental production of sarcoma of the liver of rats.
Bullock, F. D. and Curtis, M. R. (1924). *J. Cancer Res.* 8, 446–481. Reaction of rats' liver to larvae of taenia crassicollis and the histogenesis of cysticercus sarcoma.
Bullough, W. S. (1950). *Expl Cell Res.* 1, 410–420. Mitotic activity in the tissues of dead mice and in tissues kept in physiological salt solution.
Bullough, W. S. (1964). *In* "Cellular Control Mechanisms and Cancer" (P. Emmelot and O. Mühlbock, eds.), pp. 124–145. Elsevier Publishing Co., Amsterdam. Growth regulation by tissue specific factors or chalones.
Bullough, W. S. (1965). *Cancer Res.* 25, 1683–1727. Mitotic and functional homeostasis: a speculative review.
Bullough, W. S. (1967). "The Evolution of Differentiation". Academic Press, London.
Bullough, W. S. and Laurence, E. B. (1964). *Symp. zool. Soc. Lond.* 12, 1–23. The production of epidermal cells.
Burdette, W. J. (1955). *Cancer Res.* 15, 201–226. The significance of mutation in relation to the origin of tumors.
Burnet, M. (1956). "Enzyme, Antigen and Virus". Cambridge Univ. Press, Cambridge.
Burnet, M. (1957a). *Br. med. J.* i, 779–786 and 841–847. Cancer a biological approach.
Burnet, M. (1957b). *Scient. Am.* 196, 37–43. The structure of the influenza virus.

Burnet, F. M. and Stanley, W. M. (1959). "The Viruses". Academic Press, New York.
Busch, H. (1962). "Biochemistry of the Cancer Cell". Academic Press, New York.
du Buy, H. G. and Riley, F. L. (1967). *Proc. natn. Acad. Sci. U.S.A.* 57, 790–797. Hybridization between the nuclear and kinetoplast DNA's of *Leishmania enriettii* and between nuclear and mitochondrial DNA's of mouse liver.

Callan, H. G. (1963). *Int. Rev. Cytol.* 15, 1–34. The nature of lampbrush chromosomes.
Callan, H. G. and Lloyd, L. (1960). *In* "New Approaches in Cell Biology" (P. M. B. Walker, ed.), pp. 23–46. Academic Press, London. Lampbrush chromosomes.
Cameron, G. R. (1952). "Pathology of the Cell". Oliver and Boyd, Edinburgh.
Cantoni, G. L., Ishikura, H., Richards, H. H. and Tanaka, K. (1963). *Cold Spring Harb. Symp. quant. Biol.* 28, 123–132. Studies on soluble ribonucleic acid. XI. A model for the base sequence of serine sRNA.
Case, R. A. M. (1965). *Proc. R. Soc. Med.* 58, 607–609. Mortality from the cancers of childhood.
Caspar, D. L. D., Dulbecco, R., Klug, A., Lwoff, A., Stoker, M. G. P., Tournier, P. and Wildy, P. (1962). *Cold Spring Harb. Symp. quant. Biol.* 27, 49–50. Proposals.
Caspar, D. L. D. and Klug, A. (1962). *Cold Spring Harb. Symp. quant. Biol.* 27, 1–24. Physical principles in the construction of regular viruses.
Champy, Ch. and Lavedan, J. (1938). *C. r. Séanc. Soc. Biol.* 127, 1197–1199. Séminomes par régénération chez les oiseaux.
Champy, Ch. and Lavedan, J. (1939). *Bull. Ass. fr. Cancer.* 28, 503–526. Séminomes par régénération testiculaires chez les oiseaux.
Cheatle, G. L. and Cutler, L. (1931). "Tumours of the Breast". Edward Arnold & Co., London.
Chiang, K-S. and Suedka, N. (1967). *Proc. natn. Acad. Sci. U.S.A.* 57, 1506–1513. Replication of chloroplast DNA in *Chlamydomonas reinhardi* during vegetative cell cycle: its mode and regulation.
Chieco-Bianchi, L., De Benedictis, G., Tridente, G. and Fiore-Donati, L. (1963). *Br. J. Cancer* 17, 672–680. Influence of age on susceptibility to liver carcinogenesis and skin initiating action by urethane in Swiss mice.
Claude, A., and Murphy, J. B. (1933). *Phys. Rev.* 13, 246–275. Transmissible tumors of the fowl.
Clayson, D. B. (1962). "Chemical Carcinogenesis". Churchill, London.
Clever, U. and Karlson, P. (1960). *Expl Cell. Res.* 20, 623–626. Induktion von Puff veränderungen in den Speicheldrüsen. Chromosomen von Chironomus tentans durch Ecdyson.
Cohen, L. W. (1965). *Expl Cell. Res.* 37, 360–367. The basis for the circadian rhythm of mating in *Paramecium bursaria*.
Cohen, S. S. (1954). *In* "Aspects of Synthesis and Order in Growth" (D. Rudnick, ed.), pp. 127–148. Princeton Univ. Press, Princeton, N.J. Virus-induced metabolic transformations and other studies in unbalanced growth.
Cohen, S. S. (1963). *In* "General Physiology of Cell Specialization" (D. Mazia and A. Tyler, eds.), pp. 3–27. McGraw-Hill Book Co., New York. Biochemical variability and innovation.
Cohn, M. (1958). *In* "The Chemical Basis of Development" (W. D. McElroy and B. Glass, eds.), pp. 458–468. Johns Hopkins Press, Baltimore. On the differentiation of a population of *Escherichia coli* with respect to β-galactosidase formation.
Collins, D. H. and Pugh, R. C. B. (1964). *Br. J. Urol.* 36, Suppl. 2, 1–11. Classification and frequency of testicular tumours.

Colter, J. S., Bird, H. H., Moyer, A. W. and Brown, R. A. (1957). *Virology* **4**, 522–532. Infectivity of ribonucleic acid isolated from virus-infected tissues.
Colter, J. S. and Ellem, K. A. O. (1960). *Natn. Cancer Inst. Monogr.* **4**, 39–51. Viral activation of cell response.
Coman, D. R. (1953). *Cancer Res.* **13**, 397–404. Mechanisms responsible for the origin and distribution of blood-borne tumor metastases: a review.
Comfort, A. (1956). "The Biology of Senescence". Routledge, London.
Commoner, R. (1961). *Science, N.Y.* **133**, 1745–1748. In defence of biology.
Cowdry, E. V. (1942). *In* "Problems of Ageing" (E. V. Cowdry, ed.), pp. 616–663. 2nd Edition. Williams & Wilkins, Baltimore. The ageing of individual cells.
Cowen, P. N. (1950). *Br. J. Cancer.* **4**, 245–253. Strain differences in mice to the carcinogenic action of urethane and its non-carcinogenicity in chicks and guinea pigs.
Cramer, W. (1929). *Br. J. exp. Path.* **10**, 335–346. On experimental carcinogenesis: the local resistance of the skin to the development of malignancy.
Crick, F. H. C. (1957). *In* "The Chemical Basis of Heredity" (W. D. McElroy and B. Glass, eds.), pp. 532–539. Johns Hopkins Press, Baltimore. The structure of DNA.
Crick, F. H. C. (1958). *Symp. Soc. exp. Biol.* **12**, 138–163. On protein synthesis.
Crick, F. H. C. (1963). *Science, N.Y.* **139**, 461–464. On the genetic code.
Crick, F. H. C., Barnett, L., Brenner, S. and Watts-Tobin, R. J. (1961). *Nature, Lond.* **192**, 1227–1332. General nature of the genetic code for proteins.
Cutler, M. (1962). "Tumors of the Breast". J. B. Lippincott Co., Philadelphia.

Dabelstein, H. (1937). *Z. Krebsforsch.* **46**, 355–363. Wachstum und Differenzierung eines Milchgangcarcinoms in Haupttumor und Metastase.
Daland, E. M. (1927). *Surgery Gynec. Obstet.* **44**, 264–268. Untreated cancer of the breast.
Dalq, A. M. (1960). *In* "Fundamental Aspects of Normal and Neoplastic Growth" (W. W. Nowinski, ed.), pp. 305–494. Elsevier Publishing Co., Amsterdam. Germinal organization and induction phenomena.
Daniel, P. M. and Pritchard, M. M. L. (1964). *Br. J. Cancer.* **18**, 513–520. Three types of mammary tumour induced in rats by feeding with DMBA.
Daniel, P. M. and Pritchard, M. M. L. (1967). *Int. J. Cancer.* **2**, 163–177. Further studies on mammary tumours induced in rats by 7, 12-dimethylbenz (A) anthracene (DMBA).
Danielli, J. F. (1959*a*). *Expl Cell Res.* Suppl. **6**, 252–267. Some theoretical aspects of nucleocytoplasmic relationships.
Danielli, J. F. (1959*b*). *Ann. N.Y. Acad. Sci.* **78**, 675–687. The cell-to-cell transfer of nuclei in amoebae and a comprehensive cell theory.
Danielli, J. F. (1960). *In* "New Approaches in Cell Biology" (P. M. B. Walker, ed.), pp. 15–22. Academic Press, London. Cellular inheritance as studied by nuclear transfer in amoebae.
Danielli, J. F. (1963). *Harvey Lect.* **58**, 217–231. The theory of cells in relation to the study of cytoplasmic inheritance in amoebae by nuclear transplantation.
Dao, T. L. (1964). *Prog. exp. Tumor Res.* **5**, 157–216. Carcinogenesis of mammary gland in rat.
Davidson, E. H. (1963). *J. gen. Physiol.* **46**, 983–998. Heritability and control of differentiated function in cultured cells.
Davidson, E. H. (1964). *Adv. Genetics.* **12**, 143–280. Differentiation in monolayer tissue culture cells.
Davidson, E. H., Allfrey, V. G. and Mirsky, A. E. (1963). *Proc. natn. Acad. Sci. U.S.A.* **49**, 53–60. Gene expression in differentiated cells.

Davis, B. D. (1961). *Cold Spring Harb. Symp. quant. Biol.* **26**, 1–10. Opening Address: The teleonomic significance of biosynthetic control mechanisms.
Day, E. D. (1964). *Prog. exp. Tumor Res.* **4**, 57–97. Vascular relationships of tumor and host.
De Benedictis, G., Maiorano, G., Chieco-Bianchi, L. and Fiore-Donati, L. (1962). *Br. J. Cancer* **16**, 686–689. Lung carcinogenesis by urethane in newborn, suckling and adult Swiss mice.
Deelman, H. T. (1924). *Z. Krebsforsch.* **21**, 220–226. Die Entstehung des experimentellen Teerkrebs und die Bedeutung der Zellregeneration.
Delarue, J. (1947). "La Problème Biologique du Cancer". Masson et Cie, Paris.
Della Porta, G., Capitano, J., Parmi, L. and Colnaghi, M. I. (1967). *Tumori.* **53**, 81–102. Cancerogenesi da uretano in topi neonati lattenti e adulti dei ceppi $C_{57}BL$, C_3H, C_3Hf e SWR.
Denis, H. (1964*a*). *Devl Biol.* **9**, 435–457. Effets de l'actinomycine sur le developpement embryonaire. I. Étude morphologique: suppression par l'actinomysine de la compétence de l'ectoderme et du pouvoir inducteur de la lèvre blastoporale.
Denis, H. (1964*b*). *Devl Biol.* **9**, 458–472. II. Étude autoradiographique: influence de l'actinomysine sur la synthèse des acides nucléiques.
Denis, H. (1964*c*). *Devl Biol.* **9**, 473–483. III. Étude biochemique: influence de l'actinomysine sur la synthèse des proteins.
Denoix, P. F. (1954). Monographie de l'Institut National d'Hygiène, No 5, Paris. De la diversité de certains cancers.
Denoix, P. F. (ed.) (1967*a*). "Mechanisms of Invasion in Cancer". Springer-Verlag, Berlin.
Denoix, P. F. (1967*b*). *In* "Mechanisms of Invasion in Cancer" (P. F. Denoix, ed.), pp. 1–10. Springer-Verlag, Berlin. Appreciation de l'invasion dans les cancers humains.
Deringer, M. K. (1965). *J. natn. Cancer Inst.* **34**, 841–847. Response of strain DBA/2eBDe mice to treatment with urethan.
Des Ligneris, M. J. A. (1940). *Am. J. Cancer.* **40**, 1–46. Precancer and carcinogenesis.
Dettlaff, T. A. (1964). *Adv. Morphogen.* **3**, 323–362. Cell divisions, duration of interkinetic states and differentiation in early stages of development.
Di Berardino, M. A. and King, T. J. (1965). *Devl Biol.* **11**, 217–242. Transplantation of nuclei from renal adenocarcinomas. II. Chromosomal and histologic analysis of tumor nuclear-transplant embryos.
Di Paolo, J. A. (1960). *Ann. N.Y. Acad. Sci.* **89**, 408–420. Experimental evaluation of actinomycin D.
Di Paolo, J. A. and Elis, J. (1967). *Cancer Res.* **27**, 1696–1701. The comparison of teratogenic and carcinogenic effects of some carbamate compounds.
Di Paolo, J. A. and Kotin, P. (1966). *Arch. Path.* **81**, 3–23. Teratogenesis and oncogenesis: a study of possible relationships.
Dodson, J. W. (1963). *Expl Cell Res.* **31**, 233–235. On the nature of tissue interactions in embryonic skin.
Dominguez, O. V., Acevedo, H. F., Huseby, R. A. and Samuels, L. T. (1960). *J. biol. Chem.* **235**, 2608–2612. Steroid 21-hydroxylase in normal testis and malignant interstitial cell tumors.
Doncaster, L. (1920). "An Introduction to the Study of Cytology". Cambridge Univ. Press, Cambridge.
Dorfman, R. I. (1960). *In* "Biological Activities of Steroids in Relation to Cancer" (G. Pincus and E. P. Vollmer, eds.), pp. 445–456. Academic Press, New York. Steroid metabolism in endocrine tumors.

Druckrey, H., Preussmann, R., Ivankovic, S. and Schmähl, D. (1967). *Z. Krebsforsch.* **69**, 103–201. Organotrope carcinogene Wirkungen bei 65 N-Nitroso-Verbindungen.
Ducoff, H. S. and Ehret, C. F. (eds.). (1959). "Mitogenesis" Chicago Univ. Press, Chicago.
Dulbecco, R. (1960). *Cancer Res.* **20**, 751–761. A consideration of virus-host relationships in virus-induced neoplasia at the cellular level.
Dulbecco, R. (1962). *Cold Spring Harb. Symp. quant. Biol.* **27**, 519–525. Basic mechanisms in the biology of animal viruses.
Dunn, L. C. (1941). *Growth* Suppl., 147–161. Abnormal growth patterns with special reference to genetically determined deviations in early development.
Dunn, T. B. (1945). In "A Symposium on Mammary Tumors in Mice" (F. R. Moulton, ed.), pp. 13–38. American Association Advancement of Science (No. 22). Morphology and histogenesis of mammary tumors.
Dunn, T. B. (1958). In "The Physiopathology of Cancer" (F. Homburger, ed.), pp. 38–84. Hoeber Inc., New York. 2nd Edition. Morphology of mammary tumors in mice.
Dunphy, J. E. (1950). *New Engl. J. Med.* **242**, 167–176. Some observations on the natural history of cancer in man.
Dunphy, J. E. (1953). *New Engl. J. Med.* **249**, 17–25. Changing concepts in the surgery of cancer.

Eagle, H. (1963). In "General Physiology of Cell Specialization" (D. Mazia and A. Tyler, eds.), pp. 151–170. McGraw-Hill Book Co., New York. Population density and the nutrition of cultured mammalian cells.
Eagle, H. and Levintow, L. (1965). In "Cells and Tissues in Culture" (E. N. Willmer, ed.), pp. 277–296. Academic Press, London. Amino acid and protein metabolism. I. The metabolic characteristics of serially propagated cells.
Eagle, H. and Piez, K. (1962). *J. exp. Med.* **116**, 29–43. The population-dependent requirements by cultured mammalian cells for metabolites which they can synthesize.
Earle, W. R., Sanford, K. K., Evans, V. J., Waltz, H. K. and Shannon, J. E. Jr. (1951). *J. natn. Cancer Inst.* **12**, 133–153. The influence of inoculum size on proliferation in tissue cultures.
Ebert, J. D. and Kaighn, M. E. (1966). In "Major Problems in Developmental Biology" (M. Locke, ed.), pp. 29–84, Academic Press, New York. The keys to change: factors regulating differentiation.
Edmonds, H. W. (1959). *Ann. N.Y. Acad. Sci.* **80** (Art 1), 86–104. Genesis of hydatidiform mole: old and new concepts.
Edwards, C. N., Steinthorsson, E. and Nicholson, D. (1953). *Cancer* **6**, 531–554. An autopsy study of latent prostatic cancer.
Ehrlich, P. (1907). *Z. Krebsforsch.* **7**, 59–80. Experimentelle Studien an Maüse Tumoren.
Ehrlich, P. and Apolant, H. (1905). *Berl. klin. Wschr.* **42**, 871–874. Beobachtungen über maligne Mäusetumoren.
Elasser, W. M. (1958). "The Physical Foundation of Biology". Pergamon Press, London.
Elasser, W. M. (1961). *J. theoret. Biol.* **1**, 27–58. Quanta and the concept of organismic law.
Elsdale, T. R., Gurdon, J. B. and Fischberg, M. (1960). *J. Embryol. exp. Morph.* **8**, 437–444. A description of the technique for nuclear transplantation in *Xenopus laevis*.
Elsdale, T. R. and Jones, K. (1963). *Symp. Soc. exp. Biol.* **17**, 257–273. The independence of cells in the amphibian embryo.
Emerson, W. J., Kennedy, B. J., Graham, J. N. and Nathanson, I. G. (1953). *Cancer* **6**, 641–670. Pathology of primary and recurrent carcinoma of the human breast after administration of steroid hormones.

Engell, H. C. (1955). *Acta chir. scand.* Suppl. 201. Cancer cells in the circulating blood.
Ephrussi-Taylor, H. (1957). *In* "The Genetic Basis of Heredity" (W. D. McElroy and B. Glass, eds.), pp. 299–320. Johns Hopkins Press, Baltimore. X-ray inactivation studies on solutions of transforming DNA of pneumococcus.
Evans, R. W. (1966). "Histological Appearances of Tumours". E. & S. Livingstone, Edinburgh. 2nd Edition.
Evans, V. J., Earle, W. R., Wilson, E. P., Waltz, H. K. and Mackey, C. J. (1952). *J. natn. Cancer Inst.* 12, 1245–1265. The growth *in vitro* of massive cultures of liver cells.
Everson, T. C. and Cole, W. H. (1966). "Spontaneous Regression of Cancer". W. B. Saunders Co., Philadelphia.
Ewing, J. (1916). *J. Cancer Res.* 1, 71–86. Pathological aspects of some problems of experimental cancer research.
Ewing, J. (1928). "Neoplastic Diseases". W. B. Saunders Co., Philadelphia. 3rd Edition.
Ewing, J. (1936). Second *Int. Cancer Res. Congr.* Brussels. p. 215. Problems in histological tumor diagnosis.
Ezekiel, D. H. (1962). *In* "The Molecular Basis of Neoplasia", pp. 549–563. Univ. of Texas Press, Austin. Ribonucleic acid synthesis by bacterial nuclear preparations.

Falin, L. I. (1940). *Am. J. Cancer.* 38, 199–211. Experimental teratoma testis in the fowl.
Fawcett, D. W. (1961). *Expl Cell Res.* Suppl. 8, 174–187. Intercellular bridges.
Fawcett, D. W. (1966). "The Cell: An Atlas of Fine Structure". W. B. Saunders Company, Philadelphia.
Fekete, E. and Ferrigno, M. A. (1952). *Cancer Res.* 12, 438–440. Studies on a transplantable teratoma of the mouse.
Fell, H. B. (1956–57). *Proc. R. Soc.* B 146, 242–256. The effect of excess vitamin A on cultures of embryonic chicken skin explanted at different stages of differentiation.
Fell, H. B. (1957). *J. natn. Cancer Inst.* 19, 643–650. The future of tissue culture in relation to morphology.
Fell, H. B. (1964). *In* "The Epidermis" (W. Montagna and W. C. Lobitz, Jr., eds.), pp. 61–81. Academic Press, New York. The experimental study of keratinization in organ culture.
Fell, H. B. and Mellanby, E. (1953). *J. Physiol., Lond.* 119, 470–488. Metaplasia produced in cultures of chick ectoderm by high vitamin A.
Finch, B. W. and Ephrussi, B. (1967). *Proc. natn. Acad. Sci. U.S.A.* 57, 615–621. Retention of multiple developmental potentialities by cells of a mouse testicular teratocarcinoma during prolonged culture *in vitro* and their extinction upon hybridization with cells of permanent lines.
Firket, H. (1965). *In* "Cells and Tissues in Culture" (E. N. Willmer, ed.), Vol. I, pp. 203–207. Academic Press, London. Cell division.
Fischberg, M. and Blackler, A. W. (1963). *In* "Biological Organization at the Cellular and Supercellular Level" (R. J. C. Harris, ed.), pp. 111–127. Academic Press, London. Loss of nuclear potentiality in the soma versus preservation of nuclear potentiality in the germ line.
Fischberg, M., Gurdon, J. B. and Elsdale, T. D. (1958). *Expl Cell Res.* Suppl. 6, 161–178. Nuclear transfer in amphibia and the problem of the potentiality of the nuclei of differentiating tissues.
Fischer, B. (1906). *Münch. med. Wschr.* 53, 2041–2047. Die experimentelle Erzeugung atypischer Epithelwucherungen und die Entstehung bösartiger Geschwülste.
Fisher, B. and Fisher, E. R. (1959). *Science, N.Y.* 130, 918–919. Experimental evidence in support of the dormant tumor cell.

Fisher, B. and Fisher, E. R. (1967). *Cancer Res.* 27, 412–420. The organ distribution of disseminated ^{51}Cr-labelled tumor cells.
Fisher, B., Fisher, E. R. and Feduska, N. (1967). *Cancer* 20, 23–30. Trauma and the localization of tumor cells.
Fisher, E. R. and Fisher, B. (1965). *Acta cytol.* 9, 146–159. Experimental study of factors influencing development of hepatic metastases from circulating tumor cells.
Fisher, E. R. and Turnbull, R. B., Jr. (1955). *Surgery Gynec. Obstet.* 100, 102–108. Cytologic demonstration and significance of tumor cells in the mesenteric venous blood in patients with colorectal cancer.
Flickinger, R. A. (1962). *Int. Rev. Cytol.* 13, 75–98. Sequential gene action, protein synthesis and cellular differentiation.
Flickinger, R. A. (1963). *Science, N.Y.* 141, 1063–1064. Actinomycin D effects in frog embryos: sequential synthesis of DNA dependent RNA.
Flickinger, R. A., Coward, S. J., Miyagi, M., Moser, C. and Rollins, E. (1965). *Proc. natn. Acad. Sci. U.S.A.* 53, 783–790. The ability of DNA and chromatin of developing frog embryos to prime for RNA polymerase-dependent RNA synthesis.
Flickinger, R. A., Miyagi, M., Moser, C. R. and Rollins, E. (1967). *Devl Biol.* 15, 414–431. The relation of DNA synthesis to RNA synthesis in developing frog embryos.
Foulds, L. (1932). *Scient. Rep. Invest. imp. Cancer Res. Fund.* 10, 21–31. The effect of vital staining on the distribution of the Brown-Pearce rabbit tumour.
Foulds, L. (1934a). *Scient. Rep. Invest. imp. Cancer Res. Fund.* 11, 15–25. Histological studies on filterable tumours of the fowl with special reference to metastatic growths.
Foulds, L. (1934b). *Scient. Rep. Invest. imp. Cancer Res. Fund.* 9, 27–33. Autoplastic transplatation of the thymus gland in fowls.
Foulds, L. (1934c). *Scient. Rep. Invest. imp. Cancer Res. Fund.* 9, Suppl. pp. 1–41. The filterable tumours of fowls: a critical review.
Foulds, L. (1937). *J. Path. Bact.* 44, 1–18. A transplantable carcinoma of a domestic fowl with a discussion of the histogenesis of mixed tumours.
Foulds, L. (1939). *Am. J. Cancer.* 35, 363–373. The production of transplantable carcinoma and sarcoma in guinea pigs by injections of thorotrast.
Foulds, L. (1940). *Am. J. Cancer.* 39, 1–24. The histological analysis of tumours. A critical review.
Foulds, L. (1947). *Br. J. Cancer* 1, 362–370. Mammary tumours in hybrid mice: a sex factor in transplantation.
Foulds, L. (1949a). *Br. J. Cancer.* 3, 230–239. Mammary tumours in hybrid mice; the presence and transmission of the mammary tumour agent.
Foulds, L. (1949b). *Br. J. Cancer* 3, 240–246. Mammary tumours in hybrid mice: hormone responses of transplanted tumours.
Foulds, L. (1949c). *Br. J. Cancer* 3, 345–375. Mammary tumours in hybrid mice: growth and progression of spontaneous tumours.
Foulds, L. (1951). *Ann. R. Coll. Surg. Eng.* 9, 93–101. Experimental study of the course and regulation of tumour growth.
Foulds, L. (1954). *Cancer Res.* 14, 327–339. Tumor progression: a review.
Foulds, L. (1956). *J. natn. Cancer Inst.* 17, 701–801. The histological analysis of mammary tumors of mice.
I. Scope of investigations and general principles of analysis pp. 713–753. II. The histology of responsiveness and progression: The origins of tumors pp. 713–753. III. Organoid tumors pp. 755–781. IV. Secretion pp. 783–801.
Foulds, L. (1957). *In* "Cancer" (R. W. Raven, ed.), Vol. 2, pp. 27–44. Butterworth & Co., London. Biological characteristics of neoplasia.

Foulds, L. (1958a). *J. chron. Dis.* 8, 2–37. The natural history of cancer.
Foulds, L. (1958b). *In* "The Chemical Basis of Development" (W. D. McElroy and B. Glass, eds.), pp. 680–700. Johns Hopkins Press, Baltimore. Neoplastic development.
Foulds, L. (1960). *In* "Progress in the Medical Sciences in Relation to Dermatology" (A. Rook, ed.), pp. 327–342. Cambridge Univ. Press, Cambridge. Current concepts of carcinogenesis.
Foulds, L. (1961). *Acta. Unio. Intern. Contra Cancrum.* 17, 148–156. Progression and carcinogenesis.
Foulds, L. (1963). *In* "Biological Organization at the Cellular and Supercellular levels" (R. J. C. Harris, ed.), pp. 229–244. Academic Press, London. Some problems of differentiation and integration in neoplasia.
Foulds, L. (1964a). *Acta. Unio. Intern. Contra Cancrum.* 20, 663–666. Some general principles of neoplastic development.
Foulds, L. (1964b). *In* "Cellular Control Mechanisms and Cancer" (P. Emmelot and O. Mühlbock, eds.), pp. 242–258. Elsevier Publishing Co., Amsterdam. Tumour progression and neoplastic development.
Foulds, L. (1965). *Cancer Res.* 25, 1339–1347. Multiple etiological factors in neoplastic development.
Foulds, L. (1967). *Natn. Cancer Inst. Monogr.* 26, 382–383. Discussion.
Fox, F., Davidson, J. and Thomas, L. B. (1959). *Cancer* 12, 108–116. Maturation of sympathicoblastoma into ganglioneuroma.
Fraenkel-Conrat, H., Singer, B. A. and Williams, R. C. (1957). *In* "The Chemical Basis of Heredity" (W. D. McElroy and B. Glass, eds.), pp. 501–512. Johns Hopkins Press, Baltimore. The nature of the progeny of virus reconstituted from protein and nucleic acid of different strains of tobacco mosaic virus.
Franck, I. D. (1935). *Z. Krebsforsch.* 42, 381–392. Wachstum und Differenzierung rein infiltrierend wachsender Adenocarcinome.
Franklin, R. M. and Baltimore, D. (1962). *Cold Spring Harb. Symp. quant. Biol.* 27, 175–198. Patterns of macromolecular synthesis in normal and virus-infected mammalian cells.
Franks, L. M. (1954a). *J. Path. Bact.* 68, 603–616. Latent carcinoma of the prostate.
Franks, L. M. (1954b). *Ann. R. Coll. Surg.* 15, 236–249. Latent carcinoma.
Franks, L. A. (1956). *Lancet* ii, 1037–1039. Latency and progression in tumours. The natural history of prostatic cancer.
Frazell, E. L. and Foote, F. W. (1958). *Cancer* 11, 895–922. Papillary cancer of the thyroid.
Frenster, J. H. (1965). *Nature, Lond.* 206, 1269–1270. A model of specific de-repression within interphase chromatin.
Frenster, J. H., Allfrey, V. G. and Mirsky, A. E. (1963). *Proc. natn. Acad. Sci. U.S.A.* 50, 1026–1032. Repressed and active chromatin isolated from interphase lymphocytes.
Friedewald, W. F. and Rous, P. (1950). *J. exp. Med.* 91, 459–484. The pathogenesis of deferred cancer. A study of the after effects of methylcholanthrene upon rabbit skin.
Furth, J. and Kahn, M. C. (1937). *Am. J. Cancer* 31, 276–282. The transmission of leukemia in mice with a single cell.

Gaillard, P. J. (1957). *J. natn. Cancer Inst.* 19, 591–607. Morphogenesis in animal tissue cultures.
Gall, J. G. (1958). *In* "The Chemical Basis of Development". (W. D. McElroy and B. Glass, eds.), pp. 103–135. Johns Hopkins Press, Baltimore. Chromosomal differentiation.
Gallien, L., Picheral, B. and Lacroix, J. C. (1963). *C. r. hebd. Séanc. Acad. Sci., Paris* 257,

1721–1723. Modifications de l'assortiment chromosomique chez des larves hypomorphes du Triton *Pleurodeles Walthii Michah*, obtenues par transplantation de noyaux.
Gardner, W. U. (1941). *Cancer Res.* 1, 345–358. The effect of estrogen on the incidence of mammary and pituitary tumors in hybrid mice.
Gaudier, Grandclaude and Lambret, M. (1931). *Annls Anat. path.* 8, 68–70. Tumeur maligne du sein a type myoépithéliale.
Gibor, A. (1965). *Am. Nat.* 99, 229–239. Chloroplast heredity and nucleic acids.
Gibson, I. and Beale, G. H. (1962). *Genet. Res.* 3, 24–50. The mechanism whereby the genes M_1 and M_2 in *Paramecium aurelia*, stock 540, control growth of the mate-killer (mu) particles.
Gibson, I. and Beale, G. H. (1963). *Genet. Res.* 4, 42–54. The action of ribonuclease and 8-azoguanine on mate-killer paramecia.
Gibson, I. and Beale, G. H. (1964). *Genet. Res.* 5, 85–106. Infection into paramecia of metagons derived from other mate-killer paramecia.
Gibson, I. and Sonneborn, T. M. (1964). *Proc. natn. Acad. Sci. U.S.A.* 52, 869–876. Is the metagon an mRNA in Paramecium and a virus in Didinia?
Gierer, A. (1963). *J. molec. Biol.* 6, 148–157. Function of aggregated reticulocyte ribosomes in protein synthesis.
Gierke, E. (1908). *Scient. Rep. Invest. imp. Cancer Res. Fund.* 3, 115–145. The haemorrhagic mammary tumours of mice with results of research into susceptibility and resistance to inoculation.
Gilbert, W. (1963). *Cold Spring Harb. Symp. quant. Biol.* 28, 287–297. Protein synthesis in *Escherichia coli*.
Gluecksohn-Schoenheimer, S. and Rota, T. R. (1949). *Growth* 13, (Suppl.), 163–176. Causal analysis of mouse development by the study of mutational effects.
Gluecksohn-Waelsh, S. (1963). *Devl Biol.* 7, 432–444. Development in organ tissue culture of kidney rudiments from mutant mouse embryos.
Goldblatt, S. A. and Nadel, E. L. (1965). *Acta cytol.* 9, 6–20. Cancer cells in the circulating blood: a critical review. II.
Goldschmidt, R. B. (1955). "Theoretical Genetics". Univ. of California Press, Berkeley.
Goldstein, A. and Brown, B. (1961). *Biochim. biophys. Acta.* 53, 19–28. Effect of sonic oscillation upon "old" and "new" nucleic acids in *Escherichia coli*.
Goldstein, M. N., Burdman, J. A. and Journey, L. J. (1964). *J. natn. Cancer Inst.* 32, 165–199. Long-term tissue culture of neuroblastomas. II. Morphologic evidence for differentiation and maturation.
Goodall, C. M. (1969). *Internat. J. Cancer.* 4, 1–13. On para-endocrine cancer syndromes.
Goodwin, B. C. (1963). "Temporal Organization in Cells". Academic Press, London.
Goodwin, B. C. (1964). *Symp. Soc. exp. Biol.* 18, 301–326. A statistical mechanics of temporal organization in cells.
Gordon, I. (1960). *Nature, Lond.* 185, 118–119. Origin of placental trophoblast.
Gorer, P. A. (1948), *Br. J. Cancer* 2, 103–107. The significance of studies with transplanted tumours.
Goss, R. J. (1964). "Adaptive Growth". Logos Press, London.
Granick, S. (1963). *In* "Cytodifferentiation and Macromolecular Synthesis" (N. Locke, ed.), pp. 144–174. Academic Press, New York. The plastids: their morphological and chemical differentiation.
Granick, S. (1965). *Am. Nat.* 99, 193–199. Cytoplasmic units of inheritance.
Green, M. (1962). *Cold Spring Harb. Symp. quant. Biol.* 27, 219–235. Studies on the biosynthesis of viral DNA.

Greene, H. S. N. (1940). *J. exp. Med.* **71**, 305–324. Familial mammary tumors in the rabbit. IV. The evolution of autonomy in the course of tumor development as indicated by transplantation experiments.
Greene, H. S. N. (1945). *Science, N.Y.* **101**, 644–645. The production of carcinoma and sarcoma in transplanted embryonic tissues.
Greene, H. S. N. (1952). *Cancer* **5**, 24–44. Significance of heterologous transplantability of human cancer.
Greene, H. S. N. (1955). *Ann. N.Y. Acad. Sci.* **59**, 311–318. Compatibility and noncompatibility (The relation of immunology to tissue homo-transplantation).
Greene, H. S. N. (1965). *Acta cytol.* **9**, 160–168. A method of determining the presence of tumor cells in blood and organs of experimental animals and its application to the problems of metastasis and retention in organs: a review.
Greene, H. S. N. and Harvey, E. K. (1964). *Cancer Res.* **24**, 799–811. The relationship between the dissemination of tumor cells and the distribution of metastases.
Greene, H. S. N. and Harvey, E. K. (1966). *Cancer Res.* **26**, 706–714. The growth and metastasis of amelanotic melanomas in heterologous hosts.
Greenstein, J. P. (1945). *A.A.A.S. Research Conference on Cancer*. The American Association for the Advancement of Science, Washington, DC. p. 192. Enzymes in normal and neoplastic animal tissues.
Grell, K. G. (1964). *In* "The Cell" (J. Brachet and A. E. Mirsky, eds.), Vol. 6, pp. 1–79. Academic Press, New York. The protozoan nucleus.
Greulich, R. C. (1964). *In* "The Epidermis" (W. Montagna and W. C. Lobitz, Jr., eds.), pp. 116–133. Academic Press, New York. Aspects of cell individuality in the renewal of stratified squamous epithelium.
Griffith, F. (1928). *J. Hyg., Camb.* **27**, 113–159. The significance of pneumococcal types.
Griffiths, J. D. and Salsbury, A. J. (1965). "Circulating Cancer Cells". Charles C. Thomas, Springfield, Illinois.
Grobstein, C. (1954). *In* "Aspects of Synthesis and Order in Growth" (D. Rudnick, ed.), pp. 233–256. Princeton Univ. Press, Princeton, N.J. Tissue interaction in the morphogenesis of mouse embryonic rudiments *in vitro*.
Grobstein, C. (1955a). *Ann. N.Y. Acad. Sci.* **60**, 1095–1107. Tissue disaggregation in relation to determination and stability of cell type.
Grobstein, C. (1955b). *J. exp. Zool.* **130**, 319–339. Inductive interaction in the development of the mouse metanephros.
Grobstein, C. (1956a). *Expl Cell Res.* **10**, 424–440. Trans-filter induction of tubules in mouse metanephrogenic mesenchyme.
Grobstein, C. (1956b). *Adv. Cancer Res.* **4**, 187–236. Inductive tissue interaction in development.
Grobstein, C. (1959). *In* "The Cell" (J. Brachet and A. E. Mirsky, eds.), Vol. 1, pp. 437–496. Academic Press, New York. Differentiation of vertebrate cells.
Grobstein (1961a). *Expl Cell Res.* Suppl. **8**, 234–245. Cell contact in relation to embryonic induction.
Grobstein, C. (1961b). *In* "La Culture Organotypique", Colloques internationaux du C.N.R.S., Paris. No. 101 pp. 169–182. Passage of radioactivity into a membrane filter from spinal-cord pre-incubated with tritiated amino acids or nucleosides.
Grobstein, C. (1962). *Am. Scient.* **50**(1), 46–58. Levels and ontogeny.
Grobstein, C. (1963). *In* "Cytodifferentiation and Macromolecular Synthesis" (M. Locke, ed.), pp. 1–14. Academic Press, New York. Cytodifferentiation and macromolecular synthesis.
Grobstein, C. (1964). *Science, N.Y.* **143**, 643–650. Cytodifferentiation and its controls.

Grobstein, C. (1965). *In* "Cells and Tissues in Culture" (E. N. Willmer, ed.), Vol. 1, pp. 463–488. Academic Press, London. Differentiation: environmental factors, chemical and cellular.
Grobstein, C. and Parker, G. (1958). *J. natn. Cancer Inst.* **20**, 107–115. Epithelial tubule formation by mouse metanephrogenic mesechyme transplanted *in vivo*.
Gros, F. (1964). *In* "Cellular Control Mechanisms and Cancer" (P. Emmelot and O. Mühlbock, eds.), pp. 22–34. Elsevier Publishing Co., Amsterdam. The genetic code and its translation.
Gross, L. (1961). "Oncogenic Viruses". Pergamon Press, Oxford.
Grüneberg, H. (1963). "The Pathology of Development". Blackwell Scientific Publications, Oxford.
Guérin, M. (1954). "Tumeurs spontanées des animaux de laboratoire". Amedée Legrand, Paris.
Günther, R. (1937). *Virchows Arch. path. Anat. Physiol.* **300**, 449–455. Myoepitheliale Wucherung in der Brustdrüse.
Gurdon, J. B. (1960). *J. Embryol. exp. Morph.* **8**, 327–340. Factors responsible for the abnormal development of embryos obtained by nuclear transplantation in *Xenopus laevis*.
Gurdon, J. B. (1962a). *Devl Biol.* **4**, 256–273. Adult frogs from single somatic cell nuclei.
Gurdon, J. B. (1962b). *J. Embryol. exp. Morph.* **10**, 622–640. The development capacity of nuclei taken from intestinal epithelial cells of feeding tadpoles.
Gurdon, J. B. (1963). *Q. Rev. Biol.* **38**, 54–78. Nuclear transplantation in amphibia and the importance of stable nuclear changes in promoting cellular differentiation.
Gurdon, J. B. (1964). *Adv. Morphogen.* **4**, 1–43. The transplantation of living cell nuclei.
Gurdon, J. B. (1967). *In* "Cell Differentiation" (A. V. S. De Reuk, and J. Knight, eds.), pp. 65–74. J. & A. Churchill, London. Ciba Foundation Symp. Nuclear transplantation and cell differentiation.
Gurdon. J. B. and Uehlinger, V. (1966). *Nature, Lond.* **210**, 1240–1241. Fertile intestine nuclei.
Gustafson, T. and Wolpert, L. (1963). *Int. Rev. Cytol.* **15**, 139–214. The cellular basis of morphogenesis and sea urchin development.
Guthrie, J. (1962). *Expl Cell Res.* **26**, 304–311. The chromosomes and genetic sex of experimental avian testicular teratomas.
Guthrie, J. (1964). *Br. J. Cancer* **18**, 130–142. Observations on the zinc induced testicular teratomas of the fowl.

Haaland, M. (1905). *Annls Inst. Pasteur, Paris.* **19**, 165–207. Les tumeurs de la souris.
Haaland, M. (1908). *Scient. Rep. Invest. imp. Cancer Res. Fund.* **3**, 175–261. Contribution to the study of the development of sarcoma under experimental conditions.
Haaland, M. (1911). *Scient. Rep. Invest. imp. Cancer Res. Fund.* **4**, 1–113. Spontaneous tumours of mice.
Haddow, A. (1938). *J. Path. Bact.* **47**, 553–565. Biological characters of spontaneous tumours of mouse, with special reference to rate of growth.
Hadfield, G. (1954). *Br. med. J.* **ii**, 607–610. The dormant cancer cell.
Halberstaedter, L. (1923). *Z. Krebsforsch.* **19**, 104–114. Über das Röntgencarcinom.
Hammerling, J. (1953). *Int. Rev. Cytol.* **2**, 475–498. Nucleo-cytoplasmic relationships in the development of Actabularia.
Hammerling, J. (1963). *Symp. Soc. exp. Biol.* **17**, 17–137. The role of the nucleus in differentiation, especially in Acetabularia.
Hamperl, H. (1940). *Virchows Arch. path. Anat. Physiol.* **305**, 171–215. Über die Myothelien (myo-epithelialen Elemente) der Brustdrüse.

Hamperl, H. (1957). *Wien. klin. Wschr.* **69**, 201–205. Ueber die Entwicklung ("Progression") von Tumoren.
Hamperl, H. (1967). *In* "Mechanisms of Invasion in Cancer" (P. Denoix, ed.), pp. 17–25. Springer-Verlag, Berlin. Early invasive growth as seen in uterine cancer *in situ* and the role of the basal membranes.
Hanafusa, H., Hanafusa, T. and Rubin, H. (1963). *Proc. natn. Acad. Sci. U.S.A.* **49**, 572–580. The defectiveness of Rous sarcoma virus.
Hanafusa, H., Hanafusa, T. and Rubin, H. (1964). *Proc. natn. Acad. Sci. U.S.A.* **51**, 41–48. Analysis of defectiveness of Rous sarcoma virus. II. Specification of the R.S.V. antigenicity by helper virus.
v. Hansemann, D. (1902). "Die Mikroskopische Diagnose der bösartigen Geschwülste". 2 Aufl. Berlin.
v. Hansemann, D. (1907). *Z. Krebsforsch.* **5**, 510–515. Einige Bemerkungen über die Anaplasie der Geschwulstzellen.
Hartman, P. E. (1957). *In* "The Chemical Basis of Development" (W. D. McElroy and B. Glass, eds.), pp. 408–462. Johns Hopkins Press, Baltimore. Transduction: a comparative review.
Hauschka, T. S. and Levan, A. (1958). *J. natn. Cancer Inst.* **21**, 77–111. Cytologic and functional characterization of single cell clones isolated from the Krebs-2 and Ehrlich ascites tumours.
Havemann, H. V. (1936). *Z. Krebsforsch.* **44**, 365–374. Wachstum und Differenzierung des Adenocarcinoms der Mamma.
Hechter, O. and Halkerston, I. D. K. (1965). *A. Rev. Physiol.* **27**, 133–162. Effects of steroid hormones on gene regulation of cell metabolism.
Hennen, S. (1963). *Devl Biol.* **6**, 133–183. Chromosomal and embryological analyses of nuclear changes occurring in embryos derived from transfers of nuclei between *Rana pipiens* and *Rana sylvatica*.
Hennen, S. (1965). *Devl Biol.* **11**, 243–267. Nucleocytoplasmic hybrids between *Rana pipiens* and *Rana palustris*. I. Analysis of the developmental properties of the nuclei by means of nuclear transplantation.
Herriott, R. M. (1957). *In* "The Chemical Basis of Heredity" (W. D. McElroy and B. Glass, eds.), pp. 399–407. Johns Hopkins Press, Baltimore. The virulent T- even phages of *Escherichia coli* B.
Herriott, R. M. (1966). *Cancer Res.* **26**, 1971–1979. Mutagenesis.
Herrmann, H. (1960). *Science, N.Y.* **132**, 529–532. Direct metabolic interactions between animal cells.
Herrmann, H., Marchok, A. C. and Baril, E. F. (1967). *Natn. Cancer Inst. Monogr.* **26**, 1–21. Growth rate and differentiated function of cells.
Hertig, A. T. and Mansell, H. M. (1956). "Atlas of Tumor Pathology". Armed Forces Institute of Pathology, Washington. Section IX, Fasc. 33. Tumors of the female sex organs Part I. Hydatidiform mole and choriocarcinoma.
Hertz, R., Ross, G. T. and Lipsett, M. B. (1964). *Ann. N.Y. Acad. Sci.* **114**, 881–885. Chemotherapy in women with trophoblastic disease: chorioma, chorioadenoma destruens and complicated hydatidiform mole.
Heston, W. E. (1965). *Cancer Res.* **25**, 1362. Discussion in Symposium on Epidemiological Approaches to Cancer Etiology.
Heston, W. E. and Dunn, T. B. (1951). *J. natn. Cancer Inst.* **11**, 1057–1071. Tumor development in susceptible Strain A and resistant Strain L lung transplants in LAF_1 hosts.
Heston, W. E. Vlahakis, G. and Tsubura, Y. (1964). *J. natn. Cancer Inst.* **32**, 237–251. Strain DD, a new high mammary tumor strain and comparison of DD with strain C_3H.

Hilf, R. (1967). *Science, N.Y.* **155**, 826–827. Milk-like fluid in a mammary carcinoma: biochemical characterization.
Hilf, R., Michel, I., Bell, C., Freeman, J. J. and Borman, A. (1965). *Cancer Res.* **25**, 286–299. Biochemical and morphologic properties of a new lactating mammary tumor line in the rat.
Hinshelwood, C. N. (1953). *Symp. Soc. exp. Biol.* **7**, 31–42. Adaptation in microorganisms and its relation to evolution.
Hirst, A. E. and Bergman, R. T. (1954). *Cancer* **7**, 136–141. Cancer of the prostate in men 80 or more years old.
Hirst, G. K. (1962). *Cold Spring Harb. Symp. quant. Biol.* **27**, 303–309. Genetic recombination with Newcastle disease virus, polioviruses and influenza.
Holland, J. J. and Hoyer, B. H. (1962). *Cold Spring Harb. Symp. quant. Biol.* **27**, 101–112. Early stages of enterovirus infection.
Holtfreter, J. and Hamburger, V. (1955). *In* "Analysis of Development" (B. H. Willier, P. Weiss and V. Hamburger, eds.), pp. 230–296. W. B. Saunders & Co., Philadelphia. Embryogenesis: progressive differentiation. Amphibians.
Holtzer, H. (1963). *In* "General Physiology of Cell Specialization" (D. Mazia and A. Tyler, eds.), pp. 80–90. McGraw-Hill Book Co., New York. Mitosis and cell transformations.
Horava, A. and Skoryna, S. C. (1955). *Can. med. Ass. J.* **73**, 630–638. Observations on the pathogenesis of neoplasia.
Horowitz, N. H. and Metzenberg, R. L. (1965). *A. Rev. Biochem.* **34**, 527–564. Biochemical aspects of genetics.
Hotchin, J. (1962). *Cold Spring Harb. Symp. quant. Biol.* **27**, 479–499. The biology of lymphocytic choriomeningitis infection: virus-induced immune disease.
Hotchkiss, R. D. (1956). *In* "Enzymes: Units of Biological Structure and Function" (O. H. Gaebler, ed.), pp. 119–130. Academic Press, New York. The genetic organization of the deoxyribonucleate units functioning in bacterial transformations.
Hotchkiss, R. D. (1958). *J. cell. comp. Physiol.* **52**, Suppl. 1, 331–336. General discussion: molecular basis of the cause and expression of somatic cell variation.
Hou, L. T. and Azzopardi, J. G. (1967). *J. Path. Bact.* **93**, 477–481. Muco-epidermoid metaplasia and argentaffin cells in nephroblastoma.
Hou, L. T. and Holman, R. L. (1961). *J. Path. Bact.* **82**, 249–255. Bilateral nephroblastomatosis in a premature infant.
Hou, P. C. and Pang, S. C. (1956). *J. Path. Bact.* **72**, 95–104. Chorionepithelioma: an analytical study of 28 necropsied cases with special reference to the possibility of spontaneous retrogression.
Hoyle, L. (1962). *Cold Spring Harb. Symp. quant. Biol.* **27**, 113–121. The entry of myxoviruses into the cell.
Huang, R-C. C. and Bonner, J. (1962). *Proc. natn. Acad. Sci. U.S.A.* **48**, 1216–1222. Histone, a suppressor of chromosomal RNA synthesis.
Huggins, C. B. (1931). *Archs Surg.* **22**, 377–408. Formation of bone under influence of epithelium of urinary tract.
Huguenin, R. (1929). *Annls Anat. path.* **6**, 241–266. Les aspects histologiques des cancers primitifs du poumon.
Huguenin, R. (1935). *Cancer, Brux.* **12**, 213–226. Les syndromes métastatiques aigus.
Huguenin, R. (1946). "Quelques Vérités (ou soi-disant telles) sur le Cancer" Masson et Cie, Paris.
Hultin, T. (1964). *Int. Rev. Cytol.* **16**, 1–36. Ribosomal functions related to protein synthesis.

Hurwitz, J., Evans, A., Babinet, C. and Skalka, A. (1963). *Cold Spring Harb. Symp. quant. Biol.* 28, 59–65. On the copying of DNA in the RNA polymerase system.
Huxley, J. S. and de Beer, G. R. (1934). "The Elements of Experimental Embryology". Cambridge Univ. Press, London.

Ingle, D. J. (1962). *In* "On Cancer and Hormones", pp. 213–225. Chicago Univ. Press, Chicago. The search for causes of disease.
Isaacs, A. (1959). *In* "The Viruses" (F. M. Burnet and W. M. Stanley, eds.), Vol. 3 pp. 111–156. Academic Press, New York. Biological aspects of intracellular stages of virus growth.

Jackson, E. B. and Brues, A. M. (1941). *Cancer Res.* 1, 494–498. Studies on a transplantable embryoma of the mouse.
Jackson, S. F. and Fell, H. B. (1963). *Devl Biol.* 7, 394–419. Epidermal fine structure in embryonic chicken skin during atypical differentiation induced by vitamin A in culture.
Jacob, F. (1964). *In* "Cellular Control Mechanisms and Cancer" (P. Emmelot and O. Mühlbock, eds.), pp. 49–51. Elsevier Publishing Co., Amsterdam. Discussion on genetic control of protein synthesis.
Jacob, F., Brenner, S. and Cozin, F. (1963). *Cold Spring Harb. Symp. quant. Biol.* 28, 329–348. On the regulation of DNA replication in bacteria.
Jacob, F. and Monod, J. (1961). *Cold Spring Harb. Symp. quant. Biol.* 26, 193–211. On the regulation of gene activity.
Jacob, F. and Monod, J. (1963). *In* "Cytodifferentiation and Macromolecular Synthesis" (M. Locke, ed.), pp. 30–64. Genetic repression, allosteric inhibition and cellular differentiation.
Jacob, F. and Wollman, E. C. (1958). *C. r. hebd Séanc. Acad. Sci., Paris.* 247, 154–156. Les épisomes, élements génétiques ajoutées.
Jacob, F., Schaeffer, P. and Wollman, E. L. (1960). *In* "Microbial Genetics" (W. Hayes and R. C. Clowes, eds.), pp. 67–88. Cambridge Univ. Press, Cambridge. Episomic elements in bacteria.
Jensen, C. O. (1903). *Zentbl. Bakt. ParasitKde* Abt. I. Orig. Bd. 34, 28 and 122. Experimentelle Untersuchungen über Krebs bei Mäusen.
Jensen, W. A. (1963). *In* "General Physiology of Cell Specialization" (D. Mazia and A. Tyler, eds.), pp. 53–60. McGraw-Hill Book Co., New York. Specialization of the plant cell.
Joklik, W. K. (1962). *Cold Spring Harb. Symp. quant. Biol.* 27, 199–208. The multiplication of pox virus.
Jonescu, P. (1930). *Z. Krebsforsch.* 33, 264–280. Über das Vorkommen von Geschwülstzellen in stromenden Blut von Tieren mit Impftumoren.

Kacser, H. (1957). *In* "The Strategy of the Genes" (C. H. Waddington, ed.), pp. 191–249. George Allen and Unwin, London. Appendix: Some physico-chemical aspects of biological organization.
Karlson, P. and Sekeris, C. E. (1964). *In* "Comparative Biochemistry" (M. Florkin and H. S. Mason, eds.), pp. 221–243. Academic Press, New York. Biochemistry of insect metamorphosis.
Kawamata, J., Nakabayashi, N., Kawai, A. and Ushida, T. (1958). *Med. J. Osaka Univ.* 8, 753–762. Experimental production of sarcoma in mice with actinomycin.
Keck, K. (1961). *Ann. N.Y. Acad. Sci.* 94, 741–752. Nuclear and cytoplasmic factors determining the species specificity of enzyme proteins in *Acetabularia*.

REFERENCES

Keilin, D. (1959). *Proc. R. Soc.* B. **150**, 149–191. The problem of anabiosis or latent life: history and current concept.

Kellenberger, E. (1959). *Symp. Soc. gen. Microbiol.* **9**, 11–33. Growth of bacteriophage.

Kellenberger, E. (1960). *In* "Microbial Genetics" (W. Hayes and R. C. Clowes, eds.), pp. 39–66. Cambridge Univ. Press, Cambridge. The physical state of the bacterial nucleus.

Kellner, B. (1967). *In* "Mechanisms of Invasion in Cancer" (P. Denoix, ed.), pp. 11–16. Springer-Verlag, Berlin and Heidelberg. The significance of pericarcinomatous metastases.

Kelly, M. G. and O'Gara, R. W. (1961). *J. natn. Cancer Inst.* **26**, 651–679. Induction of tumors in newborn mice with dibenz (a, h) anthracene and 3-methylcholenthrone.

Kennaway, E. L. (1955). *Br. med. J.* **ii**, 749–752. The identification of a carcinogenic compound in coal tar.

Kennaway, E. L. and Hieger, I. (1930). *Br. med. J.* **i**, 1044–1046. Carcinogenic substances and their fluorescent spectra.

Kerr, I. M., Martin, E. M., Hamilton, M. G. and Work, T. S. (1962). *Cold Spring Harb. Symp. quant. Biol.* **27**, 259–269. The initiation of virus protein synthesis in Krebs ascites tumor cells infected with E.M.C. virus.

Keyes, E. L., Orrahood, M. D. and Blumenthal, H. T. (1954). *Archs Surg.* **68**, 820–828. Treated compared with untreated breast cancer.

Kidson, C. and Kirby, K. S. (1964a). *Cancer Res.* **24**, 1604–1609. Recognition of altered patterns of messenger RNA synthesis in a mouse hepatoma.

Kidson, C. and Kirby, K. S. (1964b). *Nature, Lond.* **203**, 599–603. Selective alterations of mammalian messenger-RNA synthesis: evidence for differential action of hormones on gene transcription.

Kidson, C. and Kirby, K. S. (1965). *Cancer Res.* **25**, 472–476. Selective alteration of rapidly-labelled ribonucleic acid synthesis in rat liver during azo-dye carcinogenesis.

Kim, U. (1966). *Cancer Res.* **26**, 461–464. Factors controlling metastasis of experimental breast cancer.

King, J. C. (1961). *Am. Nat.* **95**, 345–364. Inbreeding, heterosis and information theory.

Klein, G. and Klein, E. (1957). *Symp. Soc. exp. Biol.* **11**, 305–328. The evolution of independence from specific growth stimulation and inhibition in mammalian tumour-cell populations.

Klein, G. and Klein, E. (1958). *J. cell. comp. Physiol.* **52**, Suppl., 125–168. Histocompatibility changes in tumours.

Klein, M. (1951–52). *J. natn. Cancer Inst.* **12**, 1003–1010. The transplacental effect of urethan on lung tumorgenesis in mice.

Klein, M. (1954). *Cancer Res.* **14**, 438–440. Induction of lung adenomas following exposure of pregnant, newborn and immature male mice to urethan.

Klein, M. (1966). *J. natn. Cancer Inst.* **36**, 1111–1120. Influence of age on induction with urethan of hepatomas and other tumors in infant mice.

Kleinsmith, L. J. and Pierce, G. B. Jr. (1964). *Cancer Res.* **24**, 1544–1567. Multipotentiality of single embryonal carcinoma cells.

Koch, W. E. and Grobstein, C. (1963). *Devl Biol.* **7**, 303–323. Transmission of radioisotopically labelled materials during embryonic induction *in vitro*.

Koller, P. C. (1960). *In* "Cell Physiology of Neoplasia" pp. 9–48. Univ. of Texas Press, Austin. Chromosome behaviour in tumors: readjustment to Boveri's theory.

Koller, P. C. (1964). *In* "Cellular Control Mechanisms and Cancer" (P. Emmelot and O. Mühlbock, eds.), pp. 174–179. Elsevier Publishing Co. Amsterdam. Chromosomes in neoplasia.

Konigsberg, I. R. (1963). *Science, N.Y.* **140**, 1273–1284. Clonal analysis of myogenesis.

Korner, A. (1960). *J. Endocr.* **21**, 177–189. The role of the adrenal gland in the control of amino acid incorporation into protein of isolated rat liver microsomes.
Kotin, P. (1967). *In* "Carcinogenesis: a Broad Critique" (In discussion of paper by P. Shubik) pp. 738–744. The Williams and Wilkins Co., Baltimore.
Kroeger, H. (1963*a*). *J. cell. comp. Physiol.* **62**, Suppl. 1, 45–59. Experiments on the extranuclear control of gene activity in dipteran chromosomes.
Kroeger, H. (1963*b*). *Nature, Lond.* **200**, 1234–1235. Chemical nature of the system controlling gene activities in insect cells.
Kroon, A. M., Borst, P., van Bruggen, F. J. and Rottenberg, J. C. M. (1966). *Proc. natn. Acad. Sci. U.S.A.* **56**, 1836–1843. Mitochondrial DNA from sheep heart.
Kuzma, J. F. (1943). *Am. J. Path.* **19**, 473–482. Myoepithelial proliferations in the human breast.

Lacassagne, A. (1932). *C. r. hebd. Séanc. Acad. Sci. Paris.* **195**, 630–632. Apparition de cancers de la mammelle chez la souris mâle, soumise à des injections de folliculine.
Lacassagne, A. and Latarjet, R. (1946). *Cancer Res.* **6**, 183–188. Action of methylcholanthrene on certain scars of the skin of mice.
Lamerton, L. F. and Fry, R. J. M. (eds.). (1963). "Cell Proliferation". Blackwell Scientific Publications, Oxford.
Landauer, W. (1954). *J. cell. comp. Physiol.* **43**, Suppl. 1, 261–305. On the chemical production of developmental abnormalities and of phenocopies in chicken embryos.
Landauer, W. (1959). *Experentia* **15**, 409–412. The phenocopy concept: illusion or reality.
Lark, K. G. (1963). *In* "Molecular Genetics" (J. H. Taylor, ed.), Vol. I. pp. 153–206. Academic Press, New York. Cellular control of DNA biosynthesis.
Larsen, C. D. (1947). *J. natn. Cancer Inst.* **8**, 63–70. Pulmonary-tumor induction by transplacental exposure to urethane.
Lasnitzki, I. (1958). *Int. Rev. Cytol.* **7**, 79–121. The effect of carcinogens, hormones and vitamins on organ cultures.
Lathrop, A. E. C. and Loeb, L. (1916). *J. Cancer Res.* **1**, 1–20. Further investigations on the origin of tumors in mice. On the part played by hormones in the spontaneous development of tumors.
Laufer, H., Nakase, Y. and Vanderberg, J. (1964). *Devl Biol.* **9**, 367–384. Developmental studies of the dipteran salivary gland. I. The effect of actinomycin D on larval development, enzyme activity and chromosomal differentiation in *Chironomus thummi*.
Lawley, P. D. (1961). *J. Chim. phys.* **58**, 1011–1020. The action of alkylating agents on deoxyribonucleic acid.
Lawley, P. D. (1962). *In* "The Molecular Basis of Neoplasia", pp. 123–132. Univ. of Texas Press, Austin. Effects of alkylating agents on nucleic acids and their relation to other mutagens.
Lawley, P. D. and Brookes, P. (1961). *Nature, Lond.* **192**, 1081–1082. Acidic dissociation of 7:9-dialkylguanines and its possible relationship to mitogenic properties of alkylating agents.
Lawley, P. D. and Brookes, P. (1965). *Nature, Lond.* **206**, 480–483. Molecular mechanisms of the cytotoxic action of difunctional alkylating agents and of resistance to this action.
Leblond, C. P. (1965). *Am. J. Anat.* **116**, 1–28. The time dimension in histology.
Leblond, C. P., Greulich, R. C. and Pereira, J. P. M. (1964). *Adv. Biol. Skin.* **5**, 39–67. Relationship of cell formation and cell migration in the renewal of stratified squamous epithelium.
Leblond, C. P. and Walker, B. E. (1956). *Physiol. Rev.* **36**, 255–276. Renewal of cell populations.

Le Breton, E. and Moulé, Y. (1961). *In* "The Cell" (J. Brachet and A. E. Mirsky, eds.), Vol. 5. pp. 497–544. Academic Press, New York. Biochemistry and physiology of the cancer cell.
Lederberg, J. (1955). *J. cell. comp. Physiol.* **45**, Suppl. 2, 75–107. Recombination mechanisms in bacteria.
Lederberg, J. (1956). *In* "Enzymes: Units of Biological Structure and Function" (O. H. Gaebler, ed.), pp. 161–169. Academic Press, New York. Comments on gene-enzyme relationships.
Lee, A. E. (1968). *Br. J. Cancer.* **22**, 77–82. Genetic and viral influences on mammary tumours in BR6 mice.
Lefèvre, H. (1945). "Acceleration of the Development of Spontaneous Tumours in Mice" Thaning and Appels Forlag, Copenhagen.
Lehninger, A. L. (1956). *In* "Enzymes: Units of Biological Structure and Function" (O. H. Gaebler, ed.), pp. 217–233. Academic Press, New York. Physiology of mitochondria.
Leighton, J. (1957). *Cancer Res.* **17**, 929–941. Contributions of tissue culture studies to an understanding of the biology of cancer: a review.
Leighton, J. (1967). "The Spread of Cancer". Academic Press, New York.
Lindegren, C. C. and Pirie, N. W. (1963). *Nature, Lond.* **197**, 566–568. Viruses, genes and cistrons.
Leslie, I. (1961). *Nature, Lond.* **189**, 260–261. Biochemistry of heredity: a general hypothesis.
Levintow, L. and Eagle, H. (1961). *A. Rev. Biochem.* **30**, 605–640. Biochemistry of cultured mammalian cells.
Lewin, C. (1908). *Z. Krebsforsch.* **6**, 267–314. Experimentelle Beiträge zur Morphologie und Biologie bösartiger Geschwülste bei Ratten und Mäusen.
Liao, S. and Williams–Ashman, H. G. (1962). *Proc. natn. Acad. Sci. U.S.A.* **48**, 1956–1964. An effect of testosterone on amino acid incorporation of prostate ribonucleoprotein particles.
Likely, G. D., Sanford, K. K. and Earle, W. R. (1952). *J. natn. Cancer Inst.* **13**, 177–184. Further studies on the proliferation *in vitro* of single isolated tumor cells.
Linell, F. (1950). *Acta path. microbiol. scand.* **27**, 662–670. Studies on the action of carcinogenic hydrocarbons on the skin of rabbits. II. The tumour-producing effect and the increased proliferative capacity of the epithelium.
Linell, F. and Norden, J. G. (1950). *Acta path. microbiol. scand.* **27**, 394–413. Studies on the action of carcinogenic hydrocarbons on the skin of the rabbit. I. The tumour-producing effect. A comparison between 3, 4 -benzpyrene and 9, 10-dimethyl-1, 2-benzanthracene and between different areas of skin.
Linzell, J. L. (1952). *J. Anat.* **86**, 49–57. The silver staining of myoepithelial cells, particularly in the mammary gland, and their relation to the ejection of milk.
Lipschütz, B. (1924). *Z. Krebsforsch.* **21**, 50–97. Untersuchungen über die Entstehung des experimentallen Teercarcinoms der Maus.
Lipscomb, H. S., Wilson, C., Retiene, K., Matsen, F. and Ward, D. N. (1968). *Cancer Res.* **28**, 378–383. The syndrome of inappropriate secretion of antidiuretic hormone: a case report and characterization of anti-diuretic hormone-like material isolated from an oat cell carcinoma of the lung.
Lipsett, M. B. (1965). *Cancer Res.* **25**, 1068–1073. Humoral syndromes associated with cancer.
Little, C. C. (1936). *J. Am. med. Ass.* **106**, 2234–2235. Present status of our knowledge of heredity and cancer.

Little, C. C. and Strong, L. C. (1924). *J. exp. Zool.* **41**, 93–114. Genetic studies on the transplantation of two adenocarcinomata.

Little, C. C. and Tyzzer, E. E. (1915–16). *J. med. Res.* **33**, 393–453. Further experimental studies on the inheritance of susceptibility to a transplantable tumour, carcinoma (J.W.A.) of the Japanese waltzing mouse.

Loeb, L. and Fleisher, M. S. (1916). *J. Cancer Res.* **1**, 427–459. Transplantation of benign tumors.

Lorch, I. J. and Danielli, J. F. (1953). *Q. Jl microsc. Sci.* **94**, 461–480. Nuclear transplantation in amoebae. II. The immediate results of transfer of nuclei between Amoeba proteus and Amoeba discoides.

Love, R. (1959). *Ann. N.Y. Acad. Sci.* **81**, 101–117. Cytopathology of virus-infected tumor cells.

Luck, D. J. L. (1965). *Am. Nat.* **99**, 241–253. Formation of mitochondria in *Neurospora crassa*.

Lucké, B. and Schlumberger, H. C. (1957). "Atlas of Tumor Pathology" Armed Forces Institute of Pathology, Washington, D. C. Sect. VIII. Fas. 30. Tumors of the kidney, renal pelvis and ureter.

Ludford, R. J. (1925). *Proc. R. Soc.* B **98**, 557–577. The cytology of tar tumours.

Luria, S. E. (1959a). *Can. Cancer. Conf.* **3**, 261–270. Viruses as determinants of cellular functions.

Luria, S. E. (1959b). *Symp. Soc. gen. Microbiol.* **9**, 1–10. Viruses: a survey of some current problems.

Luria, S. E. (1960). *Cancer Res.* **20**, 677–688. Viruses, cancer cells and the genetic concept of virus infection.

Lwoff, A. (1950). "Problems of Morphogenesis in Ciliates". Wiley, New York.

Lwoff, A. (1953). *Bact. Rev.* **17**, 269–337. Lysogeny.

Lwoff, A. (1960). *Cancer Res.* **20**, 820–829. Tumor viruses and the cancer problem: a summation of the conference.

Lwoff, A. (1962). "Biological Order". The Massachusetts Institute of Technology Press, Cambridge, Mass.

Lwoff, A., Horne, R. and Tournier, P. (1962). *Cold Spring Harb. Symp. quant. Biol.* **27**, 51–55. A system of viruses.

Lynch, C. J. (1927). *J. exp. Med.* **46**, 917–933. Studies on the relation between tumor susceptibility and heredity. IV. The inheritance of susceptibility to tar-induced tumors in the lungs.

Maas, W. K. and Clark, A. J. (1964). *J. molec. Biol.* **8**, 365–370. Studies on the mechanism of repression of arginine synthesis in studies in *Escherichia coli*. II. Dominance of repressibility in diploids.

Maas, W. K. and McFall, E. (1964). *A. Rev. Microbiol.* **18**, 95–110. Genetic aspects of metabolic control.

McCarty, M. (1946). *Bact. Rev.* **10**, 63–71. Chemical nature and biological specificity of the substance inducing transformation of pneumococci.

McCormick, G. M. and Moon, R. C. (1965). *Br. J. Cancer.* **19**, 160–166. Effect of pregnancy and lactation on growth of mammary tumours induced by 7, 12-dimethyl (A) benzanthracene (D.M.B.A.).

Macdonald, I. (1951). *Surgery Gynec. Obstet.* **92**, 443–452. Biological predeterminism in human cancer.

McKenzie, A. (1956). *Br. J. Cancer.* **10**, 401–407. Duration of symptoms, clinical staging and survival in cancer of certain sites.

McKenzie, I. and Rous, P. (1941). *J. exp. Med.* **73**, 391–416. The experimental disclosure of latent neoplastic changes in tarred skin.
McKeown, F. (1956). *Br. J. Cancer.* **10**, 251–256. Malignant disease in old age.
McLean, A. E. M., McLean, E. and Judah, J. D. (1965). *Int. Rev. exp. Path.* **4**, 127–157. Cellular necrosis in the liver induced and modified by drugs.
McLoughlin, C. B. (1961a). *J. Embryol. exp. Morph.* **9**, 370–384. The importance of mesenchymal factors in the differentiation of chick epidermis. I. The differentiation in culture of the isolated epidermis of embryo chick and its response to vitamin A.
McLoughlin, C. B. (1961b). *J. Embryol. exp. Morph.* **9**, 385–409. II. Modifications of epidermal differentiation by contact with different types of mesenchyme.
McLoughlin, C. B. (1961c). In "Biological Approaches to Cancer Chemotherapy" (R. J. C. Harris, ed.), pp. 371–386. Academic Press, London. The connective tissue substrate.
McQuillen, K. (1965). *Symp. Soc. exp. Microbiol.* **15**, 134–158. The physical organization of nucleic acid and protein synthesis.
Madden, R. E. and Malmgren, R. A. (1962). *Cancer Res.* **22**, 62–66. Quantitative studies on circulating cancer cells in the mouse.
Magee, P. N. and Barnes, J. M. (1967). *Adv. Cancer Res.* **10**, 163–246. Carcinogenic nitroso compounds.
Makino, S. (1956). *Ann. N.Y. Acad. Sci.* **63**, Art. 5, 818–830. Further evidence favoring the concept of the stem cell in ascites tumors of rats.
Malmgren, R. A. (1967). In "Mechanisms of Invasion of Cancer" (P. Denoix, ed.), pp. 108–117. Springer-Verlag, Berlin. Studies of circulating cancer cells in cancer patients.
Malpas, J. S. Blandford, G., White, R. J. and Wrigley, P. F. M. (1968). *Proc. R. Soc. Med.* **61**, 463–464. Remote effects of non-endocrine cancer on the blood.
Marcus, P. I. (1962). *Cold Spring Harb. Symp. quant. Biol.* **27**, 351–365. Dynamics of surface modification in myxovirus-infected cells.
Marin-Padilla, M. and Benirschke, K. (1963). *Am. J. Path.* **43**, 999–1016. Thalidomide-induced alterations in the blastocyst and placenta of the armadillo Dasypus novemcinctus mexicanus, including a choriocarcinoma.
Markert, C. L. (1958). In "The Chemical Basis of Development" (W. D. McElroy and B. Glass, eds.), pp. 3–16. Johns Hopkins Press, Baltimore. Chemical concepts of cellular differentiation.
Markert, C. L. (1960). *Natn. Cancer Inst. Monogr.* **2**, 3–17. Biochemical embryology and genetics.
Markert, C. L. (1964). In "The Role of Chromosomes in Development" (M. Locke, ed.), pp. 1–9. Academic Press, New York. The role of chromosomes in development.
Markert, C. L. (1965). In "Ideas in Modern Biology" (J. A. Moore, ed.), pp. 229–258. The Natural History Press, Garden City, N.Y. Mechanisms of cellular differentiation.
Martin, E. M. (1967). *Br. med. Bull.* **23**, 192–197. Replication of small RNA viruses.
Mather, K. (1948). *Symp. Soc. exp. Biol.* **2**, 196–216. Nucleus and cytoplasm in differentiation.
Mather, K. (1961). *Nature, Lond.* **190**, 404–406. Nuclear materials and nuclear change in differentiation.
Matthaei, J. H., Jones, O. W., Martin, R. G. and Nirenberg, M. W. (1962). *Proc. natn. Acad. Sci. U.S.A.* **48**, 666–677. Characteristics and composition of RNA coding units.
Mazia, D. (1961). In "The Cell" (J. Brachet, and A. E. Mirsky, eds.), Vol. 3, pp. 77–412. Academic Press, New York, Mitosis and the physiology of cell division.
Mazia, D. (1965). *Symp. Soc. gen. Microbiol.* **15**, 379–394. The partitioning of genomes.
Melicow, W. M. M. (1965). *J. Urol.* **94**, 64–68. The new "British" classification of testicular tumors: a correlation analysis and critique.

Mendelsohn, M. L. (1963). *In* "Cell Proliferation" (L. F. Lamerton and R. J. M. Fry, eds.), pp. 190–210. Blackwell Scientific Publications, Oxford. Cell proliferation and tumour growth.

Mendelsohn, M. L. (1965). *In* "Cellular Radiation Biology" (ed.). pp. 498–513. The Williams and Wilkins Co., Baltimore. The kinetics of tumor cell proliferation.

Mendelsohn, M. L. and Dethlefsen, L. A. (1968). *Proc. Am. Ass. Cancer Res.* 9, 47. Cell proliferation and volumetric growth of fast line, slow line and spontaneous C_3H mammary tumors.

Messier, B. and Leblond, C. P. (1960). *Am. J. Ant.* 106, 247–265. Cell proliferation and migration as revealed by radioautography after injection of thymidine-H^3 into male rats and mice.

Michalowsky, I. (1928). *Virchows Arch. path. Anat. Physiol.* 267, 27–62. Eine experimentelle Erzeugung teratoider Geschwülste der Hoden beim Hahn.

Millar, M. J. and Noble, R. L. (1954). *Br. J. Cancer.* 8, 495–507. Effects of exogenous hormones on growth characteristics and morphology of transplanted mammary filroadenoma of the rat.

Miller, E. C. and Miller, J. A. (1947). *Cancer Res.* 7, 468–480. Presence and significance of bound aminoazo dyes in rats fed p-aminoazobenzene.

Miller, J. A. and Miller E. C. (1953). *Adv. Cancer Res.* 1, 339–396. The carcinogenic aminoazo dyes.

Miller, R. W. (1963). *New Engl. med. J.* 268, 393–401. Doan's syndrome (Mongolism), other congenital malformations and cancers among the sibs of leukemic children.

Miller, R. W. (1965). *Yale J. biol. Med.* 37, 487–501. Environmental agents in cancer.

Miller, R. W., Fraument, J. F. and Manning, M. D. (1964). *New Engl. J. Med.* 270, 922–927. Association of Wilms' tumor with aniridia, hemihypertrophy and other congenital malformations.

Mirsky, A. E. and Osawa, S. (1961). *In* "The Cell" (J. Brachet and A. E. Mirsky, eds.), Vol. 2, pp. 677–770. Academic Press, New York. The interphase nucleus.

Moertel, C. G. (1966). "Multiple Primary Malignant Neoplasms". Springer-Verlag, New York.

Moertel, C. G. Dockerty, M. B. and Baggenstoss, A. H. (1961*a*). *Cancer* 14, 221–230. Multiple primary malignant neoplasms. I. Introduction and presentation of data.

Moertel, C. G., Dockerty, M. B. and Baggenstoss, A. H. (1961*b*). *Cancer* 14, 231–237. Multiple primary malignant neoplasms. II. Tumors of different tissues and organs.

Moertel, C. G., Dockerty, M. B. and Baggenstoss, A. H. (1961*c*). *Cancer* 14, 238–248. Tumors of multicentric origin.

Mohr, U. and Althoff, J. (1965). *Z. Krebsforsch.* 67, 152–155. Die diaplacentare Wirkung des Cancerogenes Diathylnitrosamin bei der Maus.

Mohr, U., Althoff, J. and Authaler, A. (1966). *Cancer Res.* 26, 2349–2351. Diaplacental effect of the carcinogen diethylnitrosamine in the golden hamster.

Monod, J. (1956). *In* "Enzymes: units of Biological Structure and Function" (O. H. Graeber, ed.), pp. 7–28. Academic Press, New York. Remarks on the mechanism of enzyme induction.

Monod, J. and Jacob, F. (1961). *Cold Spring Harb. Symp. quant. Biol.* 26, 389–401. General conclusions: teleonomic mechanisms in cellular metabolism, growth and differentiation.

Montagna, W. (1962). "The Structure and Function of the Skin". Academic Press. 2nd Edition, New York.

Moore, G. E. and Sandberg, A. A. (1965). *Acta cytol.* 9, 175–184. Clinical aspects of cancer cells in the blood.

Moore, G. E., Lehner, D. F., Kikuchi, Y. and Less, L. A. (1962). *Science, N.Y.* **137**, 986–987. Continuous culture of a melanotic cell line from the golden hamster.

Moore, J. A. (1958). *Expl Cell Res.* Suppl. **6**, 179–191. The transfer of haploid nuclei between *Rana pipiens* and *Rana sylvaticus*.

Moore, J. A. (1960). *In* "New Approaches in Cell Biology" (P. M. B. Walker, ed.), pp. 1–14. Academic Press, London. Nuclear transfer of embryonic cells of the amphibia.

Moore, J. A. (1962). *J. cell. comp. Physiol.* **60**, Suppl. **1**, 19–34. Nuclear transplantation and problems of specificity in developing embryos.

Morgan, C. (1959). *Nature, Lond.* **184**, 435–436. Viral multiplications and cellular hyperplasia.

Morgan, C., Rose, H. M. and Moore, D. H. (1957). *Ann. N.Y. Acad. Sci.* **68**, 302–323. An evaluation of host cell changes accompanying viral multiplication as observed in the electron microscope.

Morris, H. P. (1963). *Prog. exp. Tumor Res.* **3**, 370–411. Some growth, morphological and biochemical characteristics of hepatoma 5123 and other new transplantable hepatomas.

Morris, H. P. (1966). *Gann Monogr.* **1**, 1–10. The development of hepatomas of different growth rate; with comments on their biology and bio-chemistry.

Morton, J. J. and Morton, J. H. (1953). *Ann. Surg.* **137**, 683–703. Cancer as a chronic disease.

Moscona, A. (1960). *In* "Developing Cell Systems and their Control". (D. Rudnick, ed.), pp. 45–70. Ronald Press, New York. Patterns and mechanisms of tissue reconstruction from dissociated cells.

Moscona, A. (1964). *In* "Cytodifferentiation" (D. Rudnick, ed.), pp. 49–50. Chicago Univ. Press, Chicago.

Moscona, A. A. and Hubby, J. L. (1963). *Devl. Biol.* **7**, 192–206. Experimentally induced changes in glutamotransferase activity in embryonic tissue.

Mühlbock, O. (1950). *J. natn. Cancer Inst.* **10**, 861–864. Mammary tumor-agent in the sperm of high-cancer-strain male mice.

Mühlbock, O. (1965). *Europ. J. Cancer.* **1**, 123–124. Note on a new inbred mouse-strain G.R./A.

Muller, H. J. (1927). *Science, N.Y.* **66**, 84–87. Artificial transmutation of the gene.

Mulligan, R. M. (1949). "Neoplasms of the Dog". The Williams and Wilkins Co., Baltimore.

Mundy, J. and Williams, P. C. (1961). *Br. J. Cancer.* **15**, 561–567. Tumour incidence and tumour-free sublines in BR6 mice.

Murray, J. A. (1908a). *Scient. Rep. Invest. imp. Cancer Res. Fund.* **3**, 69–114. Spontaneous cancer in the mouse.

Murray, J. A. (1908b). *Scient. Rep. Invest. imp. Cancer Res. Fund.* **3**, 159–174. A transplantable squamous-celled carcinoma of the mouse.

Murray, J. A. (1911). *Scient. Rep. Invest. imp. Cancer Res. Fund.* **4**, 114–130. Cancerous ancestry and the incidence of cancer in mice.

Murray, R. G. E. (1960). *In* "The Bacteria" (I. G. Gunsalus and R. Y. Stanier, eds.), Vol 1, pp. 35–96. Academic Press, New York. The internal structure of the cell.

Murray, R. G. E. (1963). *In* "General Physiology of Cell Specialization" (D. Mazia, and A. Tyler, eds.), pp. 28–52. McGraw-Hill Book Co., New York. The organelles of bacteria.

Murray, W. S. (1928). *J. Cancer Res.* **12**, 18–25. Ovarian secretion and tumor incidence.

Nadel, E. M. (ed.) (1965). *Acta cytol.* **9**, 1–188. Symposium on tumor cells in the circulating blood.

Nanney, D. L. (1957). *In* "The Chemical Basis of Heredity" (W. D. McElroy and B.

Glass, eds.), pp. 134–164. Johns Hopkins Press, Baltimore. The role of the cytoplasm in heredity.

Nanney, D. L. (1958). *Proc. natn. Acad. Sci. U.S.A.* **44**, 712–717. Epigenetic control systems.

Nanney, D. L. (1963). In "Biological Organization at the Cellular and Supercellular Level" (R. J. C. Harris, ed.), pp. 91–109. Academic Press, London. Aspects of mutual exclusion in Tetrahymena.

Nanney, D. L. (1964). In "The Role of Chromosomes in Development" (M. Locke, ed)., pp. 253–273. Academic Press, New York. Macronuclear differentiation and subnuclear assortment in ciliates.

Nanney, D. L. and Rudzinska, M. A. (1960). In "The Cell" (J. Brachet, and A. E. Mirsky, eds.), Vol. 4, pp. 109–150. Academic Press, New York. Protozoa.

Nass. M. M. K. (1966). *Proc. natn. Acad. Sci. U.S.A.* **56**, 1215–1222. The circularity of mitochondrial DNA.

Nass, M. M. K., Nass, S. and Afzelius, B. A. (1965). *Expl Cell Res.* **37**, 516–539. The general occurrence of mitochondrial DNA.

Nathanson, I. and Welch, C. (1936). *Am. J. Cancer.* **28**, 40–53. Life expectancy and insidence of malignant disease. I. Carcinoma of the breast.

Needham, A. E. (1964). "The Growth Process in Animals". Sir Isaac Pitman and Sons Ltd., London; D. van Nostrand Co. Princeton, N. J.

Needham, J. (1942). "Biochemistry and Morphogenesis". Cambridge Univ. Press, Cambridge.

Neidhart, F. C. and Frankel, D. G. (1961). *Cold Spring Harb. Symp. quant. Biol.* **26**, 63–74. Metabolic regulation of RNA synthesis in bacteria.

Nettleship, A. and Henshaw, P. S. (1943). *J. natn. Cancer Inst.* **4**, 309–319. Induction of pulmonary tumors in mice with ethyl carbamate (urethane).

Newman, W. and Cromer, J. K. (1959). *Surgery Gynec. Obstet.* **108**, 273–281. The multicentric origin of carcinomas of the female anogenital tract.

Nicholson, G. W. (1931). *J. Path. Bact.* **34**, 711–730. An embryonic tumour of the kidney in a foetus.

Nicholson, G. W. (1950). In "Studies on Tumour Formation" (R. W. Willis, ed.), Butterworth & Co., London.

van Nie, R. and Thung, P. J. (1965). *Europ. J. Cancer* **1**, 41–50. Responsiveness of mouse mammary tumours to pregnancy.

Nirenberg, M. W., Jones, O. W., Leder, P., Clark, B. P. C., Sly, W.S. and Pestka, S. (1963). *Cold Spring Harb. Symp. quant. Biol.* **28**, 549–557. On the coding of genetic information.

Noble, R. L. (1964). In "The Hormones" (G. Pincus, K. V. Thimann and E. B. Astwood, eds.), Vol 5. pp. 559–695. Academic Press, New York. Tumors and hormones.

Northrop, F. S. C. (1947). "The Logic of the Sciences and the Humanities". The Macmillan Co., New York.

Novelli, G. D., Kameyama, T. and Eisenstadt, J. M. (1961). *Cold Spring Harb. Symp. quant. Biol.* **26**, 133–143. The nature of the system catalysing the synthesis of β-galactosidase.

Novikoff, A. B. (1961). In "The Cell" (J. Brachet and A. E. Mirsky, eds.), Vol. 2, pp. 299–421. Academic Press, New York. Mitochondria (Chondriosomes).

Nowinski, W. ed. (1960). "Fundamental Aspects of Normal and Malignant Growth". Elsevier Publishing Company, Amsterdam.

Noyes, R. W. (1959). *Ann. N.Y. Acad. Sci.* **80**, 54–64. Trophoblast: problems of invasion and transport.

Ober, W. B. (ed.) (1959a). *Ann. N.Y. Acad. Sci.* **80**, Art 1, 1–284. Trophoblast and its tumors.
Ober, W. B. (1959b). *Ann. N.Y. Acad. Sci.* **80**, 1, 3–20. Historical perspectives on trophoblast and its tumors.
Ober, W. B. (1959c). *Ann. N.Y. Acad. Sci.* **80**, Art 1, 150–151. Discussion.
Oberling, Ch., Guérin, M. and Guérin, P. (1933). *Bull Ass. fr. Étude Cancer* **22**, 606–630. Recherches sure des greffes en serie de tumeurs mammaires benignes chez le rat.
Oberling, Ch., Guérin, M. and Guérin, P. (1935). *Bull Ass. fr. Étude Cancer* **24**, 232–270. A propos de la transformation sarcomateuse des fibroadénomes mammaires trasplantables du rat blanc.
Oberling, Ch., Guérin, M. and Guérin, P. (1937). *Bull Ass. fr. Étude Cancer* **26**, 483–500. Les fibro-adénomes mammaires greffable du rat blanc. Nouvelles recherches.
Oberling, Ch., Guérin, P. and Guérin, M. (1938). *Bull Ass. fr. Étude Cancer* **27**, 260–267. Plasticité morphologique et metaplasie sebacée dans les fibro-adénomes mammaires du rat.
Ochoa, S. (1962). *In* "Horizons in Biochemistry". (M. Kasha and B. Pullman, eds.), pp. 153–166. Academic Press, New York. Enzymatic mechanisms in the transmission of genetic information.
Oppenheimer, J. M. (1959a). *Q. Rev. Biol.* **34**, 271–277. Embryology and evolution: nineteenth century hopes and twentieth century realities.
Oppenheimer, J. M. (1959b). *Science, N.Y.* **130**, 686–692. Intercellular activities in vertebrate development. Problems of embryonic organization are being attacked at subcellular and supracellular levels.
Oppenheimer, J. M. (1963). *Devl Biol.* **7**, 11–21. K. E. von Baer's beginning insights into causal-analytical relationships during development.
Orr, J. W. (1958). *Br. med. Bull.* **14**, 99–100. The mechanism of chemical carcinogenesis.
Orsini, M. W. (1959). *In* Discussion of Noyes, R. W. (1959). *Ann. N.Y. Acad. Sci.* **80**, 54–64. Trophoblast: problems of invasion and transport.

Paine, C. H. (1965). *Br. J. Cancer.* **19**, 263–267. Very late recurrence of a previously excised sweat gland carcinoma: case report with a review of the literature.
Palade, G. E. (1956). *In* "Enzymes: Units of Structure and Function" (O.W. Gaebler, ed.), pp. 185–215. Academic Press, New York. Electron microscopy of mitochondria and other cytoplasmic structures.
Palm, J. (1961). *In* "Transplantation of Tissues and Cells" (R. E. Billingham and W. K. Silvers, eds.), pp. 113–132. The Wistar Institute Press, Philadelphia. Immunogenetic aspects of tissue transplantation.
Park, W. W. (1958). *J. Path. Bact.* **75**, 257–265. Experimental trophoblastic embolism of the lungs.
Park. W. W. (1959a). *Ann. N.Y. Acad. Sci.* **80**, Art 1, 152–160. Choriocarcinoma in the female.
Park, W. W. (1959b). *Ann. N.Y. Acad. Sci.* **80**, Art 1, 197–261. In discussion of Bardwil and Toy (1959).
Park, W. W. and Lees, J. C. (1950). *Archs Path.* **49**, 73–104 and 205–241. Chorioiarcinoma. A general review with an analysis of five hundred and sixteen cases.
Parsons, D. F. (1965). *Int. Rev. exp. Path.* **4**, 1–54. Recent advances correlating structure and function in mitochondria.
Pasternack, J. G. and Wirth, J. E. (1936). *Am. J. Path.* **12**, 423–435. Adeno-acanthoma sarcomatodes of the mammary gland. Report of a case with a critical review of the literature on squamous epithelium in intramammary tumors.

Paul, J. (1967). *In* "Cell Differentiation" (A. V. S. de Reuk and J. Knight, eds.), pp. 196–207. Ciba Foundation Symposium. J. & A. Churchill, London. Masking of genes in cytodifferentiation and carcinogenesis.
Pavan, C. (1959). *In* "Biological Organization" (C. H. Waddington, ed.), Chapter 3. Pergamon Press, London. Organisation of the chromosome.
Peacocke, A. R. and Drysdale, R. B. (1965). "The Molecular Basis of Heredity". Butterworths, London.
Pelc, S. R. and Fell, H. B. (1960). *Expl Cell Res.* **19**, 99–113. The effect of excess vitamin A on the uptake of labelled compounds by embryonic skin in organ culture.
Peller, S. (1960). "Cancer in Childhood and Youth". John Wright & Sons, Bristol.
Petit, N., and Peyron, A. (1927). *Bull. Ass. franc. Cancer* **16**, 510–515. Sur la coexistence de néoplasies distinctes dan la glande mammaire d'une chienne.
Peyron, A. (1924*a*). *Bull du Cancer.* **13**, 349–365. Sur la pathologie comparée des tumeurs de la mammelle.
Peyron, A. (1924*b*). *C.r. Séanc. Soc. Biol.* **90**, 1273–1276. Sur le mode de proliferation de l'assise myoépithéliale dans les tumeurs dites mixtes de la gland mammaire de la chienne.
Peyron, A. (1939). *Bull Ass. fr. Étude Cancer* **28**, 658–681. Faits nouveaux relatifs à l'origine et à l'histogénèse des embryomes.
Peyron, A., Corsy, F. and Surmont, J. (1926). *Bull. Cancer.* **15**, 21–62. Sur la pathologie comparee des tumeurs de la mamelle. Troisième demonstration.
Picheral, B. (1962). *C.r. hebd. Séanc. Acad. Sci.* **255**, 2509–2511. Capacités des noyaux de cellules endodermiques embryonaires à organiser un germe viable chez l'urodele, Pleurodeles waltlii Michah.
Picken, L. (1960). "The Organization of Cells and Other Organisms". The Clarendon Press, Oxford.
Pierce, G. B. (1961). *Can. Cancer Conf.* **4**, 119–137. Teratocarcinomas, a problem in developmental biology.
Pierce, G. B. and Beals, T. F. (1964). *Cancer Res.* **24**, 1553–1567. The ultrastructure of primordial germ cells of the fetal testes and of embryonal carcinoma cells of mice.
Pierce, G. B. and Dixon, F. J. (1959*a*). *Cancer* **12**, 573–583. Testicular teratomas. I. Demonstration of teratogenesis by metamorphosis of multipotential cells.
Pierce, G. B., and Dixon, F. J. (1959*b*). *Cancer* **12**, 584–599. Testicular teratomas. II. Teratocarcinoma as an ascitic tumor.
Pierce. G. B., Dixon, F. J. and Verney, E. L. (1960). *Lab. Invest.* **9**, 583–602. Teratocarcinogenic and tissue-forming potentials of the cell types comprising neoplastic embryoid bodies.
Pierce, G. B., Stevens, L. and Nakane, P. K. (1967). *J. natn. Cancer Inst.* **39**, 755–773. Ultrastructural analysis of the early development of teratocarcinomas.
Pietra, G., Rappaport, H. and Shubik, P. (1961). *Cancer* **14**, 308–317. The effects of carcinogenic chemicals in newborn mice.
Pietra, G., Spencer, K. and Shubik, P. (1959). *Nature, Lond.* **183**, 1689. Response of newly born mice to a chemical carcinogen.
Pitelka, D. R. (1963). "Electron-microscopic Structure of Protozoa". Pergamon Press, New York.
Pitot, H. C. (1963). *Cancer Res.* **23**, 1474–1482. Some biochemical essentials of malignancy.
Pitot, H. C. (1966). *A. Rev. Biochem.* **35**, 335–368. Some biochemical aspects of malignancy.
Pittendrigh, C. S. (1960–61). *Harvey Lect.* Series **56**, 93–125. On temporal organization in living systems.
Pollock, M. R. (1953). *Symp. Soc. gen. Microbiol.* 150–177. Adaptation in microorganisms.

Pollock, M. R. and Mandelstam, J. (1958). *Symp. Soc. exp. Biol.* 12, 195–204. Possible mechanisms by which information is conveyed to the cell in enzyme induction.
Pontecorvo, G. (1959). "Trends in Genetic Analysis". Oxford Univ. Press, London.
Pontecorvo, G. (1963). *Proc. R. Soc.* B, 158, 1–23. Microbial genetics: retrospect and prospect.
Porter, K. R. (1961). *In* "The Cell" (J. Brachet and A. E. Mirsky, eds.), Vol. 2, pp. 621–675. Academic Press, New York. The ground substance; observations from electron microscopy.
Porter, K. R. and Bonneville, M. A. (1963). "An Introduction to the Fine Structure of Cells". Lea and Febiger, Philadelphia.
Porter, K. R. and Machado, R. D. (1960). *J. biophys. biochem. Cytol.* 7, 167–180. Studies on the endoplasmic reticulum. IV. Its form and distribution during mitosis in cells of onion root tip.
Potter, V. R. (1961). *Cancer Res.* 21, 1331–1333. Transplantable animal cancer, the primary standard.
Potter, V. R. (1962). *In* "The Molecular Basis of Neoplasia". pp. 367–399. Univ. of Texas Press, Austin. Enzyme studies on the deletion hypothesis of carcinogenesis.
Potter, V. R. (1964). *In* "Cellular Control Mechanisms and Cancer" (P. Emmelot and O. Mühlbock, eds.), pp. 190–210. Elsevier Publishing Co., Amsterdam. Biochemical studies on minimal deviation hepatomas.
Prehn, R. T. and Main, J. M. (1957). *J. natn. Cancer Inst.* 18, 769–778. Immunity to methylcholanthrene-induced sarcomas.
Prescott, D. M. and Kimball, R. E. (1961). *Proc. natn. Acad. Sci. U.S.A.* 47, 686–693. Relation between RNA, DNA and protein synthesis in the replicating nucleus of Euplotes.
Pugh, R. C. B. and Smith, J. P. (1964). *Br. J. Urol.* 36, Suppl. pp. 28–44. Teratoma.

Quastler, H. (1959). *Am. Nat.* 93, 245–254. Information theory of biological integration.
Quastler, H. (1963). *In* "Cell Proliferation" (L. F. Lamerton and R. J. M. Fry, eds.), pp. 18–34. Blackwell Scientific Publications, Oxford. The analysis of cell population kinetics.

Radl, E. (1930). "The History of Biological Theories". Oxford Univ. Press, London.
Rapp, E. H. (1936). *Z. Krebsforsch.* 44, 405–414. Wachstum und Differenzierung von Lebermetastasen des Adenocarcinoms des Magens.
Raven, C. P. (1961). "Oogenesis. The Storage of Developmental Information". Pergamon Press, Oxford.
Raven, C. P. (1963). *Devl Biol.* 7, 130–143. The nature and origin of the cortical morphogenetic field in Limnaea.
Reichman, M. E. (1964). *Proc. natn. Acad. Sci. U.S.A.* 52, 1009–1017. The satellite tobacco necrosis virus: a single protein and its genetic code.
Reiskin, A. M. and Mendelsohn, M. L. (1964). *Cancer Res.* 24, 1131–1136. A comparison of the cell cycle in induced carcinomas and in their normal counterpart.
Rhodin, J. A. G. (1963). "An Atlas of Ultrastructure". W. B. Saunders Company, Philadelphia.
Rich, A. (1962). *In* "Horizons in Biochemistry" (M. Kasha and B. Pullman, eds.), pp. 103–126. Academic Press, New York. On the problems of evolution and biochemical information transfer.
Rich. A., Warner, J. R. and Goodman, H. M. (1963). *Cold Spring Harb. Symp. quant. Biol.* 28, 269–285. The structure and function of polyribosomes.

Rich, R. A. (1935). *J. Urol.* 33, 215–223. On the frequency of occurrence of occult carcinoma of the prostate.
Richardson, C. C., Schildkraut, C. L. and Kornberg. A. (1963). *Cold Spring Harb. Symp. quant. Biol.* 28, 9–19. Studies on the replication of DNA by DNA polymerases.
Richardson, K. C. (1949). *Proc. R. Soc.* B, 136, 30–45. Contractile tissues in the mammary gland with special reference to myoepithelium in the goat.
Rigler, L. G. (1964). *Ann. N.Y. Acad. Sci.* 114 (Art. 2). 755–766. The natural history of untreated lung cancer.
Ris, H. and Chandler, B. L. (1963). *Cold Spring. Harb. quant. Biol.* 28, 1–8. The ultrastructure of genetic systems in prokaryocytes and eukaryocytes.
Ritchie, A. C. and Webster, D. R. (1961). *Can. Cancer Conf.* 4, 225–236. Tumor cells in the blood.
Rivière, M., Chouroulinkov, J. and Guérin, M. (1960). *Bull. Cancer, Paris* 47, 55–87. Production de tumeurs par injections intratesticulaires, de chlorure de zinc chez le rat.
Roberts, R. B., Britten, R. J. and McCarthy, B. J. (1963). *In* "Molecular Genetics". (J. H. Taylor, ed.), Vol. 1. pp. 291–352. Academic Press, New York. Kinetic studies of the synthesis of RNA and ribosomes.
Roe, F. J. C. (1956). *Br. J. Cancer.* 10, 61–69. The development of malignant tumours of mouse skin after "initiating" and "promoting" stimuli. I. The effect of a single application of 9, 10-dimethyl-1, 2 benzanthracene (DMBA) with and without subsequent treatment with croton oil.
Roe, F. J. C., Carter, R. L. and Percival, W. H. (1967). *Br. J. Cancer* 21, 815–820. Carcinogenesis in rabbits injected at birth with 7, 12-dimethylbenz (a) anthracene.
Roe, F. J. C. and Glendenning, O. M. (1956). *Br. J. Cancer* 10, 357–362. The carcinogenicity of β-propiolactone for mouse skin.
Roe, F. J. C. and Mitchley, B. C. V. (1963). *Nature, Lond.* 200, 1016–1017. Thalidomide and neoplasia.
Roe, F. J. C., Rowson, K. E. K. and Salaman, M. H. (1961). *Br. J. Cancer* 15, 515–530. Tumours of many sites induced by injection of chemical carcinogens into newborn mice, a sensitive test for carcinogenesis: the implications for certain immunological theories.
Rogers, S. (1951). *J. exp. Med.* 93, 427–449. Age of the host and other factors affecting the production with urethane of pulmonary adenomas in mice.
Rose, H. M. and Morgan, C. (1960). *Ann. Rev. Microbiol.* 14, 217–240. Fine structure of virus-infected cells.
Rous, P. (1911). *J. exp. Med.* 13, 397–411. A sarcoma of the fowl transmissible by an agent separable from the tumor cells.
Rous, P. (1913). *J. exp. Med.* 17, 494–497. False transitions between normal and cancerous epithelium.
Rous, P. and Beard, J. W. (1934a). *J. exp. Med.* 60, 701–722. A virus-induced mammalian growth with the characters of a tumor (The Shope rabbit papilloma). I. The growth on implantation within favorable hosts.
Rous, P. and Beard, J. W. (1934b). *J. exp. Med.* 60, 741–766. A virus-induced mammalian growth with the characters of a tumor (The Shope rabbit papilloma). III. Further characters of the growth: general discussion.
Rous, P. and Beard, J. W. (1935). *J. exp. Med.* 62, 523–548. The progression to carcinoma of virus-induced papillomas (Shope).
Rous, P. and Kidd, J. G. (1939). *J. exp. Med.* 69, 399–424. A comparison of virus-induced rabbit tumors with tumors of unknown cause elicited by tarring.

Rous, P. and Kidd, J. G. (1941). *J. exp. Med.* 73, 365–389. Conditional neoplasms and subthreshold neoplastic states. A study of the tar tumours of rabbits.
Rous, P. and Smith, W. E. (1945). *J. exp. Med.* 81, 597–620. The neoplastic potentialities of mouse embryo tissues. I. The findings with skin of C strain embryos transplanted to adult animals.
Roussy, G., Guérin, M. and Guérin P. (1943). *Bull Ass. fr. Étude Cancer* 31, 150–159. Epithelioma mammaire transplantable developpé sur un adénofibrome spontanée du rat.
Roussy, G., Guérin, M. and Guérin, P. (1944). *Bull Acad. Med.* 128, 156–162. Nouvelles observations de tumeurs mammaires spontanées et transplantables chez le rat blanc.
Runnström, J., Hagström, B. E. and Perlman, P. (1959). *In* "The Cell". (J. Brachet and A. E. Mirsky, eds.). Vol. 1, 327–397. Academic Press, New York. Fertilization.
Rusch, H. P., Braun, R., Daniel, J. W., Mittermayer, C. and Sachsenmaier, W. (1964). *In* "Cellular Control Mechanisms and Cancer". (P. Emmelot and O. Mühlbock, eds.), pp. 80–85. Elsevier Publishing Co., Amsterdam. The role of DNA and RNA in mitosis and differentiation in Physarum polycephalum.
Russell, B. R. G. (1908). *Scient. Rep. Invest. imp. Cancer Res. Fund.* 3, 341–358. The nature of resistance to the inoculation of cancer.
Russell, E. S. (1930). "The Interpretation of Development and Heredity". Clarendon Press, Oxford.
Russell, E. S. (1945). "The Directiveness of Organic Activities". Cambridge Univ. Press, Cambridge.
Rutter, W. J., Wessells, N. K. and Grobstein, C. (1964). *Natn. Cancer Inst. Monogr.* 13, 51–65. Control of specific synthesis in the developing pancreas.

Sager, R. and Ryan, F. J. (1961). "Cell Heredity". John Wiley and Sons Inc. New York.
Sand, S. A. (1965). *Am. Nat.* 99, No. 904, 33–45. Position effects and the problem of coding a program.
Sanford, K. K. (1965). *Int. Rev. Cytol.* 18, 249–311. Malignant transformation of cells *in vitro*.
Sanford, K. K. (1967). *Natn. Cancer Inst. Monogr.* 26, 387–418. "Spontaneous" neoplastic transformation of cells *in vitro*: some facts and theories.
Sanford, K. K., Earle, W. R. and Likely, G. D. (1948/9). *J. natn. Cancer Inst.* 9, 229–246. The growth *in vitro* of single isolated tissue cells.
Saxen, L. and Toivonen, S. (1962). "Primary Embryonic Induction". Logos Press, Ltd., in association with Elk Books Ltd., Distributed by Academic Press, London.
Schindler, R., Day, M. and Fischer, G. A. (1959). *Cancer Res.* 19, 47–51. Culture of neoplastic mast cells and their synthesis of 5-hydroxytryptamine and histamine *in vitro*.
Schlesinger, R. W. (1959). *In* "The Viruses". (F. W. Burnet and W. M. Stanley, eds.), Vol. 3, 157–194. Academic Press, New York. Interference between animal viruses.
Schmidt, M. D. (1903). "Die verbreitunswege der Karzinome". G. Fisher, Jena.
Scholtissek, C. (1962). *Nature, Lond.* 194, 353–355. An unstable ribonucleic acid in rat liver nuclei.
Scholtissek, C., Rott, R., Hausen, P., Hausen, H. and Schäfer, W. (1962). *Cold Spring Harb. Symp. quant. Biol.* 27, 245–257. Comparative studies of RNA and protein synthesis with a myxovirus and a small polyhedral virus.
Scott, A. (1923). *Scient. Rep. Invest. imp. Cancer Res. Fund.* 8, 85–142. The occupation dermatoses of the paraffin workers of the Scottish shale oil industry with a description of the system adopted and the results obtained at the periodic examinations of these workmen.

Sengel, P. (1965). *In* "Les Cultures Organotypiques". (J. A. Thomas, ed.), pp. 283–336. Masson et Cie, Paris. Étude de l'organogenèse par la dissociation et la réassociation d'ébauches embyonnaires *in vitro*.
Serra, J. A. (1964). *J. theoret. Biol.* **6**, 371–374. The genetic concept of Treption (and brief reference to a unifying theory of cancer).
Severi, L., Biancifiori, C., Olivi, M. and Squartini, F. (1959). *Acta Un. Intn. c. Cancer.* **15**, 227–231. A microscopical study of mammary cancer in hybrid mice with particular reference to histogenesis.
Shabad, L. M. (1962). *In* "The Morphological Precursors of Cancer". (L. Severi, ed.), pp. 111–118. Division of Cancer Research, Perugia. Some aspects of the morphology of experimental pre-cancerous lesions.
Shapiro, J. A. and Kirschbaum, A. (1951). *Cancer Res.* **11**, 644–647. Intrinsic tissue response to induction of pulmonary tumors.
Sharma, A. K. and Sharma K. (1960). *Int. Rev. Cytol.* **10**, 101–136. Spontaneous and induced chromosome breaks.
Shay, H., Harris, C. and Gruenstein, M. (1952). *J. natn. Cancer Inst.* **13**, 307–331. Influence of sex hormones on the incidence and form of tumors produced in male or female rats by gastric instillation of methylcholanthrene.
Shimkin, M. B. (1951). *Cancer* **4**, 1–8. Duration of life in untreated cancer.
Shimkin, M. B. (1955a). *Adv. Cancer Res.* **3**, 223–267. Pulmonary tumors in experimental animals.
Shimkin, M. B. (1955b). *Cancer* **8**, 653–655. M. A. Novinsky: a note on the history of transplantation of tumors.
Shimkin, M. B., Lucia, E. L., Low-Beer, B. V. A. and Bell, H. G. (1954). *Cancer* **7**, 29–46. Recurrent cancer of the breast; analysis of frequency, distribution and mortality at the University of California Hospital, 1918–1947 inclusive.
Shubik, P. (1950). *Cancer Res.* **10**, 13–17. Studies on the promoting phase in the stages of carcinogenesis in mice, rats, rabbits and guinea pigs.
Shubik, P. (1961). *In* Proc. Fourth Natn. Cancer Conf. pp. 113–126. J. B. Lippincott Co., Philadelphia. Biological determination of the action of chemical carcinogens.
Shubik, P. (1966). *Can. Cancer Conf.* **6**, 244–253. Carcinogenesis: the present position.
Shubick, P. (1967a). Personal Communication.
Shubik, P. (1967b). *In* "Carcinogenesis: a Broad Critique", pp. 731–746. (Twentieth Ann. Symp. on Fundamental Cancer Res, University of Texas, M. D. Anderson Hospital and Tumor Institute). The Williams and Wilkins Co., Baltimore. Biological mechanisms in carcinogenesis.
Siegel, R. W. (1961). *Expl Cell Res.* **24**, 6–20. Nuclear differentiation and transitional cellular phenotypes in the life cycle of Paramecium.
Siekevitz, P. (1962). *In* "The Molecular Control of Cell Activity". (J. M. Allen, ed.), pp. 143–166. McGraw-Hill, New York. The relation of cell structure to metabolic activity.
Sigot, M. (1965). *In* "Les Cultures Organotropiques" (J. A. Thomas, ed.), pp. 255–282. Masson et Cie, Paris. Dissociations de cellules et réassociation *in vitro*.
Simnett, J. D. (1964). *Devl Biol.* **10**, 467–486. The development of embryos derived from the transplantation of neural ectoderm nuclei in Xenopus laevis.
Simpson, G. G. (1962). *Am. Scient.* **50**(i), 36–45. The status of the study of organisms.
Simpson, G. G. (1963). *Science, N.Y.* **139**, 81–88. Biology and the nature of science.
Sinclair, J. G. (1950). *Texas Rep. Biol. Med.* **8**, 623–632. A specific transplacental effect of urethane in mice.
Sinnott, E. W. (1939). *Science, N.Y.* **89**, 41–46. The cell and the problem of organization.
Sirlin, J. L. (1963). *Int. Rev. Cytol.* **15**, 35–96. The interacellular transfer of genetic information.

REFERENCES

Skaar, D. (1956). *Expl Cell Res.* 10, 645–656. Past history and pattern of serotype transformation in *Paramecium aurelia*.

Smith, W. E. (1947). *J. exp. Med.* 85, 459–477. The neoplastic potentialities of mouse embryo tissues. III. Tumors elicited from gastric epithelium.

Smith, W. E. (1952). *A.MA. Arch Ind. Hyg.* 5, 209–217. Lung cancer with special reference to experimental aspects.

Smith, W. E. and Rous, P. (1945). *J. exp. Med.* 81, 621–646. The neoplastic potentialities of mouse embryo tissue. II. Contributory experiments; results with skin of C_3H and Webster-Swiss embryos; general considerations.

Smith, W. E. and Rous, P. (1948). *J. exp. Med.* 88, 529–554. The neoplastic potentialities of mouse embryo tissues; lung adenomas in baby mice as result of prenatal exposure to urethane.

Smithers, D. W. (1960). "A Clinical Prospect of the Cancer Problem". E. & S. Livingstone, Edinburgh.

Smithers, D. W. (1962). *Lancet i* 493–499. Cancer: an attack on cytologism.

Smithers, D. W. (1964). "On the Nature of Cancer in Man". E. & S. Livingstone, Edinburgh.

Smithers, D. W., Rigby-Jones, P., Galton, D. A. G., and Payne, P. M. (1952). *Br J. Radiol.* Suppl. 4, 1–90. Cancer of the breast: a review.

Snell, G. D. (1953). *J. natn. Cancer Inst.* 14, 691–700. The genetics of transplantation.

Sonneborn, T. M. (1947). *Adv. Genet.* 1, 263–358. Recent advances in the genetics of Paramecium and Euplotes.

Sonneborn, T. M. (1950). *Heredity* 4, 11–36. The cytoplasm in heredity.

Sonneborn, T. M. (1951). *In* "Genetics in the 20th Century". (L. C. Dunn, ed.), pp. 291–314. The Macmillan Company, New York. The role of genes in cytoplasmic inheritance.

Sonneborn, T. M. (1964). *Proc. natn. Acad. Sci. U.S.A.* 51, 915–929. The differentiation of cells.

Sonneborn, T. M. (1965). *Am. Nat.* 99, 279–307. The metagon: RNA and cytoplasmic inheritance.

Sonneborn, T. M. (1966). *In* "The Nature of Biological Diversity". (J. M. Allen, ed.), pp. 165–221. McGraw-Hill, New York. Does preformed structure play an essential role in cell heredity?

Speyer, J. F., Lengyel, P., Basilio, C. and Ochoa, S. (1962). *Proc. natn. Acad. Sci. U.S.A.* 48, 63–68. Synthetic polynucleotides and the amino acid code.

Speyer, J. F., Lengyel, P., Basilio, C. Wahba, A. J., Gardner, R. S., and Ochoa, S. (1963). *Cold Spring Harb. Symp. quant. Biol.* 28, 559–567. Synthetic polynucleotides and the amino acid code.

Spiegelman, S. and Doi, R. H. (1963). *Cold Spring Harb. Symp. quant. Biol.* 28, 109–116. Replication and translation of RNA genomes.

Spiegelman, S. and Hayashi, M. (1963). *Cold Spring Harb. Symp. quant. Biol.* 28, 161–181. The present status of genetic information and its control.

Squartini, F. (1962). *J. natn. Cancer Inst.* 28, 911–926. Responsiveness and progression of mammary tumors of high-cancer strain mice.

Squartini, F. and Rossi, G. (1959). Lavori dell Institute di Anatomia Perugia. 19, 105–124. Accrescimente e progressione dei tumori mammari nei topi femina del substrain RIII/Dm/Se.

Squartini, F. and Rossi, G. (1962). *In* "The Morphological Precursors of Cancer". (L. Severi, ed.), pp. 319–327. Division of Cancer Research, Perugia. Responsiveness and progression of the morphological precursors of breast cancer in inbred mice: a review.

Squartini, F. and Severi, L. (1962). *In* "Tumour Viruses of Murine Origin". (G. E. W. Wolstenholme and M. O'Connor, eds.), pp. 82–106. J. & A. Churchill, London. Strain differences in the mammary tumour-inducing virus as detected by the characters and behaviour of neoplasms.

Stedman, E. and Stedman, E. (1950). *Nature, Lond.* **166**, 780–781. Cell specificity of histones.

Steele, G. G. and Lamerton, L. F. (1966). *Br. J. Cancer* **20**, 74–86. The growth rate of human tumours.

Stent, G. S. (1964). *Science, N.Y.* **144**, 816–820. The operon: on its third anniversary.

Stern, C. (1955). *In* "Analysis of Development". (B. H. Willier, P. Weiss, and V. Hamburger, eds.), pp. 151–169. W. B. Saunders and Co., Philadelphia. Gene action.

Stevens, L. C. (1958). *J. natn. Cancer Inst.* **20**, 1257–1275. Studies on transplantable testicular teratomas of strain 129 mice.

Stevens, L. C. (1959). *J. natn. Cancer Inst.* **23**, 1249–1295. Embryology of testicular teratomas in strain 129 mice.

Stevens, L. C. (1960). *Devl Biol.* **2**, 285–297. Embryonic potency of embryoid bodies derived from a transplantable testicular teratoma of the mouse.

Stevens, L. C. (1964). *Proc. natn. Acad. Sci. U.S.A.* **52**, 654–661. Experimental production of testicular teratomas in mice.

Stevens, L. C. (1966). *J. natn. Cancer Inst.* **37**, 859–867. Development of resistance to teratocarcinogenesis by primordial germ cells in mice.

Stevens, L. C. (1967). *J. natn. Cancer Inst.* **37**, 549–552. Origin of Testicular Teratomas from Primordial germ cells in Mice.

Stevens, L. C. and Hummel, K. P. (1957). *J. natn. Cancer Inst.* **18**, 719–747. A description of spontaneous congenital testicular teratomas in strain 129 mice.

Stevens, L. C. and Mackenson, J. A. (1961). *J. natn. Cancer Inst.* **27**, 443–453. Genetic and environmental influences on terato-carcinogenesis in mice.

Steward, F. C. (1963). *Scient. Am.* **209**, (4), 104–113. The control of growth in plant cells.

Stewart, F. (1952). *Texas Rep. Biol. Med.* **10**, 239–253. Experiences in spontaneous regression of neoplastic disease in man.

Stewart, H. L. (1958). *In* "The Physiopathology of Cancer". (F. Homberger, ed.), 2nd Ed. pp. 18–37. Hoeber Inc., Now York. Pulmonary tumors in mice.

Stewart, J. A. and Papaconstantinou, J. (1967). *Proc. natn. Acad. Sci. U.S.A.* **58**, 95–102. A stabilization of RNA templates in lens cell differentiation.

Stockdale, F. E. and Topper, Y. J. (1966). *Proc. natn. Acad. Sci. U.S.A.* **56**, 1283–1289. The role of DNA synthesis and mitosis in hormone-dependent differentiation.

Straus, B. S. (1962). *J. theoret. Biol.* **2**, 266–278. Some aspects of mutation.

Strong, L. C. and Smith, G. M. (1936). *Am. J. Cancer.* **28**, 112–114. Successful transplantation of hepatoma in mice.

Subtelny, S. (1965a). *J. exp. Zool.* **159**, 47–57. Single transfers of nuclei from differentiating endoderm cells into enucleated and nucleate *Rana pipiens* eggs.

Subtelny, S. (1965b). *J. exp. Zool.* **159**, 59–92. On the nature of the restricted differentiation-promoting ability of transplanted *Rana pipiens* nuclei from differentiating endoderm cells.

Swann, M. M. (1957). *Cancer Res.* **17**, 727–757. The control of cell division: a review. I. General mechanisms.

Swann, M. M. (1958). *Cancer Res.* **18**, 1118–1160. The control of cell division: a review. II. Special mechanisms.

Swift, H. (1965). *Am. Nat.* **99**, 201–227. Nucleic acids and chloroplasts.

Tannenbaum, A. (1961). *Acta Un. Intn. c. Cancer.* **17**, 72–87. Studies in urethan carcinogenesis.
Tannenbaum, A. (1962). *Cancer Res.* **22**, 1105–1112. Neoplastic response of various tissues to the administration of urethan.
Tannenbaum, A. (1964). *Natn. Cancer Inst. Monogr.* **14**, 341–356. Contribution of urethan studies to the understanding of carcinogenesis.
Tannenbaum, A. and Silverstone, H. (1958). *Cancer Res.* **18**, 1225–1231. Urethan (ethylcarbamate) as a multipotential carcinogen.
Tartar, V. (1956). *In* "Cellular Mechanisms in Differentiation and Growth". (D. Rudnick, ed.), pp. 73–100. Princeton Univ. Press, Princeton. N. J. Pattern and substance in Stentor.
Tartar, V. (1960). *J. exp. Zool.* **144**, 187–207. Reconstitution of minced Stentor coeruleus.
Tartar, V. (1961). "The Biology of Stentor". Pergamon Press, Oxford.
Tatum, E. L. (1964). *Proc. natn. Acad. Sci. U.S.A.* **51**, 908–915. Genetic determinants.
Taylor, J. H. (1963). *In* "Molecular Genetics". (J. H. Taylor, ed.), Vol. 1, pp. 65–111. Academic Press, London. The replication and organization of DNA in chromosomes.
Teutschlaender, (1926). *Z. Krebsforsch.* **23**, 209–228. Der Hornstrahlentumor (Tumeur molluscoïde Borrel-Haaland) der Maus.
Thomlinson, R. H. and Gray, L. H. (1955). *Br. J. Cancer* **9**, 539–549. The histological structure of some human lung cancers and the possible implications for radiotherapy.
Tomatis, L. (1967). Personal communication.
Toth, B. (1968). *Cancer Res.* **28**, 727–738. A critical review of experiments in chemical carcinogenesis using new born animals.
Toth, B. and Shubik, P. (1966a). *Cancer Res.* **26**, 1472–1475. Carcinogenesis in Swiss mice by isonicotinic acid hydrazide.
Toth, B. and Shubik, P. (1966b). *Science, N.Y.* **152**, 1376–1377. Mammary tumor inhibition and lung adenoma induction by isonicotinic acid hydrazide.
Toth, B., Tomatis, L. and Shubik, P. (1961). Multipotential carcinogenesis with urethan in the Syrian golden hamster.
Trager, W. (1964). *In* "The Cell". (J. Brachet and A. E. Mirsky, eds.), Vol. 6, pp. 81–137. Academic Press, New York. The cytoplasm of protozoa.
Truscott, B. McN. (1947). *Br. J. Cancer* **1**, 129–145. Carcinoma of the breast. An analysis of the symptons, factors affecting prognosis, results of treatment and recurrences in 1211 cases treated at the Middlesex Hospital.
Tuchman-Duplessis, H. and Mercier-Parot, L. (1960). *In* "Ciba Foundation Symposium on Congenital Malformations". (G. E. W. Wolstenholme and C. M. O'Connor, eds.), pp. 115–133. J & A Churchill, London. The teratogenic action of the antibiotic actinomycin D.
Tudhope, G. R. (1939). *J. Path. Bact.* **48**, 499–506. A complex malignant mammary tumour.
Twort, J. M. and Twort, C. C. (1939). *Am. J. Cancer* **35**, 80–85. Comparative activity of some carcinogenic hydrocarbons.
Tyzzer, E. E. (1909a). *J. med. Res.* **21**, 479–518. A series of spontaneous tumors in mice with observations on the influence of heredity on the frequency of their occurrence.
Tyzzer, E. E. (1909b). *J. med. Res.* **21**, 519–573. A study of inheritance in mice with reference to their susceptibility to transplantable tumors.

Various Authors (1961). *Cold Spring Harb. Symp. quant. Biol.* **26**, 1–400. Cellular regulatory mechanisms.
Various Authors (1962). *Cold Spring Harb. Symp. quant. Biol.* **27**, 1–525. Basic mechanisms in animal virus biology.

Vasiliev, J. M. (1958). *Br. J. Cancer* 12, 524–536. The role of connective tissue proliferation in invasive growth of normal and malignant tissues: a review.
Vesselinovitch, S. D. and Mihailovich, N. (1967a). *Cancer Res.* 27, 350–352. The role of periodic and interrupted treatment of newborn and infant mice with urethan on leukemogenesis.
Vesselinovitch, S. D. and Mihailovich, N. (1967b). *Cancer Res.* 27, 1422–1429. The neonatal and infant age periods as biological factors which modify carcinogenesis by urethan.
Vesselinovitch, S. D. and Mihailovich, N. (1968a). *Cancer Res.* 28, 881–887. The induction of benign and malignant liver tumors by urethan in newborn rats.
Vesselinovitch, S. D. and Mihailovich, N. (1968b). *Cancer Res.* 28, 888–897. The development of neurogenic neoplasms, embryonal kidney tumors, Harderian gland adenomas, Anitschkow-cell sarcomas of the heart and other neoplasms in urethan-treated newborn rats.
Vesselinovitch, S. D., Mihailovich, N. and Pietra, G. (1967). *Cancer Res.* 27, 2333–2337. The prenatal exposure of mice to urethan and the consequent development of tumors in various tissues.
Villee, C. A. (1963). *In* "General Physiology of Cell Specialization". (D. Mazia and A. Taylor, eds.), pp. 64–72. McGraw-Hill, New York. Enzyme specialization in mammalian cells.
Visfeldt, J. (1963). *Acta path. microbiol. scand.* 58, 414–428. Transformation of sympathicoblastoma into ganglioneuroma.

Waddington, C. H. (1932). *Phil. Trans. R. Soc. Lond.* 221, 179–230. Experiments on the development of chick and duck embryos cultivated *in vitro*.
Waddington, C. H. (1940). "Growth and Form". Cambridge Univ. Press, Cambridge.
Waddington, C. H. (1956). "Principles of Embryology". George Allen and Unwin, London.
Waddington, C. H. (1957). "The Strategy of the Genes". George Allen and Unwin, London.
Waddington, C. H. (1962). "New Patterns in Genetics and Development". Columbia Univ. Press, New York.
Wallace, A. C. (1961). *Can. Cancer Conf.* 4, 139–165. Metastasis as an aspect of cell behavior.
Wallace, D. M. (1959). *In* "Tumours of the Bladder". (D. M. Wallace, ed.), pp. 157–170. E & S Livingstone, Edinburgh. Clinico-pathological behaviour of bladder tumours.
Walther, H. E. (1948). "Krebsmetastases". Basle.
Warburg, O. (1956a). *Science, N.Y.* 123, 309–314. On the origin of cancer cells.
Warburg, O. (1956b). *Science, N.Y.* 124, 269–270. On respiratory impairment in cancer cells.
Waymouth, C. (1967). *Natn. Cancer Inst. Monogr.* 26, 1–21. Somatic cells *in vitro*: their relationship to progenitive cells and to artificial milieux.
Weinhouse, S. (1955). *Adv. Cancer Res.* 3, 269–325. Oxidative metabolism of neoplastic tissues.
Weinhouse, S. (1960). "The J. P. Greenstein Memorial Symposium on Amino Acids, Proteins and Cancer Biochemistry". (J. T. Edsall, ed.), pp. 109–119, Academic Press, New York. Enzyme activities and tumor progression.
Weinhouse, S. (1966). *Gann. Monogr.* 1, 99–115. Glycolysis, respiration and enzyme deletions in slow-growing hepatic tumors.
Weinstein, I. B. (1963). *Cold Spring Harb. Symp. quant. Biol.* 28, 579–580. Comparative studies on the genetic code.

Weiss, L. (1967). "The Cell Periphery, Metastasis and Other Contact Phenomena". North Holland Publishing Co., Amsterdam.
Weiss, P. (1950). *Quart. Rev. Biol.* **25**, 177–198. Perspectives in the field of morphogenesis.
Weiss, P. (1953). *J. Embryol. exp. Morphol.* **1**, 181–211. Some introductory remarks on the cellular basis of differentiation.
Weiss, P. (1955). *In* "Analysis of Development". (B. H. Willier, P. Weiss and V. Hamburger, eds.), pp. 346–401. N.B. Saunders, Philadelphia. Nervous system (neurogenesis).
Weiss, P. (1958a). *Int. Rev. Cytol.* **7**, 391–423. Cell contact.
Weiss, P. (1958b). *In* "Cytodifferentiation", (D. Rudnick, ed.), Chicago Univ. Press, Chicago. Evaluation and Perspectives.
Weiss, P. (1958c). *Behav. Sci.* **3**, 93–215. Concept of biology: discussion.
Weiss, P. (1959a). *In* "Biological Organization". (C. H. Waddington, ed.), pp. 224–235. Pergamon Press, London. In Discussion.
Weiss, P. (1959b). *In* "Mitogenesis". (H. S. Ducoff and C. F. Ehret, eds.), Chicago Univ. Press, Chicago. In discussion, pp. 63–64.
Weiss, P. (1962). *In* "The Molecular Control of Cellular Activity". (J. M. Allen, ed.), pp. 1–72. McGraw-Hill, New York. From cell to molecule.
Weiss, P. (1963). *J. theoret. Biol.* **5**, 389–397. The cell as unit.
Weiss, P. (1967a). *In* Discussion of Abercrombie. (1967a).
Weiss, P. (1967b). *In* "Control of Cellular Growth in Adult Organisms". (H. Teir and T. Rytöma, eds.), pp. 407–412. Academic Press, London. Concluding remarks.
Weiss, P. and James, R. (1955). *Expl Cell Res.* Suppl 3, 381–394. Skin metaplasia *in vitro* induced by exposure to vitamin A.
Weiss, P. and Kavanau, J. L. (1957). *J. gen. Physiol* **41**, 1–47. A model of growth and growth control in mathematical terms.
Weiss, P. and Taylor, A. C. (1960). *Proc. natn. Acad. Sci. U.S.A.* **46**, 1177–1185. Reconstitution of complete organs from single-cell suspensions of chick embryo in advanced stages of development.
Weisz, P. B. (1954). *Q. Rev. Biol.* **29**, 207–229. Morphogenesis in protozoa.
Wessells, N. K. (1964). *Devl Biol.* **9**, 92–114. Acquisition of actinomycin D insensitivity during differentiation of pancreas exocrine cells.
Wessells, N. K. and Cohen, J. H. (1967). *Devl Biol.* **15**, 237–270. Early pancreas organogenesis: morphogenesis, tissue interactions and mass effects.
White, C. P. (1910). *J. Path. Bact.* **14**, 450–462. Experiments on cell proliferation and metaplasia.
Wieser, C. (1934). *Arch. Gynak.* **156**, 534–549. Untersuchungen über die Mäusebrustdrüse und ihre physiologischen und pathologischen Veränderungen.
Wilde, C. E. Jr. (1959). *In* "Cell, Organism and Milieu" (D. Rudnick, ed.), pp. 3–43. Ronald Press Company, New York. Differentiation in response to biochemical environment.
Wilde, C. E. Jr. (1961a). *Adv. Morphogen.* **1**, 267–300. The differentiation of vertebrate pigment cells.
Wilde, C. E. Jr. (1961b). *In* "La Culture Organotypique". Colloques internationaux du C.N.R.S., Paris. No. 101. pp. 183–198. Factors concerning the degree of cellular differentiation in organotypic and disaggregated tissue cultures.
Wilkie, D. (1964). "The Cytoplasm in Heredity". Methuen, London.
Williams, R. C. (1956). *Proc. natn. Acad. Sci. U.S.A.* **42**, 806–810. Relations between structure and biological activity of certain viruses.
Willis, R. A. (1944). *Cancer Res.* **4**, 630–644. The mode of origin of tumors. Solitary localized squamous cell growths of the skin.

Willis, R. A. (1945). *Cancer Res.* 5, 469–479. Further studies on the mode of origin of cancer of the skin.
Willis, R. A. (1951). A.F.I.P. Atlas of Tumor Pathology, Washington, D.C. Sect. III, Fasc. 9. Teratomas.
Willis, R. A. (1952). "The Spread of Tumours in the Human Body". Butterworth & Co., London. 2nd ed.
Willis, R. A. (1958). "The Borderland of Embryology and Pathology". Butterworth & Co., London.
Willis, R. A. (1967). "The Pathology of Tumours". Butterworths, London. 4th ed.
Wilson, J. D. (1962). *J. clin. Invest.* 41, 153–161. Localization of the biochemical site of action of testosterone on protein synthesis in the seminal vesicle of the rat.
Wilson, J. G. (1965). *Ann. N.Y. Acad. Sci.* 123, (Art. 1), 219–227. Embryological considerations in teratology.
Wittman, H. G. and Wittman-Liebold, B. (1963). *Cold Spring Harb. Symp. quant. Biol.* 28, 589–595. Tobacco mosaic mutants and the genetic coding problem.
Willmer, E. N. ed. (1965). "Cells and Tissues in Culture". Academic Press, London.
Woglom, W. H. (1913). "The Study of Experimental Cancer; A Review". (George Crocker Special Research Fund No. 1) Columbia Univ. Press, New York.
Woglom, W. H. (1926). *Arch. path.* 2, 533–576, and 709–752. Experimental tar cancer.
Woglom, W. H. (1929). *Cancer Review* 4, 129–214. Immunity to transplantable tumours.
Wood, S., Holyoke, E. D. and Yardley, J. H. (1961). *Can. Cancer Conf.* 4, 167–223. Mechanisms of metastasis production by blood-borne cancer cells.
Wright, B. E. (1966). *Science, N.Y.* 153, 830–836. Multiple causes and controls in differentiation.
Wright, S. (1941). *Physiol. Rev.* 21, 487–521. The physiology of the gene.

Yamada, T. (1961). *Adv. Morphogen.* 1, 1–53. A chemical approach to the problem of the organizer.
Ycas, M. (1962). *Int. Rev. Cytol.* 13, 1–37. The coding hypothesis.
Yoshida, T. (1956). *Ann. N.Y. Acad. Sci.* 63, (Art. 5) 852–874. Contributions of the ascites hepatoma to the concept of malignancy.
Yoshida, T. (1962). In "The Morphological Precursors of Cancer". (L. Severi, ed.), pp. xxxi and xlix. On the earliest stages of cancer development.
Yoshida, T. (1966). *Gann Monogr.* 1, 21–28. Problems in the pathological evaluation of malignancy.
Young, J. S. (1959). *J. Path. Bact.* 77, 321–339. The invasive growth of malignant tumours: an experimental interpretation based on elastic-jelly models.

Zamenhof, S. (1957). In "The Chemical Basis of Heredity". (W. D. McElroy and B. Glass, eds.), pp. 351–377. Johns Hopkins Press, Baltimore. Properties of the transforming principles.
Zeidman, I. (1957). *Cancer Res.* 17, 157–162. Metastasis: a review of recent advances.
Zeidman, I. (1965a). *Cancer Res.* 25, 324–327. Fate of circulating tumor cells. III. Comparison of metastatic growth produced by tumor cell emboli in veins and in lymphatics.
Zeidman, I. (1965b). *Acta cytol.* 9, 136–140. The fate of circulating tumor cells.
Zinder, N. D. (1955). *J. cell. comp. Physiol.* 45, Suppl. 2, 23–49. Bacterial transduction.

Author Index

The numbers in *italics* indicate the pages on which names are mentioned in the reference lists.

A

Abercrombie, M., 104, 110, 303, 357, 360, 363, *387*
Acevedo, H. F., 150, *394*
Adams, W. R., 259, *387*
Afzelius, B. A., 276, *412*
Alexander, S., 150, *387*
Algire, G. H., 154, *387*
Allen, A. C., 146, *387*
Allfrey, V. G., 352, 358, 365, *387, 393, 398*
Althoff, J., 198, *410*
Ambrose, E. J., 110, *387*
Andervont, H. B., 46, 73, *387*
Anfinsen, C. B., 275, 309, *387*
Apolant, H., 25, 148, 173, *387, 395*
Apter, M. J., 310, 365, *387*
Archer, F. L., 145, 146, 147, *387*
Askanazy, M., 151, *387*
Attardi, G., 256, 258, *388*
Auerbach, C., 242, *388*
Auerbach, R., 340, *388*
Austin, M. L., 297, *388*
Authaler, A., 198, *410*
Axhausen, G., 155, *388*
Azzopardi, J. G., 150, 185, *388, 403*

B

Babinet, C., *404*
Bagg, H. J., 207, *388*
Baggenstoss, A. H., 85, *410*
Baltimore, D., 255, 256, 258, *398*
Baltus, E., 282, *388*
Barcroft, J., 312, *388*
Bardawil, W. A., 189, *388*
Baril, E. F., 347, *402*
Barnes, J. M., 196, *409*
Barnett, L., 232, *393*
Barr, G. C., 265, *388*
Barrett, M. K., 12, *388*
Barth, L. G., 325, *388*

Barth, L. J., 325, *388*
Baserga, R., 101, 102, 105, *388*
Bashford, E. F., 6, 145, 149, 152, 153, *388*
Basilio, C., 232, *419*
Bauman, R., 161, *388*
Beale, G. H., 286, 287, 289, 291, 292, 296, 297, 298, 301, *388, 389, 399*
Beals, T. F., 214, *414*
Beard, J. W., 41, 107, *389, 416*
Beer, G. R., de. 271, *404*
Beerman, W., 358, *389*
Bell, C., 144, *403*
Bell, H. G., 123, *418*
Bell, W. B., 94, *389*
Belling, J., 241, *389*
Benirschke, K., 201, *409*
Benson, C. D., 183, *389*
Benzer, S., 231, *389*
Berenblum, I., 43, 69, 86, 120, *389*
Bergman, R. T., 124, *403*
Bernal, J. D., 265, *389*
Berrill, N. J., 306, 344, *389*
Bertalanffy, F. D., 100, *389*
Bertalanffy, L., von. 97, 265, 345, *389*
Bertani, G., 251, *389*
Biancifiori, C., 68, *418*
Bielschowsky, F., 85, 129, 133, 145, *389*
Bielschowsky, M., 145, *389*
Biggs, R., 146, 147, *389*
Billingham, R. E., 13, *390*
Bird, H. H., 254, *393*
Bittner, J. J., 16, 25, 46, *390*
Black, J. W., 117, *390*
Blackler, A. W., 358, 359, 369, *396*
Blandford, G., 150, *409*
Bloom, H. J. G., 118, *390*
Blumenthal, H. T., 118, *405*
Böhmig, R., 161, *390*
Bonner, J., 365, *403*
Bonner, J. T., 231, *390*

Bonneville, M. A., 273, *415*
Bonser, G. M., 145, 148, 159, *390*
Borman, A., 144, *403*
Borst, M., 155, *390*
Borst, P., 276, *406*
Boutwell, R. K., 222, *390*
Böving, B. G., 188, *390*
Bowen, W. H., *388*
Boyd, J. S. K., 250, *390*
Boyd, W. L., 119, 125, *390*
Brachet, J., 281, 282, 321, 346, *388, 390*
Braun, A. C., 222, *390, 391*
Braun, R., 347, *417*
Brenner, S., 231, 232, 237, 261, *391, 393, 404*
Bresler, V. M., 208, *391*
Bretscher, M. S., 366, *391*
Breuer, M. E., 358, 359, *391*
Briggs, R., 281, 308, 311, 355, 358, 369, 370, 371, *391*
Brink, R. A., 371, 372, *391*
Britten, R. J., 262, *416*
Brookes, P., 242, 243, *391, 406*
Brown, B., 347, *399*
Brown, R. A., 254, *393*
Brues, A. M., 208, *404*
Bruggen, F. J., von. 276, *406*
Brunschwig, A., 87, *391*
Bullock, F. D., 22, *391*
Bullough, W. S., 97, 100, 101, 102, 104, 347, 348, 357, 365, 367, 383, 384, *391*
Burdette, W. J., 201, 242, *391*
Burdman, J. A., 182, *399*
Burnet, F. M., 253, *392*
Burnet, M., 248, 254, 257, *391*
Busch, H., 219, *392*
Butler, J. A. V., 365, *388*
Buy, H. G., du. 276, *392*

C

Callan, H. G., 358, 359, *392*
Cameron, G. R., 270, 306, *392*
Canter, H. Y., 73, *387*
Cantoni, G. L., 236, *392*
Capitano, J., 197, *394*
Carter, R. L., 196, 197, *416*
Case, R. A. M., 205, *392*
Caspar, D. L. D., 247, 257, *392*
Chalkley, H. W., 154, *387*
Champy, Ch., 208, *392*

Chandler, B. L., 261, *416*
Cheatle, G. L., 84, *392*
Chiang, K-S., 279, *392*
Chieco-Bianchi, L., 195, 198, *392, 394*
Chouroulinkov, J., 208, *416*
Clark, A. J., 262, *408*
Clark, B. P. C., 232, *412*
Claude, A., 16, *392*
Clayson, D. B., 22, *392*
Clever, U., 360, *392*
Cohen, J. H., 340, *423*
Cohen, L. W., 295, *392*
Cohen, S. S., 250, 313, *392*
Cohn, M., 238, *392*
Cole, W. H., 125, 181, 183, *389, 396*
Collins, D. H., 214, *392*
Colnaghi, M. I., 197, *394*
Colter, J. S., 254, *393*
Coman, D. R., 110, 114, *393*
Comfort, A., 311, *393*
Commoner, R., 307, 309, *393*
Corsy, F., 146, 148, *414*
Coward, S. J., 326, *397*
Cowdry, E. V., 98, *393*
Cowen, P. N., 37, *393*
Cozin, F., 231, *404*
Cramer, W., 6, 69, 153, *388, 393*
Crick, F. H. C., 230, 231, 232, *393*
Cromer, J. K., 85, *412*
Curtis, M. R., 22, *391*
Cutler, L., 84, *392*
Cutler, M., 84, *383*

D

Dabelstein, H., 162, *393*
Daland, E. M., 118,.*393*
Dalq, A. M., 318, 324, 326, 333, 343, *393*
Daniel, J. W., 347, *417*
Daniel, P. M., 144, *393*
Danielli, J. F., 239, 281, 283, *393, 408*
Dao, T. L., 52, *393*
Davidson, E. H., 348, 349, 351, 352, 353, 354, 355, *393*
Davidson, J., 181, *398*
Davis, B. D., 314, *394*
Day, E. D., 154, *394*
Day, M., 352, *417*
De Benedictis, G., 195, 198, *392, 394*
Deelman, H. T., 69, *394*
Delarue, J., 118, 119, *394*

AUTHOR INDEX

Della Porta, G., 197, *394*
Denis, H., 282, 321, *390*, *394*
Denoix, P. F., 112, 114, 118, 119, *394*
Deringer, M. K., 197, *394*
Des Ligneris, M. J. A., 69, *394*
Dethlefsen, L. A., 106, *410*
Dettlaff, T. A., 341, 343, *394*
Di Berardino, M. A., 370, *394*
Di Paolo, J. A., 199, 200, 201, 202, 203, *394*
Dixon, F. J., 211, *414*
Dockerty, M. B., 85, *410*
Dodson, J. W., 332, *394*
Doi, R. H., 255, *419*
Dominguez, O. V., 150, *394*
Doncaster, L., 306, *394*
Dorfman, R. I., 150, *394*
Druckrey, H., 198, 199, *395*
Drysdale, R. B., 231, 232, 237, 246, *414*
Ducoff, H. S., 98, 101, 104, 345, *395*
Dulbecco, R., 247, 253, 254, 257, 258, *392*, *395*
Dunn, L. C., 244, *395*
Dunn, T. B., 37, 148, 165, *395*, *402*
Dunphy, J. E., 118, *395*

E

Eagle, H., 349, 351, 353, *395*, *407*
Earle, W. R., 328, 352, *395*, *396*, *407*, *417*
Ebert, J. D., 344, *395*
Edmonds, H. W., 189, *395*
Edwards, C. N., 124, *395*
Ehret, C. F., 98, 101, 104, 345, *395*
Ehrlich, P., 148, 153, *394*
Eisenstadt, J. M., 240, *412*
Elasser, W. M., 265, 266, 267, 270, 271, 307, 310, *394*
Elis, J., 199, *394*
Ellem, K. A. O., 254, *393*
Elsdale, T. D., 369, *396*
Elsdale, T. R., 340, 348, 369, *395*
Emerson, W. J., 129, *395*
Engell, H. C., 115, *396*
Ephrussi, B., 215, *396*
Ephrussi–Taylor, H., *396*
Evans, A., *404*
Evans, R. W., 137, *396*
Evans, V. J., 328, 352, *395*, *396*
Everson, T. C., 125, 181, 183, *389*, *396*
Ewing, J., 3, 155, 179, *396*
Ezekiel, D. H., 261, *396*

F

Falin, L. I., 207, *396*
Fawcett, D. W., 273, 305, *396*
Feduska, N., 117, *397*
Fekete, E., 208, *396*
Fell, H. B., 157, 329, 330, 332, 348, *396*, *404*, *414*
Ferrigno, M. A., 208, *396*
Finch, B. W., 215, *396*
Fiore–Donati, L., 195, 198, *392*, *394*
Firket, H., 104, *396*
Fischberg, M., 358, 359, 369, *395*, *396*
Fischer, B., 20, 107, *396*
Fischer, G. A., 352, *417*
Fisher, B., 114, 117, 118, 124, *396*, *397*
Fisher, E. R., 114, 115, 117, 118, 124, *396*, *397*
Fleisher, M. S., 12, *408*
Flickinger, R. A., 314, 321, 326, 343, 361, 365, 371, *397*
Foote, F. W., 85, *398*
Foulds, L., 27, 34, 41, 44, 45, 46, 47, 49, 51, 52, 59, 68, 69, 70, 71, 72, 73, 81, 82, 93, 100, 113, 115, 119, 122, 137, 140, 148, 152, 155, 156, 157, 159, 163, 167, 172, 174, 177, 178, 221, 262, 298, 328, 332, 375, 383, 384, *397*, *398*
Fox, F., 181, *398*
Fraenkel–Conrat, H., 249, *398*
Franck, I. D., 161, *398*
Frankel, D. G., *412*
Franklin, R. M., 255, 256, 258, *398*
Franks, L. A., 124, 125, *398*
Franks, L. M., 40, 124, 125, *398*
Fraument, J. F., 184, 200, *410*
Frazel, E. L., 85, *398*
Freeman, J. J., 144, *403*
Frenster, J. H., 365, 366, *398*
Friedewald, W. F., 42, *398*
Fry, R. J. M., 101, *406*
Furth, J., 383, *398*

G

Gaillard, P. J., 335, *398*
Gall, J. G., 358, 359, *398*
Gallien, L., 370, *398*
Galton, D. A. G., 118, *419*
Gardner, R. S., 232, *419*
Gardner, W. U., 26, 52, *399*
Gaudier, N. 147, *399*

Gibor, A., 279, *399*
Gibson, I., 286, 287, *399*
Gierer, A., 237, *399*
Gierke, E., 154, *399*
Gilbert, W., 237, *399*
Glendenning, O. M., 69, *416*
Gluecksohn–Schoenheimer, S., 337, *399*
Gluecksohn–Waelsh, S., 337, *399*
Goldblatt, S. A., 116, *399*
Goldschmidt, R. B., 231, 244, 309, 324, 326, *399*
Goldstein, A., 347, *399*
Goldstein, M. N., 182, *399*
Goodall, C. M., 151, *399*
Goodman, H. M., 237, *415*
Goodwin, B. C., 240, 314, 319, 326, 365, *399*
Gordon, I., 187, *399*
Gorer, P. A., 12, *399*
Goss, R. J., 97, 99, 101, *399*
Graham, J. N., 129, *395*
Grandclaude, N., 147, *399*
Granick, S., 279, 289, 303, *399*
Gray, L. H., 105, *421*
Green, M., 253, 255, *399*
Greene, H. S. N., 41, 112, 117, 124, 193, *400*
Greenstein, J. P., *400*
Grell, K. G., 290, *400*
Greulich, R. C., 101, *400*, *406*
Griffith, F., 16, 245, *400*
Griffiths, J. D., 115, 116, *400*
Grobstein, C., 44, 223, 306, 307, 311, 316, 318, 319, 322, 324, 325, 327, 329, 333, 334, 335, 336, 337, 338, 339, 341, 344, 348, 349, 371, *400*, *401*, *405*, *417*
Gros, F., 232, 236, *401*
Gross, L., 27, *401*
Gruenstein, M., 147, *418*
Grüneberg, H., 201, 203, *401*
Guérin, M., 147, 208, *401*, *413*, *416*, *417*
Guérin, P., 147, *413*, *417*
Günthen, R., 147, *401*
Gurdon, J. B., 369, 370, *395*, *396*, *401*
Gustafson, T., 312, *401*
Guthrie, J., 207, *401*

H

Haaland, M., 10, 25, 41, 70, 138, 148, 152, 163, 170, *388*, *401*

Haddow, A., 52, 152, *401*
Hadfield, G., 123, *401*
Hagström, B. E., 315, *417*
Halberstraedter, L., 34, *401*
Halkerston, I. D. K., 232, 321, *402*
Hamburger, V., 316, 318, 324, 325, 334, 348, *403*
Hamilton, M. G., 253, 256, 258, *405*
Hammerling, J., 282, *401*
Hamperl, H., 71, 74, 110, 144, 146, 147, *401*, *402*
Hanafusa, H., 257, *402*
Hanafusa, T., 257, *402*
Hansemann, D., von. 153, 155, *402*
Harris, C., 147, *418*
Hartman, P. E., 247, *402*
Harvey, E. K., 117, *400*
Hauschka, T. S., 216, *402*
Hausen, H., 253, *417*
Hausen, P., 253, *417*
Havemann, H. V., 162, *402*
Hayashi, M., 233, 234, 256, *419*
Hechter, O., 232, 321, *402*
Hennen, S., 370, *402*
Henshaw, P. S., 36, *412*
Herriott, R. M., 241, 242, 250, *402*
Herrmann, H., 335, 347, *402*
Hertig, A. T., 188, 189, 191, *402*
Hertz, R., 189, *402*
Heston, W. E., 37, 52, 68, 159, *402*
Hieger, I., 19, *405*
Hilf, R., 144, *403*
Hinshelwood, C. N., 238, *403*
Hirst, A. E., 124, *403*
Hirst, G. K., 257, *403*
Holland, J. J., 253, *403*
Holman, R. L., 185, *403*
Holtfreter, J., 316, 318, 324, 325, 334, 348, *403*
Holtzer, H., 98, 100, 343, *403*
Holyoke, E. D., 114, 115, *424*
Horava, A., 71, 74, *403*
Horne, R., 248, *408*
Horowitz, N. H., 276, *403*
Hotchin, J., 259, *403*
Hotchkiss, R. D., 270, *403*
Hou, L. T., 185, *403*
Hou, P. C., 191, *403*
Hoyer, B. H., 253, *403*
Hoyle, L., 253, 254, *403*

Huang, R–C, C., 365, *403*
Hubby, J. L., 335, *411*
Huggins, C. B., 155, *403*
Huguenin, R., 118, 176, *403*
Hultin, T., 236, 237, 241, *403*
Hummel, K. P., 209, *420*
Humphrey, R. R., 355, 370, 371, *391*
Hurwitz, J., *404*
Huseby, R. A., 150, *394*
Huxley, J. S., 271, *404*

I

Ingle, D. J., 307, *404*
Isaacs, A., 254, *404*
Ishikura, H., 236, *392*
Ivankovic, S., 198, 199, *395*

J

Jackson, E. B., 208, *404*
Jackson, S. F., 329, 330, *404*
Jacob, F., 231, 234, 238, 239, 240, 251, 252, 268, 314, 319, 365, 367, *404, 410*
James, R., 324, 331, *423*
Jensen, C. D., 8, *404*
Jensen, W. A., 100, *404*
Joklik, W. K., 253, *404*
Jones, K., 340, 348, *395*
Jones, O. W., 232, 249, *409, 412*
Jonescu, P., 116, *404*
Journey, L. J., 182, *399*
Judah, J. D., 271, 380, *409*

K

Kacser, H., 231, 265, 268, 270, *404*
Kahn, M. C., 383, *398*
Kaighn, M. E., 344, *395*
Kameyama, T., 240, *412*
Karlson, P., 360, *392, 404*
Kavanau, J. L., 97, *423*
Kawai, A., 202, *404*
Kawamata, J., 202, *404*
Keck, K., 282, *404*
Keilin, D., 275, *405*
Kellenberger, E., *405*
Kellner, B., 110, *405*
Kelly, M. G., 195, *405*
Kennaway, E. L., 19, 27, *405*
Kennedy, B. J., 129, *395*
Kerr, I. M., 253, 256, 258, *405*
Keyes, E. L., 118, *405*

Kidd, J. G., 42, 108, 130, 131, *416, 417*
Kidson, C., 222, 323, *405*
Kikuchi, Y., 352, *411*
Kim, U., 117, *405*
Kimball, R. E., 346, *415*
King, J. C., 270, *405*
King, T. J., 281, 308, 311, 358, 369, 370, *391, 394*
Kirby, K. S., 222, 323, *405*
Kirschbaum, A., 37, *418*
Klein, E., 71, 72, *405*
Klein, G., 71, 72, *405*
Klein, M., 193, 194, 197, *405*
Kleinsmith, L. J., 142, 214, 216, *405*
Klug, A., 247, 257, *392*
Koch, W. E., 338, *405*
Koller, P. C., 14, *405*
Konigsberg, I. R., 340, 348, 350, 353, *405*
Kornberg, A., *416*
Korner, A., 241, *406*
Kotin, P., 199, 200, 201, 203, *394, 406*
Kroeger, H., 360, *406*
Kroon, A. M., 276, *406*
Kuzma, J. F., 144, *406*

L

Lacassagne, A., 25, 69, *406*
Lacroix, J. C., 370, *398*
Lambret, M., 147, *399*
Lamerton, L. F., 101, 104, *406, 420*
Landauer, W., 200, 202, 203, 244, *406*
Lark, K. G., 103, 246, *406*
Larsen, C. D., 193, 199, *406*
Lasnitzki, I., 331, *406*
Latarjet, R., 69, *406*
Lathrop, A. E. C., 24, *406*
Lau, C., 100, *389*
Laufer, H., 321, 323, *406*
Laurence, E. B., 100, *391*
Lavedan, J., 208, *392*
Lawley, P. D., 243, *406*
Leblond, C. P., 99, 101, *406, 410*
Le Breton, E., 345, *407*
Leder, P., 232, *412*
Lederberg, J., 231, *407*
Lee, A. E., 52, *407*
Lees, J. C., 23, 187, 189, 191, *413*
Lefevre, H., 35, 37, *407*
Legallais, F. Y., 154, *387*
Lenner, D. F., 352, *411*

Lehninger, A. L., 276, *407*
Leighton, J., 71, 110, 111, 114, 123, 129, 161, 216, *407*
Lengyel, P., 232, *419*
Leslie, I., 249, *407*
Less, L. A., 352, *411*
Levan, A., 216, *402*
Levintow, L., 349, *395*, *407*
Lewin, C., 146, *407*
Liao, S., 241, *407*
Likely, G. D., 352, *407*, *417*
Lindegren, C. C., 247, *407*
Lindsay, D., 145, *389*
Linell, F., 69, *407*
Linzell, J. L., 144, *407*
Lipschütz, B., 69, *407*
Lipscomb, H. S., 150, *407*
Lipsett, M. B., 150, 189, *402*, *407*
Littau, V. C., 365, *387*
Little, C. C., 11, *407*, *408*
Lloyd, L., 358, 359, *392*
Loeb, L., 12, 24, *406*, *408*
Lorch, I. J., 281, 283, *408*
Love, R., 259, *408*
Low-Beer, B. V. A., 123, *418*
Lucia, E. L., 123, *418*
Luck, D. J. L., 277, *408*
Lucké, B., 184, *408*
Ludford, R. J., 156, *408*
Luria, S. E., 250, 252, 258, *408*
Lwoff, A., 229, 247, 248, 251, 253, 258, 285, *392*, *408*
Lynch, C. J., 36, *408*

M

Maas, W. K., 238, 262, *408*
Mc Carthy, B. J., 262, *416*
McCarty, M., 245, *408*
Mc Cormick, G. M., 52, *408*
MacDonald, I., 119, *408*
McFall, E., 238, 262, *408*
Machado, R. D., 275, *415*
Mackenson, J. A., 213, *420*
McKenzie, A., 119, *408*
McKenzie, I., 108, *409*
McKeown, F., 125, *409*
Mackay, C. J., 328, *396*
McLean, A. E. M., 271, *380*, *409*
McLean, E., 271, *380*, *409*
McLoughlin, C. B., 331, *409*

McQuillen, K., 273, 276, *409*
Madden, R. E., 117, *409*
Magee, P. N., 196, *409*
Main, J. M., *415*
Maiorano, G., 195, 198, *394*
Makino, S., 216, *409*
Malmgren, R. A., 117, *409*
Malpas, J. S., 150, *409*
Mandelstam, J., 238, *415*
Manning, M. D., 184, 200, *410*
Mansell, H. M., 188, 189, 191, *402*
Marchok, A. C., 347, *402*
Marcus, P. I., 259, *409*
Marin-Padilla, M., 201, *409*
Markert, C. L., 357, 361, 362, 366, 367, *409*
Martin, E. M., 253, 255, 256, 258, *405*, *409*
Martin, R. G., 232, 249, *409*
Mather, K., 362, 373, *409*
Matsen, F., 150, *407*
Matthaei, J. H., 232, 249, *409*
Mazia, D., 98, 100, 101, 103, 104, 305, 344, 347, *409*
Melicow, W. M. M., 214, *409*
Mellanby, E., 330, *396*
Mendelsohn, M. L., 101, 105, 106, *410*, *415*
Mercier-Parot, L., 202, *421*
Messier, B., 99, *410*
Metzenberg, R. L., 276, *403*
Michalowsky, I., 207, *401*
Michel, I., 144, *403*
Mihailovich, N., 197, 199, 205, *422*
Millar, M. J., 73, *410*
Miller, E. C., *410*
Miller, J. A., *410*
Miller, R. W., 184, 200, *410*
Mirsky, A. E., 352, 359, 365, *387*, *393*, *398*, *410*
Mitchley, B. C. V., 201, *416*
Mittermayer, C., 347, *417*
Miyagi, M., 326, 343, *397*
Moertel, C. G., 85, *410*
Mohr, U., 198, *410*
Monod, J., 234, 238, 239, 240, 268, 314, 319, 367, *404*, *410*
Montagna, W., 143, *410*
Moon, R. C., 52, *408*
Moore, D. H., 259, *411*
Moore, G. E., 117, 352, *410*, *411*
Moore, J. A., 370, *411*
Morgan, C., 259, *411*, *416*

Morris, H. P., 220, *411*
Morton, J. H., 119, *411*
Morton, J. J., 119, *411*
Moscona, A., 306, 340, 350, *411*
Moscona, A. A., 335, *411*
Moser, C., 326, *397*
Moser, C. R., 343, *397*
Moulé, Y., 345, *407*
Moyer, A. W., 254, *393*
Mühlbock, O., 46, 52, *411*
Muller, H. J., *411*
Mulligan, R. M., 146, *411*
Mundy, J., 52, *411*
Murphy, J. B., 16, *392*
Murray, J. A., 6, 24, 145, 148, 152, 153, 166, *388*, *411*
Murray, R. G. E., 261, *411*
Murray, W. S., 24, *411*
Mustard, W. T., 183, *389*

N

Nadel, E. L., 116, *399*
Nadel, E. M., 115, *411*
Nakabayashi, N., 202, *404*
Nakane, P. K., 213, 214, *414*
Nakase, Y., 321, 323, *406*
Nanney, D. L., 286, 289, 290, 291, 292, 295, 299, 300, 301, 302, 303, *411*, *412*
Nass, M. M. K., 276, *412*
Nass, S., 276, *412*
Nathanson, I., 118, *412*
Nathanson, I. G., 129, *395*
Needham, A. E., 97, 271, 328, 344, 381, *412*
Needham, J., 328, *412*
Neidhart, F. C., *412*
Nettleship, A., 36, *412*
Newman, W., 85, *412*
Nicholson, D., 124, *395*
Nicholson, G. W., 94, 144, 161, 162, 184, 186, 224, *412*
Nie, R., van. 52, 55, *412*
Nirenberg, M. W., 232, 249, *409*, *412*
Noble, R. L., 52, 73, *410*, *412*
Nordon, J. G., 69, *407*
Northrop, F. S. C., 266, 269, *412*
Novelli, G. D., 240, *412*
Novikoff, A. B., 275, 276, *412*
Nowinski, W., 97, *412*
Noyes, R. W., 188, *412*

O

Ober, W. B., 188, 189, 191, *413*
Oberling, Ch., 147, *413*
Ochoa, S., 232, 233, *413*, *419*
O'Gara, R. W., 195, *405*
Olivi, M., 68, *418*
Oppenheimer, J. M., 308, 315, *413*
Orlando, R. A., 145, 146, 147, *387*
Orr, J. W., 86, *413*
Orrahood, M. D., 118, *405*
Orsini, M. W., 188, *413*
Osawa, S., 365, *410*

P

Paine, C. H., 119, *413*
Palade, G. E., 274, 275, *413*
Palm, J., 11, *413*
Pang, S. C., 191, *403*
Papaconstantinou, J., 364, *420*
Park, H. D., 154, *387*
Park, W. W., 23, 187, 189, 191, *413*
Parker, G., 325, 338, *401*
Parmi, L., 197, *394*
Parsons, D. F., 275, *413*
Pasternack, J. G., 143, 145, *413*
Paul, J., 323, 366, 383, *414*
Pavan, C., 358, 359, *391*, *414*
Payne, P. M., 118, *419*
Peacocke, A. R., 231, 232, 237, 246, *414*
Pelc, S. R., 329, *414*
Peller, S., 204, *414*
Percival, W. H., 196, 197, *416*
Pereira, J. P. M., 101, *406*
Perlmann, P., 315, *417*
Pestka, S., 232, *412*
Petit, N., 146, *414*
Peyron, A., 146, 148, 211, *414*
Picheral, B., 370, *398*, *414*
Picken, L., 231, 247, 251, 268, 273, 276, 277, 282, 283, 314, *414*
Pierce, G. B., 211, 213, 214, 216, *414*
Pierce, G. B., Jr., 142, 214, 216, *405*
Pietra, G., 195, 197, 199, *414*, *422*
Piez, K., 353, *395*
Pirie, N. W., 247, *407*
Pitelka, D. R., 285, 290, *414*
Pitot, H. C., 220, 221, 318, 384, *414*
Pittendrigh, C. S., 313, *414*
Pollock, M. R., 238, 240, *414*, *415*

Pontecorvo, G., 231, 233, 240, 241, 261, 278, 309, *415*
Porter, K. R., 273, 274, 275, *415*
Potter, V. R., 71, 72, 220, *415*
Prehn, R. T., *415*
Prescott, D. M., 346, *415*
Preussmann, R., 198, 199, *395*
Prince, A. M., 259, *387*
Pritchard, M. M. L., 144, *393*
Pugh, R. C. B., 214, *392, 415*

Q
Quastler, H., 101, 309, *415*

R
Radl, E., 307, *415*
Rapp, E. H., 162, *415*
Rappaport, H., 195, *414*
Raven, C. P., 310, 314, 315, *415*
Ravitch, M. M., 183, *389*
Reichman, M. E., 257, *415*
Reiskin, A. M., 105, *415*
Retiene, K., 150, *407*
Rhodin, J. A. G., 273, *415*
Rich, A., 237, 309, *415*
Rich, R. A., 40, 124, *416*
Richards, H. H., 236, *392*
Richardson, C. C., *416*
Richardson, K. C., 144, *416*
Rigby–Jones, P., 118, *419*
Rigler, L. G., 118, *416*
Riley, F. L., 276, *392*
Ris, H., 261, *416*
Ritchie, A. C., 115, 116, *416*
Riviere, M., 208, *416*
Roberts, R. B., 262, *416*
Roe, F. J. C., 69, 195, 196, 197, 201, *416*
Rogers, S., 38, *416*
Rollins, E., 326, 343, *397*
Rose, H. M., 259, *411, 416*
Ross, G. T., 189, *402*
Rossi, G., 52, *419*
Rota, T. R., 337, *399*
Rott, R., 253, *417*
Rottenberg, J. C. M., 276, *406*
Rous, P., 15, 37, 41, 42, 88, 107, 108, 130, 131, 193, 199, *389, 398, 409, 416, 417, 419*
Roussy, G., 147, *417*
Rowson, K. E. K., 195, *416*

Rubin, H., 257, *402*
Rudzinska, M. A., 286, 289, 291, 292, 299, 301, 303, *412*
Runnström, J., 315, *417*
Rusch, H. P., 347, *417*
Russell, B. R. G., 153, *417*
Russell, E. S., 305, 307, 308, 312, 314, *417*
Rutter, W. J., 322, 339, *417*
Ryan, F. J., 231, *417*

S
Sachsenmaier, W., 347, *417*
Sager, R., 231, *417*
Salaman, M. H., 195, *416*
Salsbury, A. J., 115, 116, *400*
Samuels, L. T., 150, *394*
Sand, S. A., 307, 310, 365, *417*
Sandberg, A. A., 117, *410*
Sandford, K. K., 22, 132, 349, 352, 354, *395, 407, 417*
Saxen, L., 318, 323, 324, 325, 326, *417*
Schaeffer, P., 252, *404*
Schäfer, W., 253, *417*
Schildkraut, C. L., *416*
Schindler, R., 352, *417*
Schlesinger, R. W., 254, *417*
Schlumberger, H. C., 184, *408*
Schmähl, D., 198, 199, *395*
Schmidt, M. D., 113, *417*
Scholtissek, C., 235, 237, 253, 255, *417*
Scott, A., 34, *417*
Sekeris, C. E., 360, *404*
Sengel, P., 332, *418*
Serra, J. A., 371, *418*
Severi, L., 52, 68, *418, 420*
Shabad, L. M., 74, *418*
Shannon, J. E., Jr., 352, *395*
Shapiro, J. A., 37, *418*
Sharma, A. K., 242, *418*
Sharma, K., 242, *418*
Shay, H., 147, *418*
Shimkin, M. B., 7, 36, 37, 73, 118, 119, 123, *387, 418*
Shubik, P., 39, 43, 86, 120, 121, 195, 196, 198, 205, *389, 414, 418, 421*
Siegel, R. W., 302, *418*
Siekevitz, P., 276, *418*
Signoret, J., 355, 370, 371, *391*
Sigot, M., 306,
Silverstone, H., 197, *421*

Simnett, J. D., 370, *418*
Simpson, G. G., 307, 314, *418*
Sinclair, J. G., 199, *418*
Singer, B. A., 249, *398*
Sinnott, E. W., 270, *418*
Sirlin, J. L., 235, *418*
Skaar, D., 297, 301, 302, *419*
Skalka, A., *404*
Skoryna, S. C., 71, 74, *403*
Sly, W. S., 232, *412*
Smith, G. M., 152, *420*
Smith, J., 256, 258, *388*
Smith, J. P., 214, *415*
Smith, W. E., 37, 38, 193, 199, *417*, *419*
Smithers, D. W., 85, 93, 95, 118, 122, 125, 180, 381, *419*
Snell, G. D., 11, *419*
Snyder, W. H., Jr., 183, *389*
Sonneborn, T. M., 278, 286, 287, 288, 289, 291, 292, *399*, *419*
Spencer, K., 195, *414*
Speyer, J. F., 232, *419*
Spiegelman, S., 233, 234, 255, 256, *419*
Squartini, F., 52, 68, *418*, *419*, *420*
Stanley, W. M., 253, *392*
Stedman, E., 365, *420*
Steele, G. G., 104, *420*
Steinthorsson, E., 124, *395*
Stent, G. S., 262, *420*
Stern, C., 244, *420*
Stevens, L., 213, 214, *414*
Stevens, L. C., 209, 210, 211, 213, *420*
Steward, F. C., 362, *420*
Stewart, F., 37, 94, *420*
Stewart, H. L., 37, *420*
Stewart, J. A., 364, *420*
Stockdale, F. E., 343, 344, *420*
Stoker, M. G. P., 247, *392*
Straus, B. S., 242, *420*
Strong, L. C., 11, 152, *408*, *420*
Subtelny, S., 370, *420*
Suedka, N., 279, *392*
Surmount, J., 146, 148, *414*
Swann, M. M., 98, 101, 103, 346, 347, *420*
Swift, H., 278, 279, *420*

T

Tanaka, K., 236, *392*
Tannenbaum, A., 38, 197, *421*
Tartar, V., 285, *421*

Tatum, E. L., 233, 236, 237, *421*
Taylor, A. C., *423*
Taylor, J. H., 103, *421*
Teutschlaender, 171, *421*
Thomas, L. B., 181, *398*
Thomlinson, R. H., 105, *421*
Thung, P. J., 52, 55, *412*
Toivonen, S., 318, 323, 324, 325, 326, *417*
Tomatis, L., 195, 313, *421*
Topper, Y. J., 343, 344, *420*
Toth, B., 195, 196, 198, *421*
Tournier, P., 247, 248, *392*, *408*
Toy, B. L., 189, *388*
Trager, W., 285, *421*
Tridente, G., 195, 198, *392*
Truscott, B., McN., 123, *421*
Tschetter, D., 87, *391*
Tsubura, Y., 52, 68, *402*
Tuchman–Duplessis, H., 202, *421*
Tudhope, G. R., 145, 177, *421*
Turnbull, R. B., Jr., 115, *397*
Twort, C. C., 69, *421*
Twort, J. M., 69, *421*
Tyzzer, E. E., 36, *408*, *421*

U

Uehlinger, V., 370, *401*
Ushida, T., 202, *404*

V

Vanderberg, J., 321, 323, *406*
Various Authors, 238, 253, *421*
Vasiliev, J. M., 107, 111, *422*
Verney, E. L., 211, *414*
Vesselinovitch, S. D., 197, 199, 205, *422*
Villee, C. A., 313, *422*
Visfeldt, J., 182, *422*
Vitry, F., de. 282, 321, *390*
Vlahakis, G., 52, 68, *402*

W

Waddington, C. H., 231, 240, 261, 266, 269, 278, 306, 309, 311, 312, 313, 318, 319, 322, 324, 325, 326, 327, 333, 334, 361, 375, *422*
Wahba, A. J., 232, *419*
Walker, B. E., 101, *406*
Walker, L. M., 297, *388*
Wallace, A. C., 104, 109, 110, 113, 114, 115, *422*

Wallace, D. M., 136, *422*
Walther, H. E., 113, *422*
Waltz, H. K., 328, 352, *395*, *396*
Warburg, O., *422*
Ward, D. N., 150, *407*
Warner, J. R., 237, *415*
Watts–Tobin, R. J., 232, *393*
Waymouth, C., 351, 354, *422*
Webster, D. R., 115, 116, *416*
Weinhouse, S., 72, 219, *422*
Weinstein, I. B., 232, *422*
Weiss, L., 110, 114, 118, *423*
Weiss, P., 97, 265, 266, 275, 278, 305, 307, 313, 317, 318, 319, 320, 324, 327, 329, 331, 332, 334, 336, 345, 346, 379, 380, *423*
Weisz, P. B., 285, *423*
Welch, C., 118, *412*
Welch, J. K., 183, *389*
Wessells, N. K., 322, 339, 340, *417*, *423*
White, C. P., 143, *423*
White, R. J., 150, *409*
Widmayer, D., 297, *388*
Wieser, C., 143, *423*
Wilde, C. E., Jr., 306, 322, 328, 335, 340, 346, 348, *423*
Wildy, P., 247, *392*
Wilkie, D., 275, 276, 287, 279, 282, 286, *423*
Williams, E. D., 150, *388*
Williams, P. C., 52, *411*
Williams, R. C., 249, *398*, *423*
Williams,–Ashman, H. G., 241, *407*

Willis, R. A., 84, 85, 86, 87, 88, 113, 119, 123, 132, 137, 152, 155, 160, 179, 180, 185, 188, 204, 206, 330, 344, *423*, *424*
Willmer, E. N., 350, *424*
Wilson, C., 150, *407*
Wilson, E. P., 328, *396*
Wilson, J. D., 241, *424*
Wilson, J. G., 203, *424*
Wirth, J. E., 143, 145, *413*
Wittman, H. G., 232, *424*
Wittman–Liebold, B., 232, *424*
Woglom, W. H., 10, 18, 27, 154, *424*
Wollman, E. C., 251, *404*
Wollman, E. L., 252, *404*
Wolpert, L., 310, 312, 365, *387*, *401*
Wood, S., 114, 115, *424*
Work, T. S., 253, 256, 258, *405*
Wrigley, P. F. M., 150, *409*
Wright, B. E., 271, 381, *424*
Wright, S., 314, *424*

Y

Yamada, T., 318, *424*
Yardley, J. H., 114, 115, *424*
Ycas, M., 232, 249, *424*
Yoshida, T., 71, 93, 94, 216, *424*
Young, J. S., 110, *424*

Z

Zamenhof, S., *424*
Zeidman, I., 114, 115, 117, *424*
Zinder, N. D., 247, *424*

Subject Index

A

Aetiological factors in neoplasia, 31–39
Anaplasia, 150
Anomalous associations of independently variable characters, 50, 71, 73, 94, 121–125, 192, 221, 223, 328, 329, 384
Anomalies of clinical behaviour, 121–125
Ascites tumours, 14, 160

B

Bacterial transduction, 246, 247
Bacterial transformation, 245, 246
Benign neoplasia, 3, 4, 12, 17, 26, 72, 73, 91, 93, 94, 105, 109, 112
  Transplantation of, 12
Biochemical theories of neoplasia, 219–222
Biological Organization,
  Cell theory of, 305–307
  Complexity of, 229, 271
  Mutual conformance of genetic and biotoxic systems in, 284, 294, 316, 341, 364, 366, 384
  Network organization of, 271, 380
  Static and dynamic components of, 229, 263, 265, 360, 379–381
Biological problems of neoplasia, 222–225
Biotonic systems, 270, 271, 303

C

Cancer, 93, 94, 381, 383
Capacity (Developmental), 44, 300, 317, 319, 324–329
  (See also Competence and Facultative genome)
Capacity (Neoplastic), 42, 44, 78, 79, 140, 234, 382
Carcinogenesis (experimental),
  by enhancement, 21, 35–40
  by induction, 17–20, 31–35
Carcinogenesis and Teratogenesis, 200–205
  relationship between, 200–203
Carcinogens,
  Chemical, 17–22
  Physical, 22
  Metazoan parasites, 22
  Effective carcinogens, 32
  Incomplete carcinogens, 43
  Effective exposure to, 33
  Carcinogenicity, 33
Cell proliferation (Normal),
  Cell division, 101
  Differential mitoses, 100
  Mitotic (cell) cycle, 101–104
  Preparatory role in differentiation, 344
Cell proliferation (Neoplastic),
  Mitotic cycles, 105–106
Cell proliferation and differentiation,
  Bullough's hypothesis, 347–348
  Inverse relationship, 343, 349–355
Ciliated protozoa, 284–304
  Cortical organization, 284, 285
  Functions of micronuclei and macronuclei, 291
  Mating types, 292
  Metagons in, 286–289
  Mutual exclusion of types, 297–300
  Organization of nuclear materials, 289
  Serotypes, 295
Cistrons, 231
Classifications and definitions of neoplasms, 4, 72, 91, 95, 96
Clinical course of neoplasia in man, 118–127
Collateral hyperplasia, 155–157, 185, 187
Competence, 39, 140, 324–339, 375, 376
  Autonomous ageing of, 325
  Genetic basis of, see Facultative genome
  (See also Capacity)
Complex tumours, 176–179

Components of Neoplasms, 139
  Invaded tissues, 1
  Parenchyma, 141–153
  Stroma, 153–155
Conditional tumours, 44, 46–50, 52–65, 108, 128, 133, 134
Confused cells, 328, 329, 331
Creodes, 327
Cytoplasmic inheritance, 286, 304
Cytoplasmic organization,
  Network organization, 269–271, 380, 381
  Operative cytosome, 320
  Ultrastructure. Endoplasmic reticulum, 273–275
    Ribosomes, 273
    Golgi apparatus, 274
    Mitochondria, 275–279
    Chloroplasts, 279

D

Decision in advance of performance, 102, 302, 334
De-differentiation,
  in cultures *in vitro*, 349–355
  in neoplasms, 150
Deflection of development, 317, 329, 361, 375, 376
  (See also Metaplasia)
Determination, 333–334, 337, 339, 341, 376
  (See also Preselection of effective genomes)
Development of metazoa, 305–341
  Directiveness of, 312, 334, 381
  Early vertebrate development, 314–317
  Genetic basis of, 310
  preformation and epigenesis, 308–310
  Programme of development, 310, 311, 361
Developmental bias, 332, 339, 363
Developmental imminence, 39, 69, 325, 339, 361
Differential utilization of the genome, 140, 225, 262, 292
  General principles of, 140, 357, 358, 360–365
  Mechanisms of, 365–373
  (See also Effective genome and Facultative genome)
Differentiation in normal development, 139–141, 318–324, 375–377

Definitions of, 318–319
Discreteness of, 327, 384
Exclusivity of, 327, 384
Genetic limitation of, 327, 384
Strain differentiation, 319, 323, 345, 361, 368
Successive steps in, 320
  (See also Expression of effective genomes)
Differentiation in neoplasms, 194, 383–384
  Type of differentiation, 141–151
  Degree of differentiation, 151–153
  Divergent differentiation, 145, 192, 217
Differentiation *in vitro*, 349–355
Dissociations of independently variable characters, 71, 94, 122, 189, 328, 329
Dormant tumours, 123–124
Dynamic chemical equations, 267–269, 379
Dynamic organization, 266–272, 378–380
  Biotonic systems, 270, 271, 303
  Epigenetic systems, 303
  Open systems, 266, 267

E

Effective genome, 140, 262, 291, 299, 311, 319, 360, 362, 363
  (See also Expression of genetic information)
Embryonic tumours,
  In man, 179–192
  In experimental animals, 192–205
Endocrine neoplasia, 22–24, 85
Endocrinoma, 23, 191
Enhancement of neoplasia, 21, 35–38, 193, 195, 198
  Genetic factor in, 36–38
Enzyme synthesis, 237
Eukaryocytes, 261
Expression of genetic information, 140, 345, 354, 376–377, 382
  Preselection of effective genome, 302–326
  Selection of effective genome, 295, 323, 324, 376
  Transcription of effective genome, 141, 260
  Translation of effective genome, 141
Extra-nuclear genetic mechanisms, 377, 378
  In ciliated protozoa, 284–285, 286–289, 303
  In chloroplasts, 279

SUBJECT INDEX

In mitochondria, 276
Morphogenetic substances in plants, 282
Stable informational RNA, 378

F

Facultative genome, 140, 262, 298, 311, 357, 362, 363
Field theory of Willis, 87–89

G

Gastrulation, 316
Genes,
  Structural genes, 233, 262, 263
  Regulatory genes, 233, 262, 263
Genetic code, 231–233
  Transcription of, 233
  Translation of, 236–241, 260
Genetic materials and genetic actions, 229–263
  DNA: structure, 230
    replication, 230, 260
    transcription, 233, 260
  RNA: messenger RNA, 234
    transfer RNA, 235–236
    ribosomal RNA, 234–235
    stable informational RNA, 322
Group A lesions, 79, 92
Group B lesions, 79–86, 121, 122
  Possible fates of, 80–81
Group C lesions, 81
Growth (normal), 97–101
  Expanding, renewing and static cell populations, 99–101
Growth (neoplastic), 104–106

I

"Inappropriate" secretion, 150–151, 384
Incipient neoplasia, 44, 45, 78, 79, 82–84, 86, 92, 130, 382
Independently variable characters of neoplasma, 50, 71, 72, 94, 119, 128, 166, 170, 189, 192, 384
Individuality of tumours, 72, 119, 221, 384
Induction (embryonic), 317, 334–341
  Primary induction, 317–323
  in amphibian eye, 335, 336
  in higher vertebrates, 317, 334–341

Induction (neoplastic), 31–35, 199
Initiation of neoplasia, 20, 42, 43, 44, 45, 78, 79, 233
Invasion,
  by non-neoplastic tissues, 107, 188
  by benign tumours, 107–109
  by malignant tumours, 109–112

J

Jacob-Monod regulatory circuits, 238–241, 379, 384

L

Latent carcinoma, 124–125
Latent neoplastic potentialities, 42
Latent tumour cells, 42
Localized neoplasia, 86–89
Lysogeny, 251

M

Maintenance systems, 283, 318, 321, 377–378
Malignancy, 12, 72, 73, 91–94, 106, 193, 223
Malignant tumours, 3, 4, 91–93, 109–121
Mammary hyperplastic nodules, 26, 68
Mammary neoplasia,
  in mice, 24–46, 46–75
  in rats, 26, 27
  in women, 84
Mammary plaques, 26, 59–61, 68
Metaplasia, 329–334, 361, 363, 376
  (See also Deflection and Facultative genome)
Metastasis, 5, 112–117
  Steps in, 113–114
  Experimental study of, 113–115
  Neoplastic cells in circulating blood, 115–117
  Fate of liberated cells, 117
  Patterns of distribution of secondary tumours, 117, 118
Minimal deviation tumours, 220, 384
Modulation, 13, 142, 318
Morphogenesis, 277–279, 282, 284, 285, 316, 318, 319, 323, 361
Morphogenesis in neoplasia, 155–160, 175, 187, 192, 206, 224
  (See also Organoid tumours and Teratoma)

SUBJECT INDEX

"Morphogenetic Memory" in neoplasms, 175, 216, 217, 383
Multifocal origin of neoplasms, 85–86
Mutation, 241–243
Mutation theory of cancer, 14, 91
Mutual conformance of genetic and biotonic patterns, 281, 294, 316, 341, 464, 366, 379

N

Neoplasms,
  General pathology and classifications, 3–7
  Definitions, classifications and terminologies, 91–96
Nephroblastoma, 184–187
Network organizations, 271, 380, 381, 385
Neuroblastoma and ganglioneuroma, 181–184
Nuclear transplantation,
  In unicellular organisms, 283, 284
  In amphibia, 369–371

O

Occult primary tumours, 122, 123
Operative cytosome, 320, 344
Operon concept (Jacob-Monod), 233, 234, 238–240, 261–262
Organizer, 317
Organoid tumours, 161–175, 187, 206

P

Para-endocrine cancer syndromes, 151
Paramutation, 371–373
Patterns of neoplastic development, 75–89
  A generalized schema, 76–81
  Group A lesions, 79
  Group B lesions, 79–81
  Group C lesions, 81
Patterns of transcription, 323, 375
Phenocopy, 200, 202, 243, 244
Physiological functions in neoplasms, 48, 144, 145, 172, 191
Polysomes, 236
Precancerous Lesions, 72, 79, 80, 92, 190
  (See also Group B lesions)
Preselection of effective genomes, 302, 326, 333, 340, 341, 375
Programme of development, 310, 311, 361, 364

Programmes of neoplastic development, 120, 130–132, 134, 175, 183, 216, 217, 224, 225, 383
Progression, 41, 45, 51, 57–59, 65–75, 86, 87, 94, 182, 223,
  Definition of, 45
  General principles of, 69–75
Prokaryocytes, 261
Promotion of neoplasia, 42–44

R

Regional neoplasia, 33–35, 77, 81–86
  of epidermis, 33–35
  of mammary gland, 84
  of urinary bladder, 83, 84
  of uterine cervix, 82, 83
Regulatory genes, 221, 233, 262, 263, 384
Regulons, 261
Residual neoplasia, 45, 56, 65, 129, 131, 133, 135, 382
Retinoblastoma, 180
Retrogressive neoplasia, 125–134, 180, 181, 190
Reversibility of neoplasia, 131, 135, 222, 382, 383
Reversion to embryonic type, 179, 383
Rous sarcoma, 15

S

Specific stroma reaction, 9, 112, 139, 153, 177
Sporadic neoplasia, 23, 38–40
Spurious malignancy, 106–109, 188
Stem cells,
  normal, 100, 106, 330, 343, 361
  neoplastic, 1, 129, 142, 144, 185, 207, 210, 215–217
Strain differentiation, 319, 323, 345, 361, 376, 382
Stroma of neoplasma, 153–155
Structural genes, 233, 262, 263, 381
Structural organization of neoplasms, 160–175
Sub-threshold neoplastic state, 42

T

Tar cancer, 17–18
Teleonomic processes, 314

Teratomas, 205–217
  In man, 205–207
  In mice, 208–209
  Experimental induction of, 207–208
Teratomas of mouse testis, 208–217
  Transplantation of, 209, 216
  Cloning experiments with, 214–216
  Embryoid bodies in, 211, 215
  Histogenesis of, 212–214
  Programme of development of, 216
  Stem-cell concept of, 216
Tissue-resistance to neoplasia, 135–136
Transplacental carcinogenesis, 195, 198, 199, 203–205
Transplantable tumours,
  Composition of, 9
  Immunity to, 10
  Long-established strains, 9
  Modulation in, 13
  Progression in, 13
  Properties of, 13, 14
  Stability of, 9–10
Transplantation of tumours, 7–14
  Early experiments, 7–9
  Genetic theory of, 11
  Autotransplantation, 10
  Homotransplantation, 8
  Iso-transplantation, 11
  Heterotransplantation, 8
  as a test of "malignancy", 12
  as "artificial metastasis", 10
Trophoblast and its neoplastic derivatives, 187–192

V

Viruses, 247–260
  Nature of, 247, 248
  Tobacco mosaic virus, 248–250
  Bacteriophage, 250–252
  Animal viruses, 252–260
Virus-tumours, 15–17
  Filterable tumours of fowls, 15
  Mammalian virus-tumours, 16–17

DATE DUE			
MAY 0 0 ~~~			

DEMCO 38-297